Advances in Robotics and Automatic Control: Reviews

Book Series, Volume 1

S.Yurish
Editor

Advances in Robotics and Automatic Control: Reviews

Book Series, Volume 1

IFSA International Frequency Sensor Association Publishing

S. Yurish, *Editor*
Advances in Robotics and Automatic Control: Reviews, Book Series, Vol. 1

Published by IFSA Publishing, S. L., 2018,
E-mail (for print book orders and customer service enquires):
ifsa.books@sensorsportal.com

Visit our Home Page on http://www.sensorsportal.com

ISBN: 978-84-09-02449-0
BN-20180530-XX
BIC: TJFM1

Contents

Preface

By 2020 the International Federation of Robotics (IFR) estimates that more than 1.7 million new industrial robots will be installed in factories worldwide and robots for domestic could reach almost 32 million units in the period 2018-2020, with an estimated value of about €10 bn ($11.7 bn).

Industrial robots offer many benefits, including cost reduction, increased rate of operation and improving quality, along with improved manufacturing efficiency and flexibility. The demand for industrial robotics is majorly observed in industries such as automotive, electrical & electronics, chemical, rubber & plastics, machinery, metals, food & beverages, precision & optics, and others. In its turn, industrial automation control market will witness considerable growth during the same period with the growing demand of products such as sensors, drives and various robots.

The first volume of the *Advances in Robotics and Automatic Control: Reviews*, Book Series started by IFSA Publishing in 2018 contains ten chapters written by 32 contributors from 9 countries: Belgium, China, Germany, India, Ireland, Japan, Serbia, Tunisia and USA.

Chapter 1 discusses the electrostatic inchworm motors with low energy consumption using a small size power source. The leg of the microrobot is designed to allow reciprocal motions and powered by Si photovoltaic (hereafter PV) cells.

Chapter 2 describes an adaptive trajectory tracking control for nonholonomic mobile manipulators under modeling uncertainties and external disturbances. One feature of the proposed controller is its model-independent control scheme that can avoid the knowledge of the dynamic parameters and the bound of the external disturbances. Furthermore, the control law is formulated in task space and the redundancy problem is resolved by an extended approach.

Chapter 3 presents a fast approximate nearest neighbor search tree based novelty filter for mobile robotic and video surveillance applications.

Chapter 4 describes control algorithms for the centrifuge flight simulator/spatial disorientation trainer, calculate their kinematic and dynamic parameters in each interpolation period to predict their dynamic behaviour.

Chapter 5 presents state-of-the-art review in the area of continuous hard turning. The various wear mechanisms of polycrystalline cubic boron nitride tool materials are discussed with a view to identifying the critical factors that determine their behaviour in application.

Chapter 6 discusses an approach, which involves factorization of the SLAM posterior over the robot's path, in which each individual particle follows a constant time stereo SLAM approach and the particle distribution is harnessed by the algorithm to estimate the optimal trajectory.

Chapter 7 reports topics in Systems, Control and Optimisation and their evolution through recently funded projects, since about 2013, as well as the EU (e.g. H2020, ERC), National and other Programmes vis-à-vis broader developments.

Chapter 8 summarized a real time switching-model detection Innovation Squared Mismatch (ISM) strategy is presented to enable closed loop control of the switched systems.

Chapter 9 reports H_∞ tracking adaptive fuzzy sliding mode design controller for a class of non square nonlinear systems.

Chapter 10 discusses two formulations of the optimal control problem associated with the optimization of the energy consumed by the induction motor under vector control. The emphasis was placed on the advantage of limiting the control quantities during a real application in order to protect the actuators and the machine.

I hope that readers will enjoy this book and it can be a valuable tool for those who involved in research and development of various robots and automatic control systems.

Sergey Y. Yurish

Editor   Barcelona, Spain

Contributors

Riadh Abdelati
University of Monastir, National School of Engineers of Monastir, Tunisia, E-mail: riaabdelati@yahoo.fr

S. Aloui
National Engineering School of Sfax, Laboratory of Sciences and Techniques of Automatic Control & Computer Engineering (Lab-STA), University of Sfax; BP 1173, 3038, Sfax, Tunisia, E-mail: aloui_sinda@yahoo.fr

Mohamed Boukattaya
Laboratory of Sciences and Techniques of Automatic Control & Computer Engineering (Lab-STA), National School of Engineering of Sfax, University of Sfax, Sfax, Tunisia

D. Chatterjee
Indian Statistical Institute, Kolkata, India

Daniel S. Contreras
University of California, Berkeley, USA

M. Elloumi
National Engineering School of Sfax, Laboratory of Sciences and Techniques of Automatic control & computer engineering (Lab-STA), University of Sfax; BP 1173, 3038, Sfax, Tunisia, E-mail: mourad.elloumi@yahoo.fr

Sayan Ghosal
Seagate Technologies LLC, Shakopee, MN, USA

Seamus Gordon
University of Limerick, Plassey, Limerick, Ireland

Minami Kaneko
Nihon University, Tokyo, Japan

13

Satoshi Kawamura
Nihon University, Tokyo, Japan

Alkis Konstantellos
European Commission, Complex Systems and Advanced Computing
Unit Brussels, Belgium

Y. Koubaa
National Engineering School of Sfax, Laboratory of Sciences and
Techniques of Automatic Control & Computer Engineering
(Lab-STA), University of Sfax; BP 1173, 3038, Sfax, Tunisia,
E-mail: Yassine.Koubaa@enis.run.tn

Vladimir Kvrgić
Lola Institute, Kneza Viseslava 70 a, 11030 Belgrade, Serbia

Cora Lahiff
University of Limerick, Plassey, Limerick, Ireland

Liwei Lin
University of California, Berkeley, USA

Yoshio Mita
The University of Tokyo, Tokyo, Japan

Isao Mori
The University of Tokyo, Tokyo, Japan

P. Mukherjee
Department of Computer Science, IIT Kharagpur, Kharagpur, India

S. Patranabis
Department of Computer Science, IIT Kharagpur, Kharagpur, India

Pat Phelan
University of Limerick, Plassey, Limerick, Ireland

Kristofer S. J. Pister
University of California, Berkeley, USA

R. Reiger
Airbus Group HQ HWD1, Germany

Ken Saito
Nihon University, Tokyo, Japan

Murti Salapaka
Department of Electrical and Computer Engineering, University of Minnesota, Twin Cities, USA

A. Sarkar
Department of Mathematics, IIT Kharagpur, Kharagpur, India
Techno India University, Kolkata, India

H. Singh
Department of Computer Science, IIT Kharagpur, Kharagpur, India

Daisuke Tanaka
Nihon University, Tokyo, Japan

Taisuke Tanaka
Nihon University, Tokyo, Japan

Fumio Uchikoba
Nihon University, Tokyo, Japan

Xia Li Wang
Changan University, Xi'an, Shaanxi, China

Xiaochun Wang
Xi'an Jiaotong University, Xi'an, Shaanxi, China

D. Mitchell Wilkes
Vanderbilt University, Nashville, TN, USA

Chapter 1

Electrostatic Inchworm Motors Driven by High-Voltage Si Photovoltaic Cells for Millimeter Scale Multi-Legged Microrobots

Ken Saito, Daniel S. Contreras, Isao Mori, Daisuke Tanaka, Satoshi Kawamura, Taisuke Tanaka, Minami Kaneko, Fumio Uchikoba, Yoshio Mita, Liwei Lin and Kristofer S. J. Pister

1.1. Introduction

Several microrobot systems from the micrometer to centimeter scale have been demonstrated [1-12]. Among these demonstrations, the micrometer scale ones have potential usages in special environments such as surgery inside the narrow blood vessel of a human brain or micro assembly for the small size mechanical system [4, 8] but it is difficult to add power sources and controllers into the microscale system. Therefore, passive control schemes by external electrical or magnetic forces are commonly implemented. On the other hand, a lot of centimeter-size robots have been constructed by the miniaturizations of electrical components with integrated sensors, actuators, power sources and controllers [6, 9]. Despite the fact that multiple bio-inspired robots have been proposed, millimeter scale robots do not perform like insects due to the difficulty in integrating power sources and actuators onto the robot [13-14]. In particular, the locomotion mechanisms of insects attract the attention of researchers [5, 7]. In seeking further miniaturization, some researchers use micro fabrication technology to fabricate small sized actuators [15-16]. For example, piezoelectric actuators, shape memory alloy actuators, electrostatic actuators, ion-exchange polymer actuators, and so on are a few examples. These actuators have different strengths, such as power consumption, switching speed, force generation, displacement, and fabrication difficulty. In general, an actuator can only

Ken Saito
Nihon University, Tokyo, Japan

generate either rotary or linear motion and mechanical mechanisms are necessary to convert the movements generated by the actuators to locomotion.

Previously, the authors have shown a millimeter scale hexapod-type microrobot to perform the tripod gait locomotion of an ant [17], and a quadruped-type microrobot to replicate the quadrupedal gait locomotion of an animal [18] by using shape memory alloy actuators for large deformation and large force. This chapter discusses the electrostatic inchworm motors [19-21] with low energy consumption using a small size power source. The leg of the microrobot is designed to allow reciprocal motions and powered by Si photovoltaic (hereafter PV) cells [22].

1.2. Multi-Legged Microrobot

Fig. 1.1 (a) shows the previous multi-legged microrobot using shape memory alloy type actuator [18]. A previous multi-legged microrobot using shape memory alloy actuator is changed to electrostatic inchworm motors in this work, where each leg of the robot can perform the stepping motion via a single actuator. The leg is fixed on both sides of the body and the microrobot can increase the number of the legs easily. In this chapter, the actuator connection part has been redesigned to accommodate the electrostatic inchworm motors. Fig. 1.1 (b) shows the mechanical parts of the leg made from a silicon wafer except for the shaft and the steady pin. The shapes of the mechanical parts are machined by the inductively coupled plasma dry etching process with photolithography technology. The authors have manual assembled the mechanical parts of the robot because microfabrication technology is hard to construct the complicated three-dimensional structure. In the process, 200 μm-thick silicon wafers were used for the mechanical parts except for the washer which used 100 μm-thick silicon wafers. The shaft was constructed by using 0.1 ± 0.002 mm in diameter cemented carbide. The washer was mounted to the end of a shaft to fix the silicon parts. To keep the parts rigidly connected, the washer and the shaft were glued using cyanoacrylate. All silicon parts have a clearance of a 10 μm gap with respect to the other fitted parts. Since these actuators can only generate the rotary motion or linear motion, linkage assemblies are needed for a microrobot to move using the stepping pattern. The stepping pattern realized by two sets of four-bar linkages. Bar 1, bar 2, bar 5 and bar 6 are the primary (top) four-bar linkage. Bar 3, bar 4, bar 5 and bar 6 are the secondly (bottom) four-bar linkage. The primarily four-bar linkage and secondly four-bar linkage are combined with each other with bar 5 and bar 6 (Fig. 1.1 (c)).

Fig. 1.2 shows the leg motion and trajectory of the leg. The inflection point of the trajectory has four points such as (x_1, y_1), (x_2, y_2), (x_3, y_3) and (x_4, y_4). The steady pin and the hole of bar 5 cause the inflection of the trajectory. The four points can be expressed by the difference of angles of θ_A and θ_{Foot}. The difference of θ_A and θ_{Foot} can perform the reciprocal movement of point P. In other words, Fig. 1.2 shows that the designed leg can perform the stepping motion by the reciprocal movement of point P.

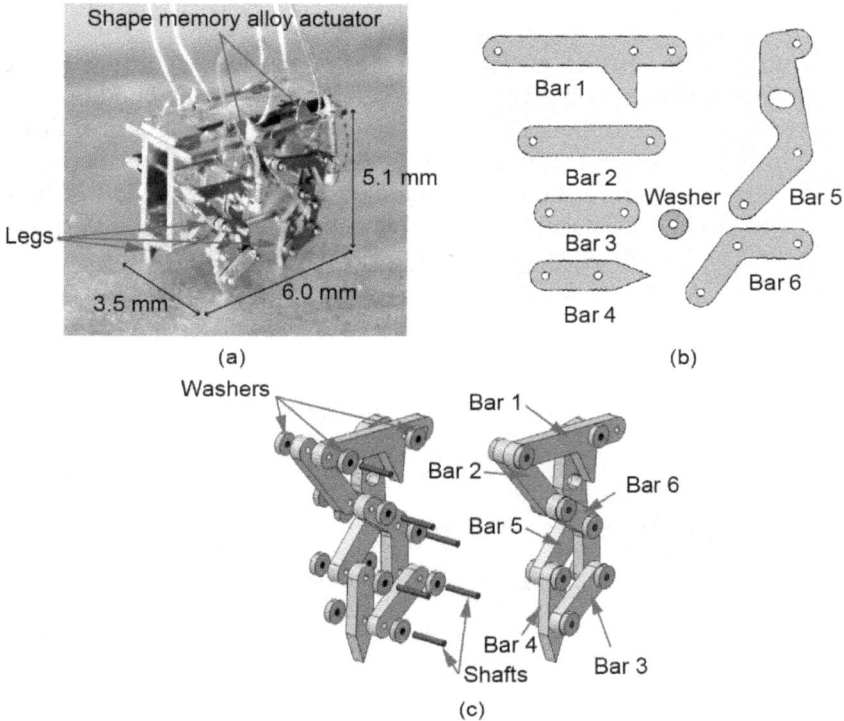

Fig. 1.1. (a) Previous multi-legged microrobot using shape memory alloy type actuator [18]. Mechanical parts of the leg for microrobot with (b) individual parts; (c) assembled structure.

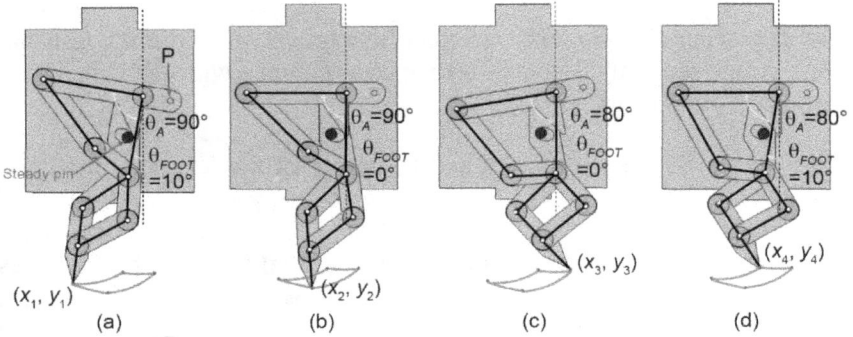

Fig. 1.2. Leg motion and trajectory of the leg. (a) (x_1, y_1) at θ_A=90°, θ_{FOOT}=10°; (b) (x_2, y_2) at θ_A=90°, θ_{FOOT}=0°; (c) (x_3, y_3) at θ_A=80°, θ_{FOOT}=0°; (d) (x_4, y_4) at θ_A=80°, θ_{FOOT}=10°.

The authors design the mechanical parts of the leg according to the mathematical equations. Fig. 1.3 and Table 1.1 show the conditions to describe the point of the leg (x_n, y_n). The (x_0, y_0) is the origin coordinate which is the only fixed point of the robot. The upper case alphabet A, B, C, D, E, F, G, H and I show the name of each lengths. L1 and L3 show the auxiliary lines from (x_0, y_0) to bar 4 which is the bar contains the point of the leg. θ_3, θ_5, θ_7, θ_B and θ_{A0} are described as Fig. 1.3 (b).

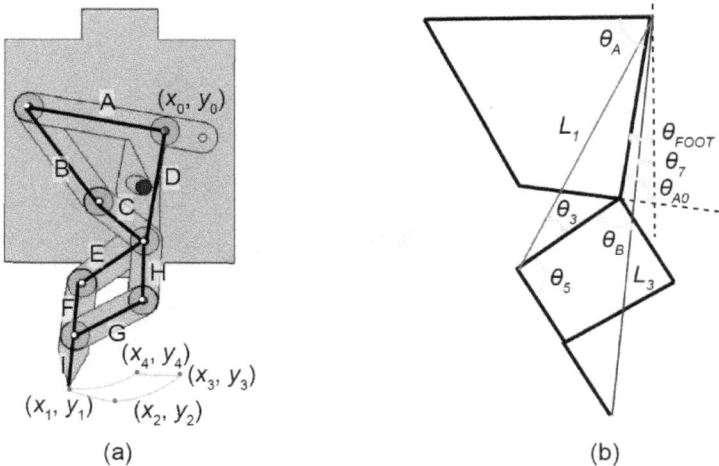

Fig. 1.3. Name of each bars, coordinates and angles of the leg.
(a) Length and coordinates; (b) Angles and auxiliary lines.

Table 1.1. Length between the node points.

Name of bar	Name	Length (μm)
Bar 1	A	1800
Bar 2	B	1556
Bar 6	C	800
Bar 5	D	1500
Bar 5	E	1000
Bar 4	F	700
Bar 3	G	1000
Bar 6	H	800
Bar 4	I	700

The (x_n, y_n) (n=1, 2, 3, 4) can describe by the Equation (1.1):

$$(x_n = L_3 \cos(-90° - \theta_7 - \theta_{Foot}), y_n = L_3 \sin(-90° - \theta_7 - \theta_{Foot})), \quad (1.1)$$

where θ_7, L_3, L_1, θ_5, θ_3, θ_B and θ_{A0} are the following Equation (1.2), (1.3), (1.4), (1.5), (1.6), (1.7) and (1.8), respectively.

$$\theta_7 = \cos^{-1} \frac{L_1{}^2 + D^2 - E^2}{2L_1 D} - \cos^{-1} \frac{L_1{}^2 + L_3{}^2 - (F+I)^2}{2L_1 L_3}, \quad (1.2)$$

$$L_3 = \sqrt{L_1{}^2 + (F+I)^2 - 2L_1(F+I)\cos(\theta_3 + \theta_5)}, \quad (1.3)$$

$$L_1 = \sqrt{D^2 + E^2 - 2D^2 \cos 135°}, \quad (1.4)$$

$$\theta_5 = \cos^{-1}\left(\frac{E - H\cos\theta_B}{\sqrt{E^2 + H^2 - 2EH\cos\theta_B}}\right) + \\ + \cos^{-1}\left(\frac{E^2 + H^2 - G^2 + F^2 - 2EH\cos\theta_B}{2F\sqrt{E^2 + H^2 - 2E\cos\theta_B}}\right), \quad (1.5)$$

$$\theta_3 = 180° - \left(135° + \cos^{-1}\frac{L_1{}^2 + D^2 - E^2}{2L_1 D}\right), \quad (1.6)$$

$$\theta_B = 360° - (\theta_{A0} + 135° + 50°), \quad (1.7)$$

$$\theta_{A0} = 180° - \left(\cos^{-1}\left(\frac{D - A\cos\theta_A}{\sqrt{D^2 + A^2 - 2DA\cos\theta_A}}\right) + \\ + \cos^{-1}\left(\frac{D^2 + A^2 - B^2 + C^2 - 2DA\cos\theta_A}{2C\sqrt{D^2 + A^2 - 2DA\cos\theta_A}}\right)\right), \quad (1.8)$$

Table 1.2 shows the derived coordinates of the each foot point using the above equations and conditions. This result shows the designed leg can perform the stepping motion which is needed to move the multi-legged microrobot.

Table 1.2. Coordinates of each foot point.

Foot point	θ_A and θ_{Foot}	Coordinates
(x_1, y_1)	90° and 10°	(-1248.6, -3440.3)
(x_2, y_2)	90° and 0°	(-632.3, -3604.9)
(x_3, y_3)	80° and 0°	(227.2, -3249.4)
(x_4, y_4)	80° and 10°	(-340.5, -3239.5)

Fig. 1.4 shows the measurement method for the required force F_S to actuate the leg. The required force can describe by the Equation (1.9):

$$F_s = M_W\, g,　\qquad (1.9)$$

where M_W is the mass of the weight and g is the gravity acceleration. The authors vary the mass of the weight to find the minimum weight to actuate the leg. Fig. 1.4(a) is the required force for the push motion.

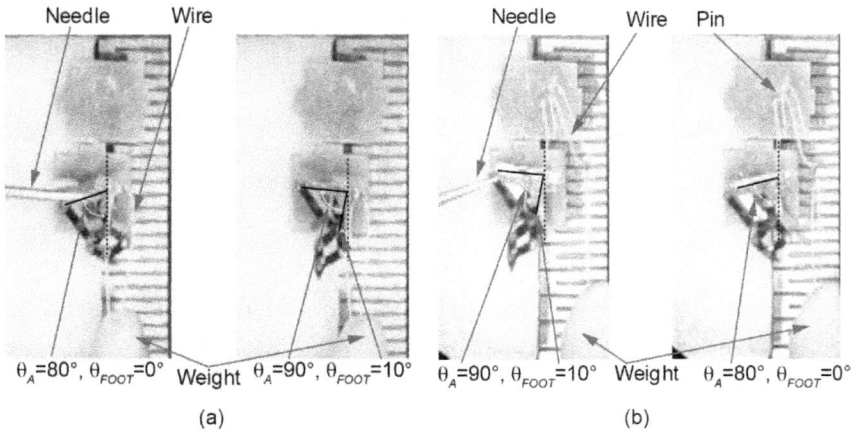

(a)　　　　　　　　(b)

Fig. 1.4. Measurement method for the required force to actuate the leg
for (a) push motion; (b) pull motion.

The weight was attached to point P of bar 1 using a wire. The required force of 542 μN is measured in order to move the leg to the regular position under the lightest weight of 55.3 mg. Fig. 1.4 (b) shows the required force for the pull motion. The weight was attached to the point P of bar 1 using a wire through the pin and the lightest weight was 36.2 mg, while the force for this pull motion was 355 μN.

1.3. Electrostatic Inchworm Motors

As an alternative low-power means of actuation, electrostatic inchworm motors can be used to drive the legs of the microrobot. MEMS electrostatic inchworm motors are based on capacitively driven gap-closing actuators (GCA) working in tandem to displace a shuttle linearly at over 100 μN force output without any static current [19].

The authors used an angled-arm design based on work from [20]. In this design, the GCAs use an attached angled-arm to impact a central shuttle and move it in a preferential direction. The motors have a gap size of 2.1 μm and each step of the motor moves the shuttle by 1 μm. Each GCA has 70 fingers, totalling 140 fingers for each actuation step. The inchworm motor chiplet measures a total area of approximately 2.2 mm × 2.5 mm. The electrostatic inchworm motors are fabricated in a 3-mask silicon-on-insulator (SOI) process. The SOI wafers had a 40 μm device layer, 2 μm buried oxide and 550 μm handle wafer. A layer of 100 nm-thick aluminium is deposited on the device-layer silicon to define the contact pads. The device layer silicon is etched to form the structure of the motors using DRIE. A backside etch is then performed to reduce the mass and release the singulated chiplets from the substrate.

Fig. 1.5 shows the force output of an electrostatic inchworm motor. Force measurements are taken using a serpentine spring assembly attached to the motor shuttle. The serpentine assembly has a spring constant of 18.5 N/m. By measuring the displacement of the inchworm shuttle, we can relate this to the force output of the motor. The solid line highlights the analytical calculation of the force output. We can see that at 60 V we get an average force output of over 1 mN from 5 measured devices. The original angled-arm inchworm motors shown in [20] were able to generate 1.88 mN at 110 V. Previous work has shown 500 μN of force at 60 V [21] while the newly fabricated devices have demonstrated 1 mN of force at 60 V. Discrepancies between the analytical model and the measured values can be attributed to unaccounted lateral etching of the silicon sidewalls. This can increase the effective finger gap size and change the spring constants of the springs.

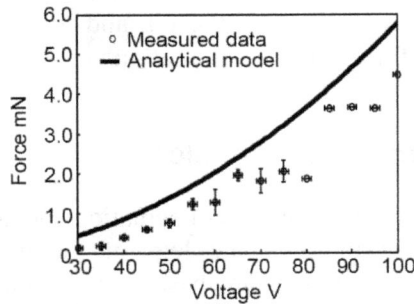

Fig. 1.5. The raw force output of the inchworm motor used in these experiments.

1.4. High-Voltage Si PV Cells

Fig. 1.6 shows the fabricated high-voltage Si PV cell array. The PV cell array was designed in an area of about 3 mm square. The device was made by CMOS post-process dry release and device isolation method. The array consists of 125 PV cells connected in series and each cell has a p-diffusion layer on n-well. The details of the design and process method are shown in reference [22].

In the reference [22], the light source of the PV cell array was a red LED with 30 mA current. The open circuit voltage (VOC) was 57.9 V, from which we can deduce that the open circuit voltage of each cell was about 0.46 V on average. The short circuit current (ISC) was 976 nA. The maximum power (Pmax) was 43.3 µW, where the voltage was 53.2 V and the current was 683 µA. The fill factor (FF; FF = Pmax/VOC ISC) was 76.7 %. The FF generally indicates the quality of the pn-junctions (a high fill factor means a high quality of the junction) and the value 76.7 % is relatively high. This high value was achieved by using a commercial CMOS process performed by a foundry. However, the maximum power in the reference [22] was not high enough to actuate the electrostatic inchworm motors. The authors changed the light source to a xenon lamp with 5 A current to achieve VOC=60.0 V and ISC=105 µA. Fig. 1.7 shows the I-V characteristics of the PV cell array lighted with the xenon lamp. The xenon lamp irradiated the PV cell from a distance of 10 cm. This result shows that the xenon lamp can produce 30 times the power shown in the reference [22], large enough to actuate the electrostatic inchworm motors.

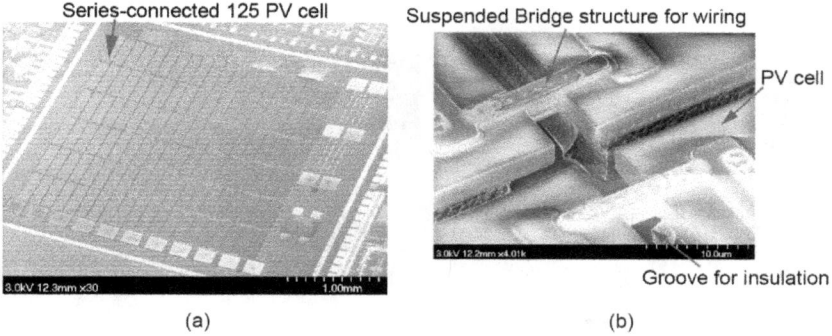

Fig. 1.6. Fabricated PV cell array [22]. (a) Whole view; (b) Magnified view.

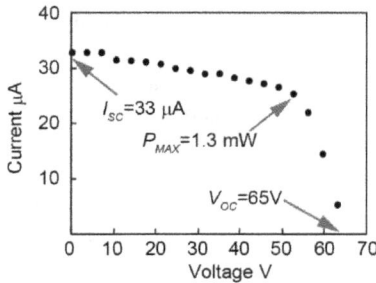

Fig. 1.7. I-V curve of 125-cell PV array (Light source: xenon lamp).

1.5. Experimental Results

Fig. 1.8(a) shows an inchworm motor chip. The image highlights the ring meant to engage to a complimentary post on the leg, the gap closing actuators, the reset spring, and the signal pads that receive signals from the drive circuit. The inchworm motor is fabricated in the 3-mask process described in Section 1.3. This motor has the force profile shown in Fig. 1.5. Fig. 1.8 (b) highlights the methodology of integration of the inchworm chip with the leg. The motor is taped onto a platform off of a micromanipulator stage. This is because aligning the ring with the post and interfacing the parts require careful precision. Once the leg engagement ring is in place on the leg post, a set of probes are dropped onto the motor contact pads to provide the electrical signals from the circuit needed to drive the motor. This circuit is powered by the solar cell.

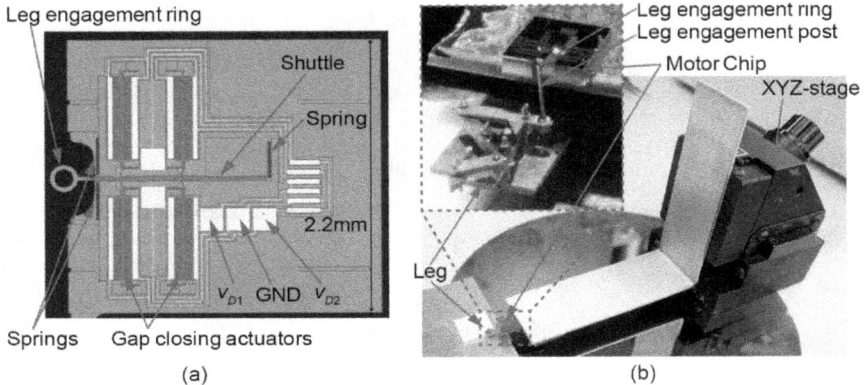

Fig. 1.8. Details of the inchworm motor chip integration (a) A micrograph of the motor chip highlighting the leg engagement ring, gap closing actuators, shuttle, spring, and the electrical contact pads. The pads are driven with probes that are connected to the circuit that is driven by the solar cells; (b) A diagram of the experimental setup showing the leg engagement post, meant to interface with the leg engagement ring. The motor chip is held on a platform on a micromanipulator stage and the ring is maneuvered around the post. Once the motor is in place, the probes are dropped onto the motor chip to drive the leg.

Fig. 1.9 shows the actuation experimental setup of the electrostatic inchworm motors using PV cell array (Fig. 1.9 (a)). The anode-side of PV cell array was connected to the solid resistor at the collector of the transistor. In other words, the generated voltage by the PV cell, V_{PV}, was used as the voltage source of the circuit. The Arduino was used for switching the transistor for generating the driving waveform v_{D1} and v_{D2} for the electrostatic inchworm motors (Fig. 1.9(b)). The driving waveforms were two offset 60 V amplitude 500 Hz square waves, one for each of the GCAs of the motor.

Fig. 1.10 shows the generated force of the electrostatic inchworm motors. The force gauge system was attached on the shuttle and the scale of the force gauge was characterized as 1 dot = 0 μN, 2 dot = 370 μN, 3 dot = 740 μN. The guideline is attached to the shuttle to point the dot. The result in Fig. 1.11 shows that the guideline points the 3 dot. This result shows the generated force was 740 μN according to the gauge system, which is high enough to actuate the leg of the microrobot.

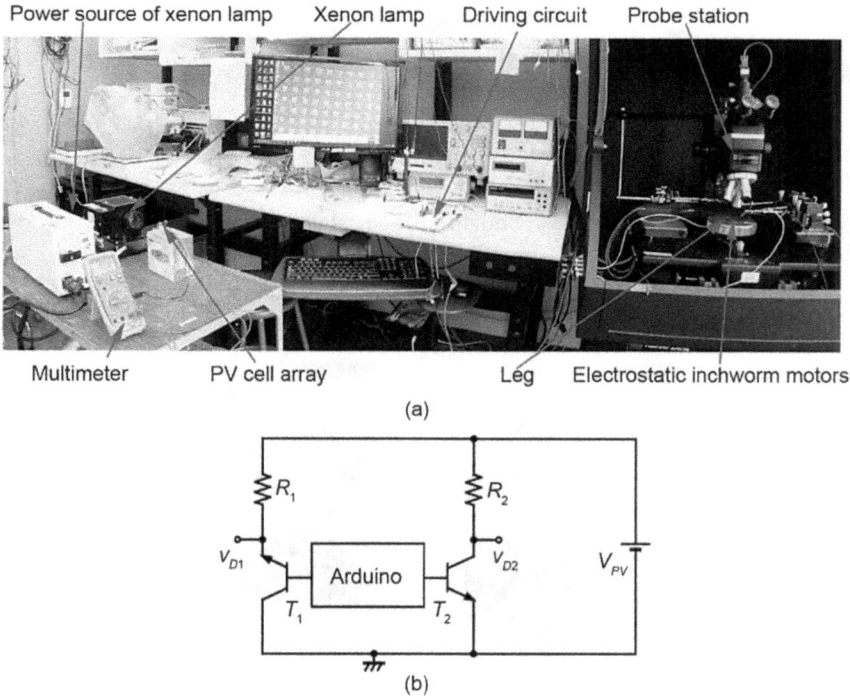

Fig. 1.9. Actuation experimental setup: (a) Whole setup; (b) Circuit diagram of driver circuit.

Fig. 1.11 shows the actuation of the leg using electrostatic inchworm motors. The ring structure was attached to the shuttle of the electrostatic inchworm motors using the method described above. The electrostatic inchworm motors was connected to the leg through the shaft of point P. The result in Fig. 1.11 shows that the electrostatic inchworm motors produced about 250 μm in displacement to move the leg of the microrobot. However, the pull motion was not enough to actuate the leg from (x_4, y_4) to (x_1, y_1). This is because spring was designed to generate the 250 μN pull motion. The pull motion needs 355 μN to complete the motion. The strength of the spring needs for the future examination.

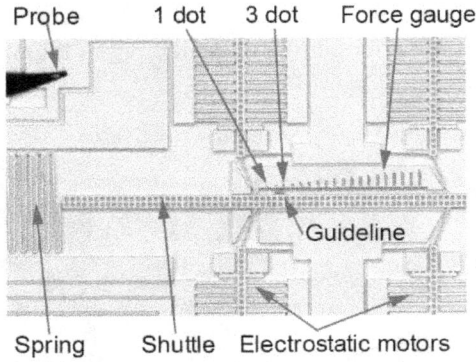

Fig. 1.10. Generated force measurement of the electrostatic inchworm motors.

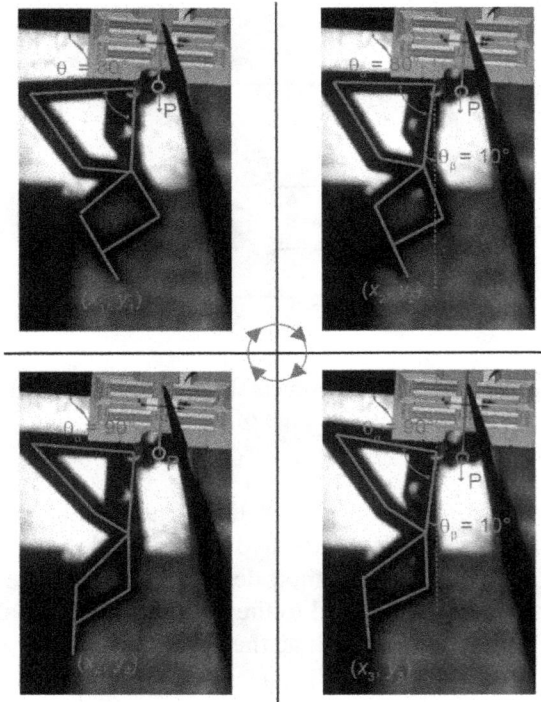

Fig. 1.11. Actuation of leg: (a) Pull motion; (b) Push motion.

1.6. Conclusions

In this chapter, the electrostatic actuator with low energy consumption is powered by a 3 mm × 3 mm Si photovoltaic cells with an output voltage of 60 Volts. The generated force of the electrostatic inchworm motors was 740 µN to actuate the leg of the microrobot. The leg of the microrobot could move using the electrostatic inchworm motors with proper driving waveforms for large displacements. In the future, the authors will design the millimeter scale locomotive robot with Si PV cell driven electrostatic inchworm motors.

Acknowledgements

The fabrication of the microrobot was supported by Research Center for Micro Functional Devices, Nihon University. Fabrication of the inchworm motors was supported by the UC Berkeley Marvell Nanofabrication Laboratory. The authors would like to acknowledge the Berkeley Sensor and Actuator Center and the UC Berkeley Swarm Lab for their continued support. VLSI Design and Education Center (VDEC), the University of Tokyo (UTokyo) and Phenitec Semiconductor are acknowledged for CMOS-SOI wafer fabrication. Japanese Ministry of Education, Sports, Culture, Science and Technology (MEXT) is acknowledged for financial support through Nanotechnology Platform to UTokyo VDEC used for PV cell post-process.

References

[1]. Ebefors T., Mattsson J. U., Kälvesten E., Stemme G., A Walking Silicon Micro-robot, in *Proceedings of the 10th Int. Conference on Solid-State Sensors and Actuators (TRANSDUCERS' 99)*, Sendai, Japan, 1999, pp. 1202-1205.

[2]. Hollar S., Flynn A., Bellew C., Pister K. S. J., Solar powered 10 mg silicon robot, in *Proceedings of the IEEE Sixteenth Annual International Conference on Micro Electro Mechanical Systems*, Kyoto, Japan, 2002, pp. 706-711.

[3]. Ryu J., Jeong Y., Tak Y., Kim B., Kim B., Park J., A ciliary motion based 8-legged walking micro robot using cast IPMC actuators, in *Proceedings of the International Symposium on Micromechatronics and Human Science*, 2002, pp. 85-91.

[4]. Donald B. R., Levey C. G., McGray C. D., Paprotny I., Rus D., An Untethered, Electrostatic, Globally Controllable MEMS Micro-Robot, *Journal of Microelectromechanical Systems*, Vol. 15, No. 1, 2006, pp. 1-15.

[5]. Hoover M. A., Steltz E., Fearing S. R., RoACH: An autonomous 2.4 g crawling hexapod robot, in *Proceedings of the IEEE/RSJ International Conference on Intelligent Robots and Systems*, Nice, France, 22–26 September 2008, pp. 26–33.

[6]. Kernbach S., Kernbach O., Collective energy homeostasis in a large-scale microrobotic swarm, *Robotics and Autonomous Systems*, Vol. 59, 2011, pp. 1090-1101.

[7]. Wood R. J., Finio B., Karpelson M., Ma K., Pérez-Arancibia N. O., Sreetharan P. S., Tanaka H., Whitney J. P., Progress on "Pico" Air Vehicles, *The International Journal of Robotics*, Vol. 31, No. 11, 2012, pp. 1292-1302.

[8]. Donald B. R., Levey C. G., Paprotny I., Rus D., Planning and control for microassembly of structures composed of stress-engineered MEMS microrobots, *The International Journal of Robotics Research*, Vol. 32, No. 2, 2013, pp. 218–246.

[9]. Rubenstein M., Cornejo A., Nagpal R., Programmable self-assembly in a thousand-robot swarm, *Science*, Vol. 345, No. 6198, 15 Aug. 2014, pp. 795-799.

[10]. Jinhong Qu J., Oldham K. R., Multiple-Mode Dynamic Model for Piezoelectric Micro-Robot Walking, in *Proceedings of the 21st Design for Manufacturing and the Life Cycle Conference and 10th International Conference on Micro- and Nanosystems*, Vol. 4, 2016, Paper No. DETC2016-59621.

[11]. Vogtmann D., Pierre R. S., Bergbreiter S., A 25 MG Magnetically Actuated Microrobot Walking at > 5 Body Lengths/sec, in *Proceedings of the IEEE 30th International Conference on Micro Electro Mechanical Systems*, Las Vegas, NV, USA, 2017, pp. 179-182.

[12]. Rahmer J., Stehning C., Gleich B., Spatially selective remote magnetic actuation of identical helical micromachines, *Sci. Robot.*, Vol. 2, 2017.

[13]. Abbott J. J., Nagy Z., Beyeler F., Nelson B. J., Robotics in the Small, Part I: Microbotics, *IEEE Robotics & Automation Magazine*, Vol. 14, No. 2, 2007, pp. 92-103.

[14]. Cho K., Wood R., Biomimetic robots, *Springer International Publishing*, Cham, Switzerland, Chap. 23, 2016.

[15]. Fearing R. S., Powering 3 Dimensional Microrobots: Power Density Limitations, in *Proceedings of the IEEE International Conference on Robotics and Automation, Tutorial on Micro Mechatronics and Micro Robotics*, 1998.

[16]. Bell D. J., Lu T. J., Fleck N. A., Spearing S. M., MEMS actuators and sensors: observations on their performance and selection for purpose, *Journal of Micromechanics and Microengineering*, Vol. 15, No. 7, 2005, pp. S153-S164.

[17]. Saito K., Maezumi K., Naito Y., Hidaka T., Iwata K., Okane Y., Oku H., Takato M., Uchikoba F., Neural Networks Integrated Circuit for Biomimetics MEMS Microrobot, *Robotics*, Vol. 3, 2014, pp. 235-246.

[18]. Tanaka D., Uchiumi Y., Kawamura S., Takato M., Saito K., Uchikoba F., Four-leg independent mechanism for MEMS microrobot, *Artificial Life and Robotics*, Vol. 22, No. 3, September 2017, pp 380–384.

[19]. Yeh R., Hollar S., Pister K. S. J., Single mask, large force and large displacement electrostatic linear inchworm motors, *Journal of Microelectromechanical Systems*, Vol. 11, No. 4, 2002, pp. 330-336.

[20]. Penskiy I., Bergbreiter S., Optimized electrostatic inchworm motors using a flexible driving arm, *Journal of Micromechanics and Microengineering*, Vol. 23, No. 1, 2012, pp. 1-12.

[21]. Contreras D. S., Drew D. S., Pister K. S. J., First steps of a millimeter-scale walking silicon robot, in *Proceedings of the 19th Int. Conference on Solid-State Sensors, Actuators and Microsystems*, Kaohsiung, Taiwan, 2017.

[22]. Mori I., Kubota M., Lebrasseur E., Mita Y., Remote power feed and control of MEMS with 58 V silicon photovoltaic cell made by a CMOS post-process dry release and device isolation method, in *Proceedings of the Symposium on Design, Test, Integration & Packaging of MEMS/MOEMS (DTIP'14)*, Cannes, France, 1-4 April 2014.

Chapter 2

Adaptive Trajectory Tracking Control and Dynamic Redundancy Resolution of Nonholonomic Mobile Manipulators

Mohamed Boukattaya

2.1. Introduction

A mobile manipulators consists of a mobile platform carrying a robotic manipulators. Such system combines the locomotion and manipulation abilities to perform complex tasks. To solve the tracking control problems, many nonlinear control algorithms have been proposed, such as nonlinear feedback control [1-2], input-output decoupling control [3-4], task space and null space decoupling control [5-6], computed torque control [7], etc... The drawback of the aforementioned schemes is their dependence on the precise knowledge of the complex dynamics of the system which is difficult to obtain is practical applications. In order to overcome the effects of uncertainties and external disturbances, many advanced nonlinear controllers are considered such as, neural network control [8-10], fuzzy control [11-13], backstepping control [14] and sliding- mode control [15-16]. In [8], an adaptive Neural Network (NN) based controls for the arm and the base is proposed for the joint-space position control of a mobile manipulator. Each NN control output comprises a linear control term and a compensation term for parameter uncertainty and disturbances. In [9], a control strategy for duct cleaning robot in the presence of uncertainties and various disturbances is proposed which combines the advantages of neural network technique and advanced adaptive robust theory. In [10], an adaptive recurrent neural network controller is proposed to deal with the unmodelled system dynamics. The proposed control strategy guarantees that the system

Mohamed Boukattaya
Laboratory of Sciences and Techniques of Automatic Control & Computer
Engineering (Lab-STA), Sfax, Tunisia

motion asymptotically converges to the desired manifold while the constraint force remains bounded. In [11], a trajectory tracking controller for mobile manipulators with model uncertainties and external disturbances is presented. It is composed by an adaptive estimator for the trajectory tracking and a fuzzy controller to avoid collision with obstacles. In [12], an adaptive position tracking system and a force control strategy for nonholonomic mobile manipulator robot, which combine the merits of Recurrent Fuzzy Wavelet Neural Networks (RFWNNs). The design of the adaptive online learning algorithms approximates unknown dynamics without the requirement of prior controlled system information. In [14], an adaptive sliding mode controller based on the backstepping method is applied to the robust trajectory tracking of the wheeled mobile manipulator. The control algorithm rests on adopting the backstepping method to improve the global ultimate asymptotic stability and applying the sliding mode control to obtain a high response and invariability to uncertainties. In [15], the authors propose a sliding mode adaptive neural network controller for the trajectory following of nonholonomic mobile manipulators in task space. Multi-layered perceptron (MLP) is used to approximate the dynamic model and a sliding-mode control with a direct adaptive technique are combined together to suppress bounded disturbances and modelling errors caused by parameter uncertainties. In [16], a robust tracking controller is proposed for the trajectory tracking problem of a dual-arm wheeled mobile manipulator. The design of the controller is divided into two levels, a backstepping controller for the kinematic level and a sliding-mode equivalent controller, composed of neural network control, robust scheme and proportional control, for the dynamic level. The proposed control law can deals with inadequate modelling and parameter uncertainties.

However, the most of the previous research works have designed a complicated and computationally expensive control algorithm. Moreover, the mobile manipulator is controlled in generalized coordinates and the problem of redundancy is not treated or resolved. In order to overcome these difficulties, a simple adaptive control law will be proposed for the trajectory tracking of mobile manipulator in task space. The redundancy is resolved by using a suitable extended task space dynamics and the unknown robot parameters and external disturbances are estimated and compensated using adaptive update laws. The control design and the stability analysis are performed via Lyapunov theory. The main contributions of this work are listed as follows:

1) Adaptive control law is presented for nonholonomic mobile manipulators operating in task space, under external disturbances and dynamics uncertainties.

2) The proposed controller does not rely on precise knowledge of dynamic parameters and can suppress the effects of external disturbances.

3) Estimation and compensation of unknown dynamics and external disturbances is realized using only joint position and joint velocity measurements which are available by simple encoders.

The rest of this chapter is organized as follows. The mobile manipulators subject to nonholonomic constraints is described in Section 2.2. Section 2.3 presents the extended formulation for resolving the problem of redundancy. The main results of control design are presented in Section 2.4. Simulation studies that illustrate the effectiveness of the proposed task control are presented in Section 2.5. Concluding remarks are given in Section 2.6.

2.2. System Description

Nonholonomic mobile manipulators (Fig. 2.1) is composed by an n DOF robotic manipulators mounted on a nonholonomic differential mobile robot. The navigation is achieved by two driving wheels mounted on the same axis.

Using Euler-Lagrange formulation, the dynamics equation of a n-DOF nonholonomic mobile manipulators can be expressed in the generalized form as

$$M(q)\ddot{q} + C(q,\dot{q})\dot{q} + G(q) + A^T(q)\lambda = B(q)(\tau + \tau_d),$$
$$A(q)\dot{q} = 0 \qquad (2.1)$$

where q, $\dot{q}$ and $\ddot{q} \in \mathfrak{R}^n$ are the vector of generalized position, generalized velocity and the generalized acceleration, respectively, $M(q) \in \mathfrak{R}^{n \times n}$ is the inertia matrix which is symmetric and positive definite, $C(q,\dot{q}) \in \mathfrak{R}^{n \times n}$ represents the centripetal- Coriolis matrix, $G(q) \in \mathfrak{R}^n$ is the gravitational force vector, $A(q) \in \mathfrak{R}^{m \times n}$ in the constrained matrix, $\lambda \in \mathfrak{R}^m$ is the vector of the constraint force,

35

$B(q) \in \Re^{n\times(n-m)}$ is the input transformation matrix, $\tau_d \in \Re^{n-m}$ is the vector of external disturbance, and $\tau \in \Re^{n-m}$ denotes the input torque vector.

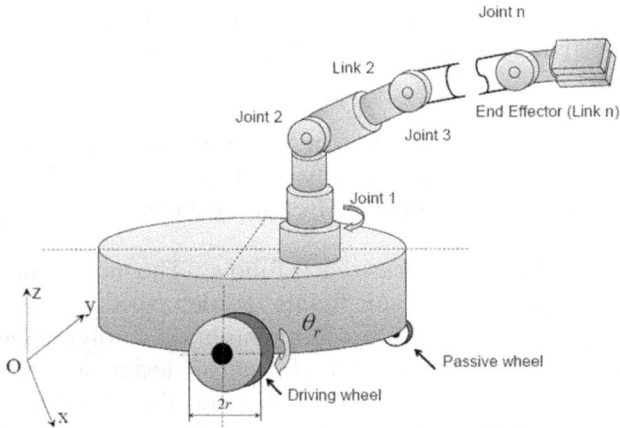

Fig. 2.1. Description of the nonholonomic mobile manipulators.

It is possible to find a smooth and linearly independent vector $S(q) \in \Re^{n-m}$ satisfying the following relation

$$S^T(q)A^T(q) = 0 \qquad (2.2)$$

Moreover, Equation (2.2) implies the existence of joint velocity vector $\dot{\theta} \in \Re^{n-m}$, such that:

$$\dot{q} = S(q)\dot{\theta} \qquad (2.3)$$

The derivative of (2.3) can be written as

$$\ddot{q} = S(q)\ddot{\theta} + \dot{S}(q)\dot{\theta} \qquad (2.4)$$

Equation (2.3) represents the kinematic model of the nonholonomic mobile manipulators in which the vector of joint velocity can be expressed as

$$\dot{\theta} = \begin{bmatrix} \dot{\theta}_r & \dot{\theta}_l & \dot{\theta}_1 & \cdots & \dot{\theta}_{n_b} \end{bmatrix}^T, \qquad (2.5)$$

where $\dot{\theta}_r$ and $\dot{\theta}_l$ are the angular velocities of driving right and the left wheel, respectively, and $\dot{\theta}_{n_j}$ is the angular velocity of the jth joint manipulators for $j = 1, ..., n_b$.

By substituting (2.3) and (2.4) into (2.1) and then left multiplying S^T, the dynamic of the nonholonomic mobile manipulators can be expressed in joint space as follows:

$$M_\theta(q)\ddot{\theta} + C_\theta(q,\dot{q})\dot{\theta} + G_\theta(q) = B_\theta(q)(\tau + \tau_d), \qquad (2.6)$$

where $M_\theta = S^T M S$, $C_\theta = S^T(CS + M\dot{S})$, $G_\theta = S^T G$ and $B_\theta(q) = S^T B$.

Remark 2.1:

The dynamic equation in joint space (2.6) reflects only the feasible motions allowable by satisfying the constraints.

Remark 2.2:

Generally, the DOF of the robot is superior to the DOF of the task performed by the end-effector. In this situation, the robot is called kinematically redundant and there exists an infinity of configuration for the same task.

2.3. Redundancy Resolution by Extended Formulation

For a redundant nonholomic mobile manipulators, we assume that the task x performed by the end-effector is described by the following a nonlinear function

$$x = f(q) \qquad (2.7)$$

Taking the derivative of (2.7) with respect to time and using (2.3), one has

$$\dot{x} = \frac{\partial f(q)}{\partial q}\dot{q} = \frac{\partial f(q)}{\partial q}S(q)\dot{\theta} = J_1(q)\dot{\theta}, \qquad (2.8)$$

where $J_1(q) \in R^{s \times r}$ is the non-invertible Jacobian matrix which describes the relation between r-dimensional joint velocity vector $\dot{\theta} \in \Re^r$

($r = n - m$) and the s-dimensional end-effector task velocity vector $\dot{x} \in \mathfrak{R}^s$.

For the purpose of redundancy resolution, we use a unified dynamic extended approach. It uses a set of p-desired additional task $y \in \mathfrak{R}^p$ that will be performed due to the redundancy, one obtains

$$y = g(q) \qquad (2.9)$$

Taking the time derivative of (2.9), it follows

$$\dot{y} = \frac{\partial g(q)}{\partial q}\dot{q} = \frac{\partial g(q)}{\partial q}S(q)\dot{\theta} = J_2(q)\dot{\theta} \qquad (2.10)$$

By combining (2.8) and (2.10), we obtain the extended differential kinematic model of the redundant mobile manipulators as follows

$$\dot{X} = \begin{bmatrix} \dot{x} \\ \dot{y} \end{bmatrix} = \begin{bmatrix} J_1(q) \\ J_2(q) \end{bmatrix}\dot{\theta} = J(q)\dot{\theta}, \qquad (2.11)$$

where $J(q) \in \mathfrak{R}^{m \times m}$ is the extended Jacobian matrix which is supposed to be square and invertible.

Hence, from (2.11), we can obtain

$$\begin{aligned} \dot{\theta} &= J^{-1}\dot{X} \\ \ddot{\theta} &= J^{-1}\ddot{X} - J^{-1}\dot{J}J^{-1}\dot{X} \end{aligned} \qquad (2.12)$$

Now, substituting (2.12) into (2.6) and left multiplying J^{-T}, one can obtain the dynamic of the nonholomic mobile manipulators in the extended task space as

$$M_x(q)\ddot{X} + C_x(q,\dot{q})\dot{X} + G_x(q) = B_x(q)(\tau + \tau_d), \qquad (2.13)$$

where

$$M_x(q) = J^{-T}(q)M_\theta(q)J^{-1}(q);$$
$$C_x(q,\dot{q}) = J^{-T}(q)\left[C_\theta(q,\dot{q}) - M_\theta(q)J^{-1}(q)\dot{J}(q)\right]J^{-1}(q);$$

$$G_x(q) = J^{-T}(q)G_\theta(q);$$
$$B_x(q) = J^{-T}(q)B_\theta(q);$$

Remark 2.3:

The control law with an extended formulation for the redundancy resolution of the nonholonomic mobile manipulators is developed in dynamic level which is suitable for practical applications.

The dynamic model of the mobile manipulators in the extended task space has the following beneficial properties:

Property 2.1: $M_x(q)$ is a symmetric and positive definite matrix, which is upper and lower bounded, that is

$$m_1 \le \|M_x(q)\| \le m_2, \tag{2.14}$$

where m_1 and m_2 are the positive scalar constants and $\|.\|$ is the standard Euclidian norm.

Property 2.2: $\dot{M}_x - 2C_x$ is skew-symmetric, that is to say

$$x^T(\dot{M}_x - 2C_x)x = 0, \quad \forall x \in \mathfrak{R}^{n-m} \tag{2.15}$$

Property 2.3: For any differential vector χ:

$$M_x(q)\ddot{\chi} + C_x(q,\dot{q})\dot{\chi} + G_x(q) = Y(q,\dot{q},\dot{\chi},\ddot{\chi})\Psi, \quad \forall \dot{\chi}, \ddot{\chi} \in \mathfrak{R}^{n-m}, \tag{2.16}$$

where $\Psi \in \mathfrak{R}^l$ is the vector of unknown parameters and $Y \in \mathfrak{R}^{(n-m) \times l}$ is so-called "regressor" matrix and contains known parameters.

2.4. Control Design

In this section, two strategies are proposed for the trajectory tacking control of nonholonomic mobile manipulators: a) Passive Control (PC) under the assumption that the dynamics of the mobile manipulators are known and the external disturbances are neglected; b) Adaptive Passive Control (APC) under the assumption that the dynamics parameters of the mobile manipulators are totally unknown and external disturbances may occur.

Let $X_d(t)$ be the extended desired trajectory in the task space and $\dot{X}_d(t)$ and $\ddot{X}_d(t)$ be the desired extended velocity and acceleration. Define the tracking errors as

$$
\begin{aligned}
e(t) &= X_d(t) - X(t) \\
\dot{e}(t) &= \dot{X}_d(t) - \dot{X}(t)
\end{aligned}
\tag{2.17}
$$

For the controller design, define the following new variables:

$$
\begin{aligned}
\dot{X}_r &= \dot{X}_d + \Lambda e \\
\ddot{X}_r &= \ddot{X}_d + \Lambda \dot{e}
\end{aligned}
\tag{2.18}
$$

where Λ is the positive definite matrix.

Define also the filtered tracking error as

$$
s = \dot{X}_r - \dot{X} = \dot{e} + \Lambda e
\tag{2.19}
$$

Before the controller design, we make the following assumptions:

Assumption 2.1:

The desired extended trajectory $X_d(t)$ and their first and second derivatives $\dot{X}_d(t)$, $\ddot{X}_d(t)$ are bounded.

Assumption 2.2:

The parameters of the mobile manipulators are in known compact sets. Thus, the matrices M_x, C_x, G_x, B_x and E_x are all bounded.

Assumption 2.3:

The external disturbances $\tau_d(t)$ is bounded, that is $\|\tau_d(t)\| \le d$, where d is a positive constant.

Assumption 2.4:

We suppose that the unknown dynamic parameters $\Psi(t)$ and the external disturbances $\tau_d(t)$ change slowly, that is $\lim_{t\to\infty}\|\dot{\Psi}(t)\|=0$ and $\lim_{t\to\infty}\|\dot{\tau}_d(t)\|=0$.

2.4.1. Passive Control (PC) Design

The design procedure of PC law can be summarized in the following theorem:

Theorem 2.1:

Consider a nonholonomic mobile manipulators described by (2.1), (2.3), (2.11) and (2.13). Assume that the dynamic parameters are known and the external disturbances are neglected $(\tau_d(t) = 0)$. If the following PC law is applied:

$$\tau(t) = B_x^{-1}(q)\left(M_x(q)\ddot{X}_r + C_x(q,\dot{q})\dot{X}_r + G_x(q) - Ks\right), \quad (2.20)$$

where K is the positive gain matrix.

Then, the tracking errors e and $\dot{e}$ asymptotically converge to zero as $t \to \infty$. Furthermore, all the closed-loop signals in the system are bounded.

Proof:

By substituting the control law (2.20) into the dynamics (2.13), one can obtain the closed-loop dynamic as follows:

$$M_x\dot{s} + (C_x + K)s = 0 \qquad (2.21)$$

Now, consider the Lyapunov candidate function as:

$$V = \frac{1}{2}s^T M_x s \qquad (2.22)$$

The time derivative of V can be obtained as:

$$\dot{V} = s^T (M_x \dot{s} + \frac{1}{2} \dot{M}_x s) \qquad (2.23)$$

If we substitute the closed-loop dynamic (2.21) in the above equation, we have:

$$\dot{V} = s^T (-Ks + \frac{1}{2} s(\dot{M}_x - 2C_x)) \qquad (2.24)$$

The skew-symmetric property 2.2, yields:

$$\dot{V} = -s^T Ks \leq 0 \qquad (2.25)$$

We can conclude from the Lyapunov theory that $V(t)$ is bounded. Hence, s is bounded and it follows that e, $\dot{e}$ are also bounded. From assumption 2.1 it is easy to show that the signals X, $\dot{X}$, $\dot{X}_r$, $\ddot{X}_r$ are all bounded. Moreover, assumption 2.2 can be utilized to show that the control law (2.20) is also bounded. That is to say, all the signals in the system are bounded. Furthermore, from the closed loop dynamics (2.21), we can conclude that $\dot{s}$ is bounded. Finally, we can apply Barbalat's Lemma to conclude that $s \rightarrow 0$ as $t \rightarrow \infty$. It follows that e and $\dot{e} \rightarrow 0$ as $t \rightarrow \infty$. That is all for the proof.

2.4.2. Adaptive Passive Control (APC) Design

The design procedure of APC law can be summarized in the following theorem:

Theorem 2.2:

Consider a nonholonomic mobile manipulators described by (2.1), (2.3), (2.11) and (2.13). Assume that the dynamic parameters are unknown and external disturbances are applied to the system $(\tau_d(t) \neq 0)$. If the following APC law is applied

$$\tau(t) = B_x^{-1}(q)\left(Y(q, \dot{q}, \dot{x}_r, \ddot{x}_r)\hat{\Psi} - Ks - B_x(q)\hat{\tau}_d\right), \qquad (2.26)$$

where $\hat{\Psi}$ and $\hat{\tau}_d$ are the estimates of the dynamic uncertainties Ψ and external disturbance τ_d, respectively which are updated by the following adaptive laws

$$\dot{\hat{\Psi}} = -K_u^{-T} Y^T s, \qquad (2.27)$$

$$\dot{\hat{\tau}}_d = K_d^{-T} B_x^T s, \qquad (2.28)$$

where K, K_u and K_d are the positive gain matrixes.

Then, the tracking errors e and $\dot{e}$ asymptotically converge to zero as $t \to \infty$. Furthermore, all the closed-loop signals in the system are bounded.

Proof:

In practice applications, the dynamic parameters of a mobile manipulator are partially known or totally unknown. Moreover, external disturbances may occur. For this case, consider $\hat{M}_x$, $\hat{C}_x$ and $\hat{G}_x$ the estimated values of M_x, C_x and G_x respectively. Hence, the control law in (2.20) is modified as

$$\tau(t) = \hat{B}_x^{-1}(q)\left(\hat{M}_x(q)\ddot{X}_r + \hat{C}_x(q,\dot{q})\dot{X}_r + \hat{G}_x(q) - Ks - B_x(q)\hat{\tau}_d\right) \quad (2.29)$$

Using the regression property 2.3, the above control law is transformed as

$$\tau_x = \hat{B}_x^{-1}(q)\left(Y(q,\dot{q},\dot{X}_r,\ddot{X}_r)\hat{\Psi} - Ks - B_x(q)\hat{\tau}_d\right), \qquad (2.30)$$

where $\hat{\Psi}$ is the estimate vector of the parameter uncertainties Ψ. The closed-loop dynamic is obtained by substituting the control law (2.26) into the dynamic model (2.13), we have

$$M_x\dot{s} + V_x s + Ks = Y(q,\dot{q},\dot{X}_r,\ddot{X}_r)\tilde{\Psi} - B_x(q)\tilde{\tau}_d, \qquad (2.31)$$

where $\tilde{\Psi} = \hat{\Psi} - \Psi$ is the dynamic uncertainties error, which is the difference between the estimated and the actual value.

Now, consider the Lyapunov candidate function as:

$$V = \frac{1}{2}s^T M_x s + \frac{1}{2}\tilde{\tau}_d^T K_d \tilde{\tau}_d + \frac{1}{2}\tilde{\Psi}^T K_u \tilde{\Psi} \qquad (2.32)$$

The time derivative of V can be obtained as:

$$\dot{V} = s^T(M_x\dot{s} + \frac{1}{2}\dot{M}_x s) + \tilde{\dot{\tau}}_d^T K_d \tilde{\tau}_d + \tilde{\dot{\Psi}}^T K_u \tilde{\Psi} \qquad (2.33)$$

Substituting the closed-loop dynamic (2.31) in the above equation, we have:

$$\dot{V} = s^T(-Ks + \frac{1}{2}s(\dot{M}_x - 2C_x)) + (\tilde{\dot{\tau}}_d^T K_d - s^T B_x)\tilde{\tau}_d + (\tilde{\dot{\Psi}}^T K_u + s^T Y)\tilde{\Psi} \quad (2.34)$$

Applying, the skew-symmetric property 2.2, one can obtain

$$\dot{V} = -s^T Ks + (\tilde{\dot{\tau}}_d^T K_d - s^T B_x)\tilde{\tau}_d + (\tilde{\dot{\Psi}}^T K_u + s^T Y)\tilde{\Psi} \qquad (2.35)$$

If we substitute the adaptation laws (2.27) and (2.28) with the consideration of assumption 2.4, we have:

$$\dot{V} = -s^T Ks \leq 0 \qquad (2.36)$$

We can conclude from the Lyapunov theory that $V(t)$ is bounded. Hence, s is bounded and it follows that e, $\dot{e}$ are also bounded. From assumption 2.1 it is easy to show that the signals X, $\dot{X}$, $\dot{X}_r$, $\ddot{X}_r$ are all bounded. Furthermore, due to the boundedness of the external disturbances (assumption 2.3), we can deduce that $\hat{\tau}_d$ is also bounded. Moreover, from assumption 2.2, it is trivial to show that the regression matrix Ψ is bounded and similarly it estimates value $\hat{\Psi}$ is also bounded. Now, assumption 2.2 can be utilized to show that the control law (2.26) with adaptive update laws (2.27) and (2.28) are all bounded. That is to say, all the signals in the system are bounded. Furthermore, from the closed loop dynamics (2.31), we can conclude that $\dot{s}$ is bounded. Finally, we can apply Barbalat's Lemma to conclude that $s \to 0$ as $t \to \infty$. It follows that e and $\dot{e} \to 0$ as $t \to \infty$. That is all for the proof.

Remark 2.4:

The estimation of the parameters uncertainties and the external disturbances are achieved using only joint position and velocity measurement which are available by simple encoders.

2.5. Simulation Results

To verify the effectiveness of the proposed adaptive control, let us consider a planar 2 DOF mobile manipulators system shown in Fig. 2.2.

The parameters of the robot are assumed to be:

$b = 0.182\,m,\ r = 0.0508\,m,\ d = 0.116\,m,\ L_a = 0.1\,m,\ L_1 = 0.514\,m,$
$L_{11} = 0.252\,m,\ L_2 = 0.362\,m,\ L_{22} = 0.243\,m,\ m_b = 17.25\,kg,$
$m_w = 0.159\,kg,\ m_1 = 2.56\,kg,\ m_2 = 1.07\,kg,\ I_b = 0.297\,kg.m^2,$
$I_w = 0.0002\,kg.m^2,\ I_1 = 0.148\,kg.m^2,\ I_2 = 0.0228\,kg.m^2.$

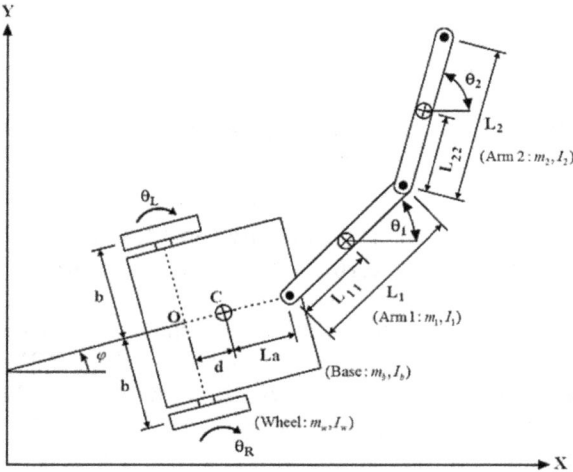

Fig. 2.2. Schematic of a planar 2 DOF mobile manipulators.

The desired trajectories are selected as:

$$X_d(t) = \begin{bmatrix} 0.2t + 0.3 \\ 0.5 + 0.25\sin(0.2\pi t) \\ 0.2t \\ 0 \end{bmatrix}$$

It consists of a sinusoidal path for the end-effector and a straight line path for the mobile platform.

We also define $q = \begin{bmatrix} x_C & y_C & \varphi & \theta_R & \theta_L & \theta_1 & \theta_2 \end{bmatrix}^T$ as a vector of the generalized coordinate and $\dot{\theta} = \begin{bmatrix} \dot{\theta}_R & \dot{\theta}_L & \dot{\theta}_1 & \dot{\theta}_2 \end{bmatrix}^T$ as a vector of the joints velocities.

Hence, the initial conditions are given as follow:

$$q(0) = \begin{bmatrix} 0 & 0 & \dfrac{\pi}{2} & 0 & 0 & \dfrac{\pi}{4} & \dfrac{-\pi}{4} \end{bmatrix}^T \text{ and } \dot{\theta}(0) = \begin{bmatrix} 0 & 0 & 0 & 0 \end{bmatrix}^T.$$

To illustrate the robustness of the proposed method, two differents simulations were carried out. In the first simulation, a parametric variant of 20 % and a time varying sinusoidal disturbances are introduced into the system. In the second simulation, a parametric variant of 20 % and a time varying rectangular pulse disturbances are introduced into the system.

In the two cases, a parametric variant of 20 % indicates that the nominal physical parameters used in the APC are 20 % less than used in the model. For the two experiments, the gains *are* selected as: $K = 100$, $\Lambda = 2$, $K_d = 0.001$ and $K_u = 2$.

The tracking performances are illustrated in Figs. 2.3–2.6 for the first simulation and in Figs. 2.7–2.10 for the second simulation.

It can be seen that under uncertainties and sinusoidal disturbances, the robot tracked the desired trajectories correctly (Fig. 2.3), and after a transient due to errors in the initial condition, both the end-effector and the platform tracking errors tend to zero (Fig. 2.4). This good performance can be observed for the case of rectangular pulse disturbances (Fig. 2.7 and Fig. 2.8). Therefore, the proposed APC scheme is robust to both the plant uncertainty and time varying external disturbance. Moreover, the APC produces smooth control commands with slow variation (Fig. 2.6 and Fig. 2.10). One can deduce that the input profiles are mainly related to the applied disturbances, to reject its undesirable effects. From Fig. 2.5 and Fig. 2.9, it is clear that the estimation process using APC is performed very well where the deviation of estimated disturbance from the actual disturbance is very small. Then the estimated disturbance vector is utilized to compensate for the external disturbance effect in order to achieve a better trajectory tracking performance.

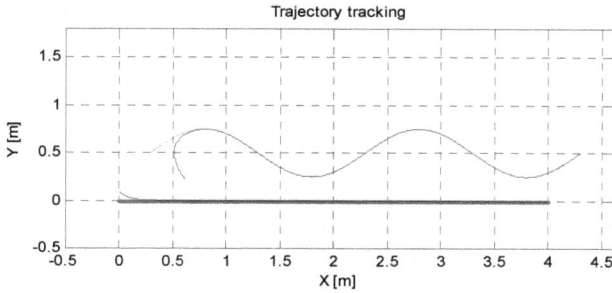

Fig. 2.3. Trajectory tracking control using APC under uncertainties and sinusoidal disturbances.

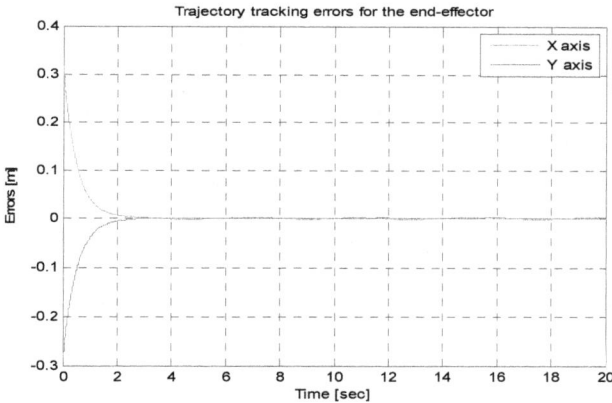

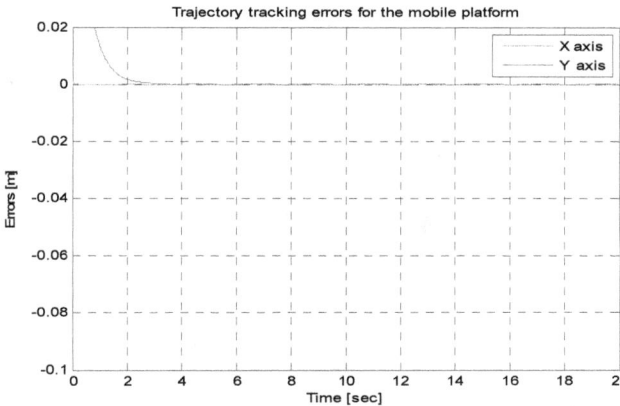

Fig. 2.4. Tracking errors using APC under uncertainties and sinusoidal disturbances (a) for the end-effector; (b) for the mobile platform.

47

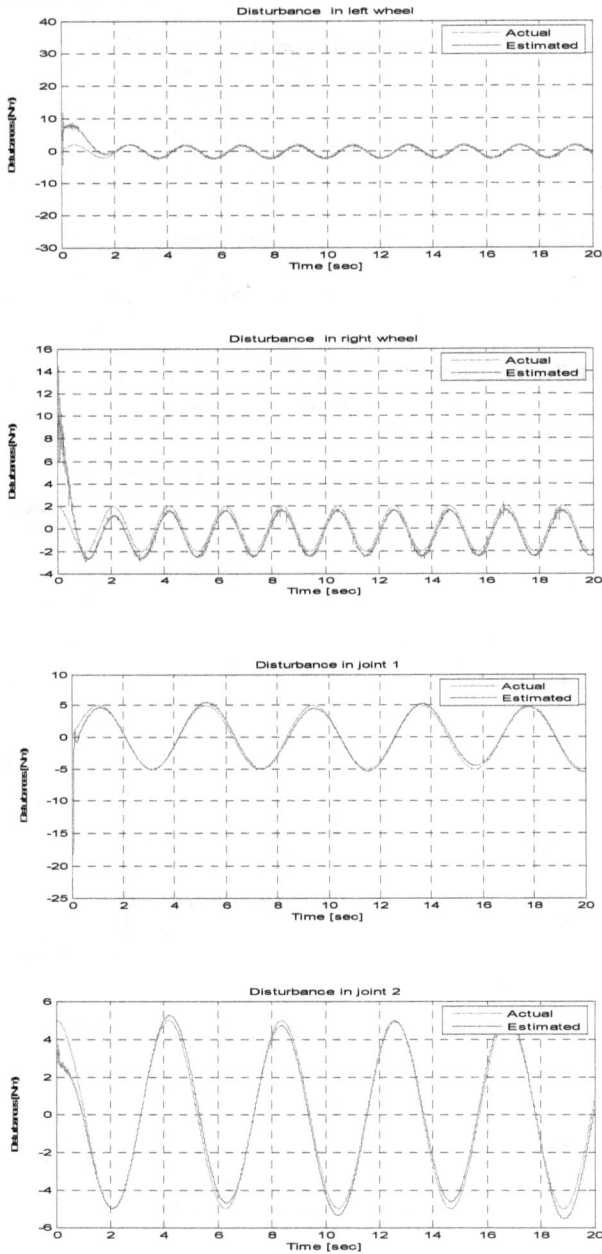

Fig. 2.5. Estimation of sinusoidal disturbances using APC (a) in the left wheel; (b) in the right wheel; (c) in joint 1 and (d) in joint 2.

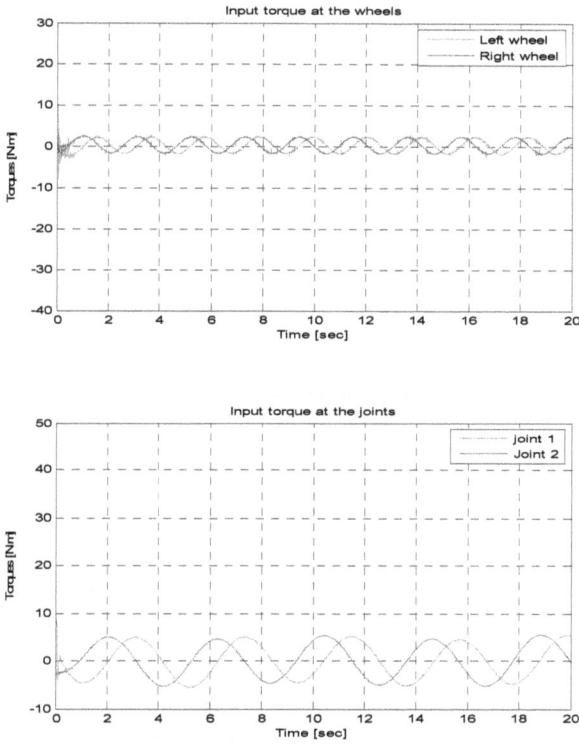

Fig. 2.6. Input torques using APC under uncertainties and sinusoidal disturbances (a) at the wheels, (b) at the joints.

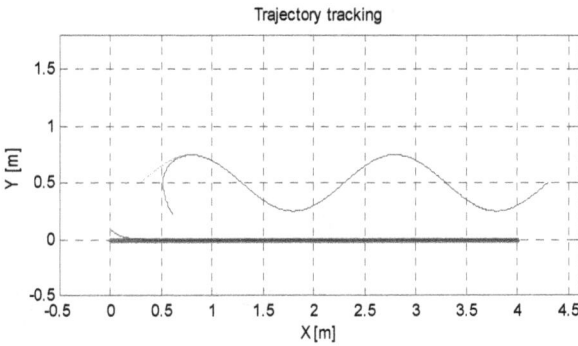

Fig. 2.7. Trajectory tracking control using APC under uncertainties and rectangular pulse disturbances.

49

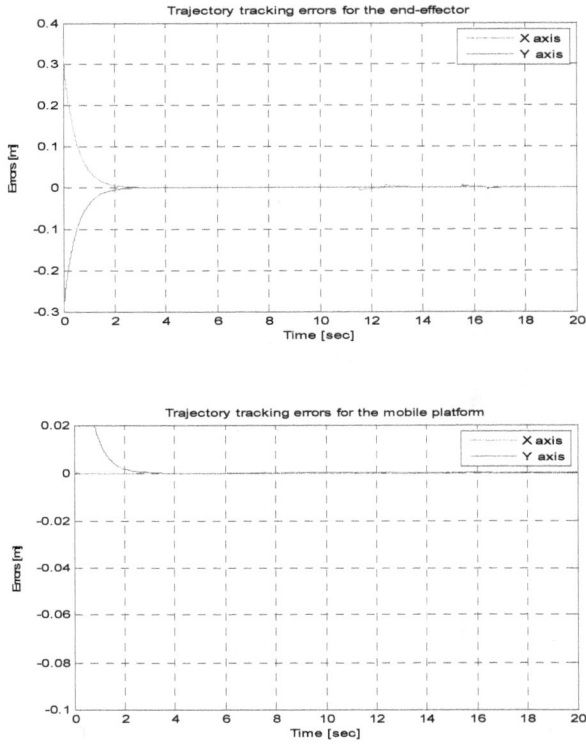

Fig. 2.8. Tracking errors using APC under uncertainties and rectangular pulse disturbances (a) for the end-effector, (b) for the mobile platform.

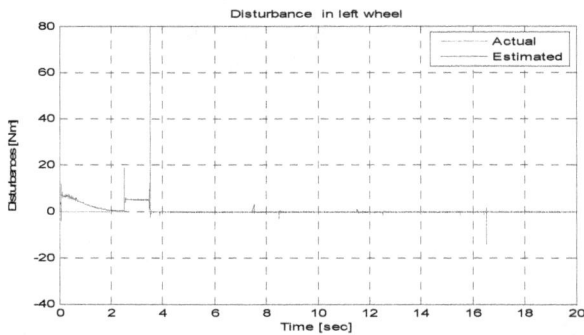

Fig. 2.9 (a). Estimation of rectangular pulse disturbances using APC (a) in left wheel.

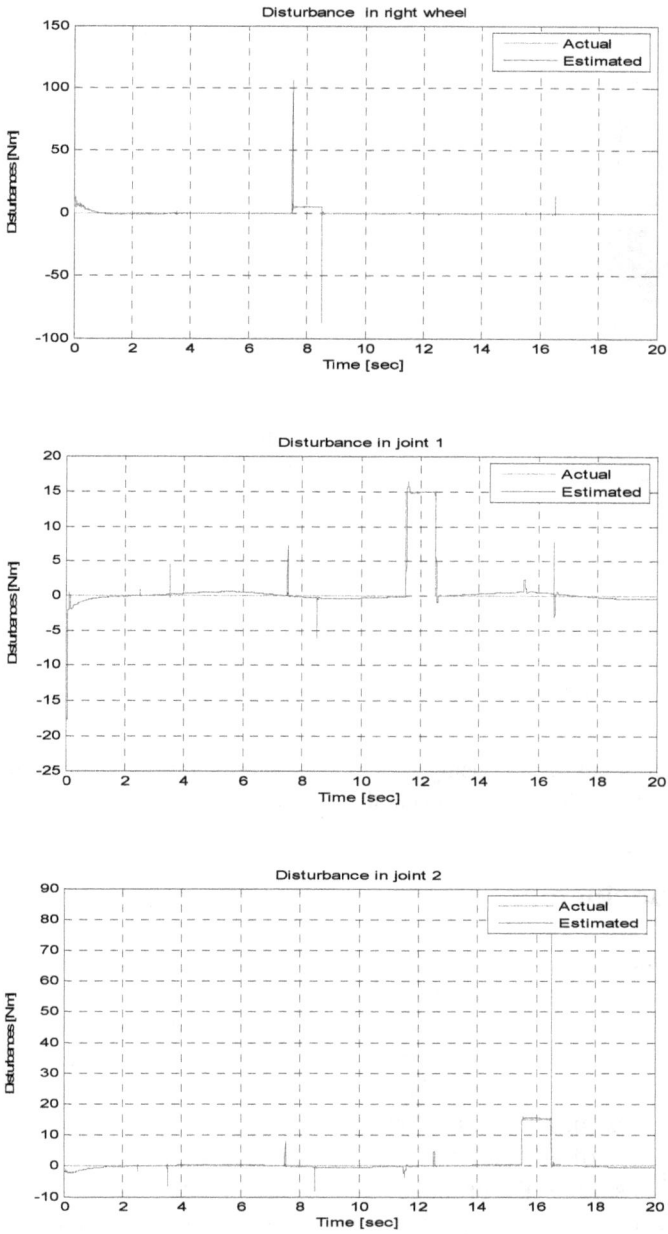

Fig. 2.9 (b-d). Estimation of rectangular pulse disturbances using APC (b) in right wheel, (c) in joint 1, (d) in joint 2.

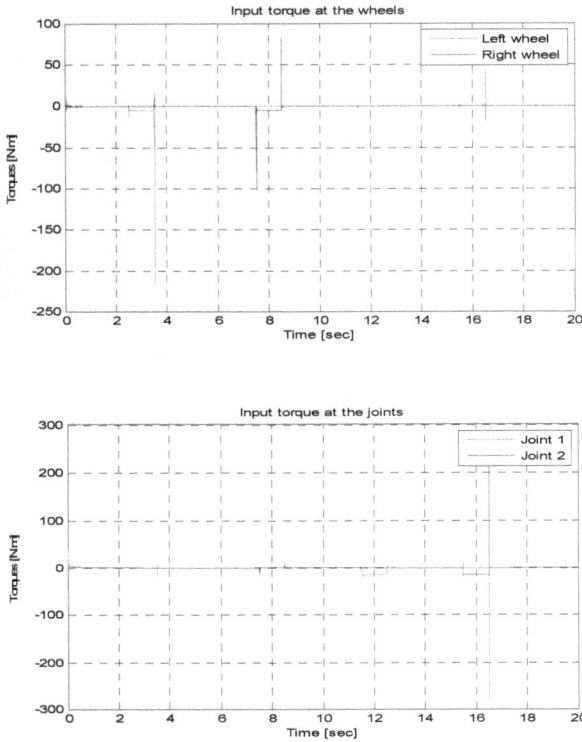

Fig. 2.10. Input torques using APC under uncertainties and rectangular pulse
disturbances (a) at the wheels, (b) at the joints.

2.6. Conclusions

In this chapter, effective adaptive control strategy has been presented to
control a noholonomic mobile manipulators in the presence of
uncertainties and disturbances. By considering an extended approach,
the desired trajectories are formulated in task space. Moreover, adaptive
updates laws can effectively estimate and compensate the effects of
uncertainties and external disturbances. The system stability and the
boundedness of all the signals are proved using Lyapunov synthesis. In
simulation studies, the proposed control method demonstrated robust and
effective control performance with good disturbance rejection. As a
future work, the proposed adaptive robust controller will be proven
experimentally and will be also applied to other kinds of electro-
mechanical systems.

Acknowledgements

We thank the ministry of higher education and scientific research of Tunisia for funding this work.

References

[1]. Y. Yamamoto, X. Yun, Effect of the Dynamic Interaction on Coordinated Control of Mobile Manipulators, *IEEE Transaction on Automatic Control*, Vol. 12, No. 5, 1996, pp. 816-824.

[2]. X. Yun, Y. Yamamoto, Stability Analysis of the Internal Dynamics of a Wheeled Mobile Robot, *Journal of Robotic Systems*, Vol. 14, No. 10, 1997, pp. 697-709.

[3]. A. Mazur, New approach to designing input-output decoupling controllers for mobile manipulators, *Bulletin of the Polish Academy of Science Technical Sciences*, Vol. 53, No. 1, 2005, pp. 31-37.

[4]. A. Mazur, D. Szakiel, On path following control of nonholonomic mobile manipulators, *International Journal of Applied Mathematics and Computer Science*, Vol. 19, No. 4, 2009, pp. 561–574.

[5]. G. White, R. Bhatt, V. Krovi, Dynamic redundancy Resolution in a Nonholonomic Wheeled mobile manipulator, *Robotica*, Vol. 25, No. 2, March 2007, pp. 147-156.

[6]. G. White, R. Bhatt, C. Tang, V. Krovi, Experimental Evaluation of Dynamic Redundancy Resolution in a Nonholonomic Wheeled Mobile Manipulator, *IEEE/ASME Transactions on Mechatronics*, Vol. 14, No. 3, 2009, pp. 349-357.

[7]. E. Papadopoulos, J. Poulakakis, Planning and Model Based Control for Mobile Manipulators, in *Proceedings of the IEEE/RSJ International Conference on Intelligent Robots and Systems (IROS'2000)*, Vol. 3, October 2000, pp. 1810-1815.

[8]. S. Lin, A. A. Goldenberg, Neural Network Control of Mobile Manipulator, *IEEE Transactions on Neural Networks*, Vol. 12, No. 5, 2001, pp. 1121-1133.

[9]. B. Wei, Y. Hongshan, W. Cong, Z. Hui, Adaptive Robust Control based on RBF Neural Networks for Duct Cleaning Robot, *International Journal of Control, Automation, and Systems*, Vol. 13, No. 2, 2015, pp. 475-487.

[10]. Z. P. Wang, T. Zhou, Y. Mao, Q. J. Chen, Adaptive recurrent neural network control of uncertain constrained nonholonomic mobile manipulators, *International Journal of Systems Science*, Vol. 45, 2014, pp. 133-134.

[11]. A. Karray, M. Njah, M. Feki, M. Jallouli, Intelligent mobile manipulator navigation using hybrid adaptive-fuzzy controller, *Computers and Electrical Engineering*, Vol. 56, 2016, pp. 773-783.

[12]. Y. Wang, T. Mai, J. Mao, Adaptive motion/force control strategy for non-holonomic mobile manipulator robot using recurrent fuzzy wavelet neural networks, *Engineering Applications of Artificial Intelligence*, Vol. 34, 2014, pp. 137-153.

[13]. M. Long, W. Nan, Adaptive Position Tracking System and Force Control Strategy for Mobile Robot Manipulators Using Fuzzy Wavelet Neural Networks, *Journal of Intelligent and Robotic Systems*, Vol. 79, No. 2, 2015, pp. 175-195.

[14]. N. Chen, F. Song, G. Li, X. Sun, C. Ai, An adaptive sliding mode backstepping control for the mobile manipulator with nonholonomic constraints, *Communications in Nonlinear Science and Numerical Simulation*, Vol. 18, No. 10, 2013, pp. 2885-2899.

[15]. Y. Lui, Y. Li, Sliding Mode Adaptive Neural Network Control for Nonholonomic Mobile Modular Manipulators, *Journal of Intelligent and Robotic Systems*, Vol. 44, No. 3, November 2005, pp. 203-224.

[16]. C. Tsai, M. Cheng, S. Li, Robust Tracking Control for Wheeled Mobile Manipulator with Dual Arms using Hybrid Sliding Mode Neural Network, *Asian Journal of Control*, Vol. 9, No. 4, December 2007, pp. 377-389.

Chapter 3

An Automated On-line Novel Visual Percept Detection Method for Mobile Robot and Video Surveillance

Xiaochun Wang, Xia Li Wang and D. Mitchell Wilkes

3.1. Overview

In this chapter, we present a fast approximate nearest neighbor search tree based novelty filter for mobile robotic and video surveillance applications. We begin with an introduction to the concept of novelty detection in general and an image-patch-based perceptual learning system as a basis for visual novelty detection in specific. The proposed on-line novel visual percept detection method is next presented. Finally, the performance of the proposed filter is compared with that of the well-known Grow-When-Required neural network approach for a novelty detection task in an indoor environment, and with that of efficient support vector data description method for a novelty detection task in an outdoor environment.

3.2. Introduction

Because of our persistent interest in novelty, the human species has been able to succeed in this world. Novelty detection can be defined as the task of classifying test data that differ in some respect from the data that are available during training. Being one of the fundamental requirements of a good classification or identification system, novelty detection techniques can identify new or unknown data or signals that a machine learning system is not aware of at the time of training the model.

In the novelty detection problems, a set of data indicative of normal system behaviours is given and the goal is to learn a description of

Xiaochun Wang
Xi'an Jiaotong University, Xi'an, Shaanxi, China

normality by constructing a model of it, $M(\theta)$ (where θ represents the free parameters of the model), from the data under the assumption that the data used to train the learning system constitute the basis to build a model of normality. A relatively few number of previously unseen patterns x are then compared with the model of normality and each often assigned a novelty score $z(x)$, which may or may not be probabilistic. Large novelty scores $z(x)$ should correspond to increased "abnormality" with respect to the model of normality. Deviations from normality are then detected according to a decision boundary that is usually referred to as the novelty threshold $z(x) = T$. By comparing the scores to the novelty threshold, those test data for which the threshold is exceeded are then deemed to be "abnormal". In other words, a novelty threshold $z(x) = T$ is defined such that x is classified "normal" if $z(x)<T$, or "abnormal" otherwise.

Different types of models for constructing M, methods for setting their parameters θ, and methods for determining novelty thresholds T have been proposed in the literature. Markou and Singh distinguished between two main categories of novelty detection techniques: statistical approaches [1] and neural network based approaches [2]. In statistical approaches, the statistical properties of data are used to estimate whether a new test data point comes from the same distribution as the training data points or not. Generally, there are two main statistical novelty detection techniques, parametric modeling, which assumes that the data come from a family of known distributions, and nonparametric modeling, in which no assumption on the form of the data distribution is made. The former is somewhat efficient in practice, but may often be limited by the assumptions about the distributions, while the latter is more flexible but often more computationally expensive, especially for large data sets. Neural network-based approaches come from a wide range of flexible non-linear regression and classification models, data reduction models, and non-linear dynamical models that make no a priori assumptions on the properties of data, demand a relatively small number of parameters for optimization, and therefore, have been extensively used for novelty detection [3-4]. However, neural networks have increased computational complexity in training and typically cannot be retrained to learn new units with computational efficiency in both time and space comparable to that of statistical models, particularly when the dimensionality of the data is high.

Novelty detection has gained much research attention in application domains, including the detection of mass-like structures in mammograms [5] and other medical diagnostic problems [6-7], faults and failure detection in complex industrial systems [8], structural damage [9], intrusions in electronic security systems, credit card or mobile phone fraud detection [10-11], video surveillance [12-13], mobile robotics [14-15], sensor networks [16], astronomy catalogues [17-18] and text mining [19]. In mobile robotic and video surveillance applications, novelty detection has been extensively employed to recognizing novel objects in images and video streams. It is deemed a very important task and gaining attention because of the availability of large amounts of video data and because of the lack of automated methods for extracting important details from such media.

Early work in neural network based on-line novelty detection for mobile robots was conducted by Marsland et al. through a Grow-When-Required (GWR) neural network [20]. The GWR network is a self-organizing learning mechanism which is able to determine whether an input is novel or not through the use of a model of habituation. Capable of continuous learning, this unsupervised approach was used in mobile robots having sonar readings [21]. In 2004, this GWR network was successfully applied to visual novelty detection based on local color histograms [22-23]. To scale to high-dimensional visual data (such as visual inputs acquired from a moving platform), more recently in 2007, Neto and Nehmzow proposed to use an incremental Principle Component Analysis to reduce the dimensionality of input to the novelty filter while trying to preserve discriminability between different classes of features as much as possible [24]. Their approach combined both unsupervised continuous learning and color vision with no explicit image segmentation or restrictions in field of view.

Being another popular neural network-based approach, support vector data description (SVDD) [25] uses a minimum-volume sphere to enclose all or most of the data, either in the feature space directly or by replacing the dot products between patterns with the corresponding kernel functions. The novelty detection is performed by using the obtained boundary as the decision boundary to determine whether a test sample is novel or not. Although it is specially designed for handling large datasets, the computation cost scales linearly with the number of support vectors. To improve the prediction speed, a fast SVDD (F-SVDD) method was proposed in 2010 [26] to adopt a novel preimage method to approximate the sphere center of SVDD as a single vector instead of a linear combination of support vectors. Noticing that the decision hypersphere of F-SVDD becomes a sphere in the input space using the given Gaussian kernel, which is not suitable for many real world cases, Peng and Xu proposed an efficient SVDD (E-SVDD) [27]. Based on the observation that most real datasets contain at most a few clusters, E-SVDD assigns each cluster a unique preimage and then uses these preimages to represent the sphere center of SVDD, improving the prediction performance of SVDD while maintaining the computation efficiency.

Different metrics are used to evaluate the effectiveness and efficiency of novelty detection methods. The effectiveness of novelty detection techniques can be evaluated according to how many novel data points are correctly identified as well as according to how many normal data are incorrectly classified as novel data. The latter is also known as the false alarm rate. Novelty detection techniques should aim to have a high detection rate while keeping the false alarm rate low. Generally speaking, it is in this sense that the novelty threshold is regarded as one of the most critical parameters in novelty detection tasks and directly affects the false negative rate. The efficiency of novelty detection approaches is evaluated according to both time and space complexity. In addition, depending on the specific novelty detection task, the amount of memory required to implement the technique is typically considered to be an important performance evaluation metric. Efficient novelty detection techniques should be scalable to large and high-dimensional datasets.

In this chapter, we present a fast approximate nearest neighbor search tree based novelty filter for mobile robotic and video surveillance applications. First, it identifies novel samples using statistically obtained thresholds. Next, these samples are accumulated and evaluated whether they represent truly novel patterns and form dense clusters in both the feature space and the image space. By this way, the novelty detection problem becomes the identification of novel percepts in the segmented images. Finally, the proposed on-line novelty detection mechanism completes the learning process by inserting the verified novel samples into the search tree of image vocabulary percepts. One important contribution of the proposed method is its significant efficiency in the tree construction over Vocabulary tree (VT) [28] for image patch-based highly sparse high dimensional data while maintaining reasonably good search capability. Another important contribution of the proposed method is the automated strategy for setting the novelty threshold. Finally, the proposed novelty detection strategy incorporates a capability for automatically learning multiple newly discovered objects by using spatial information to cluster the detected novel samples into different categories and by inserting them into the tree on-line to allow for fast incremental updating of the knowledge base.

To demonstrate the soundness of this manuscript, the performance of the proposed filter is compared with that of the well-known GWR neural network approach, for a novelty detection task in an indoor environment, and with that of the E-SVDD method for a novelty detection task in an outdoor environment. The ultimate goal of the research presented here is to relieve human operators by equipping autonomous robots with such automatic inspection capability.

3.3. A Percept Learning System

Novel object detection is an important application of novelty detection in both robotics and computer vision. Due to the lack of examples from the novel targets, model selection in such scenarios is a very challenging research topic [29] and requires a synthesis of techniques from computer vision, machine learning and pattern recognition. In the following, an image-patch-based perceptual learning system is first described.

A primary goal of human perceptual system is to recognize and locate objects and to keep track of them as new images are captured. Perceptual learning is the ability to construct compact representations of sensory events based on their statistical properties in the perceptual level as opposed to the behaviour or cognitive level [30-31]. However, not having human's sophisticated visual system, the robot perceives the world differently and there is no guarantee that the robot vision system can easily detect or recognize objects defined by human designers. As a result, to operate in realistic environments and to recognize objects in a given image, a robot must have the ability to form percepts on its own by natural association among features of sensory information, and, based on that, to segment the image into nonoverlapping but meaningful regions, denoting reliable objects [32]. With high spatial frequencies being filtered out, image segmentation reduces the amount of storage and, when done well, makes behaviour resistant to some loss of information. The approach proposed in this chapter is based on a computer vision system inspired loosely by neuro-biological evidences [33-34], a simplified description of which is given in the following.

3.3.1. Feature Generation

Color is an identifying feature that is local and largely independent of view and resolution. However, the color of a single pixel has little information by itself and may even contain only noise. If the color pixels in a simple $N \times N$ patch of an image are considered, the histogram of these N^2 colors in the patch can describe a color pattern which may not only contain considerable more information but also potentially be less sensitive to noise. Also, this color histogram description remains relatively independent of translation and rotation, and is also somewhat independent of scale such as the change of distance to the object [35]. For texture-based measure, Gabor filters have been frequently used in image processing due to their match with the receptive fields of primate's V1 simple cells. As a result, the combination of color histogram and Gabor texture measure is a good appearance feature to use for segmenting images and identifying objects.

To obtain feature vectors from small patches of an image in this research, as shown in Fig. 3.1, a moving window of size $N \times N$ (pixels) is used which is shifted by M pixels in the row and column directions. The window shift controls the resolution of the result and should not exceed the border of the image.

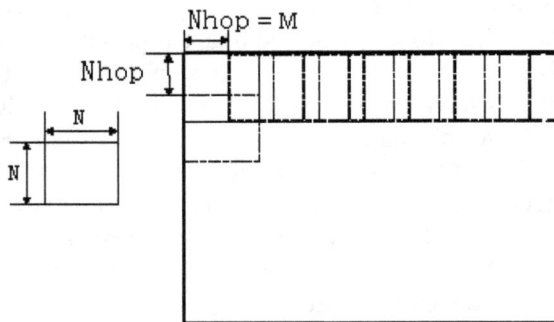

Fig. 3.1. Image segmentation model.

The images and videos taken from a camcorder are stored in RGB color space. To be more robust to lighting conditions, the RGB color images are converted to the HSV (Hue, Saturation, Value) color space for use [36]. The conversion formula is given by,

$$H = \begin{cases} (0+\dfrac{G-B}{MAX-MIN})\times 60 & if \quad R=MAX, \\ (2+\dfrac{B-R}{MAX-MIN})\times 60 & if \quad G=MAX, \\ (4+\dfrac{R-G}{MAX-MIN})\times 60 & if \quad B=MAX, \end{cases} \quad S=\dfrac{MAX-MIN}{MAX}, \quad V=MAX \qquad (3.1)$$

where MAX is the maximum value of (R, G, B), and MIN is the minimum. Hue describes each color by a normalized number in the range [0.0 ... 1.0], starting at red and cycling through yellow, green, cyan, blue, magenta, and back to red. Saturation represents the purity of a color such as the "redness" of red. Value describes the brightness of the color. To construct color features, a histogram of color measurements in the HSV space for each moving window (i.e., each image patch) in an image is computed as follows. The hues are broken evenly into 100 bins. The saturations and values are evenly distributed into 10 bins each ranging from 0.0 to 1.0. Each color can be represented by combining the three bins, one hue bin, one saturation bin and one value bin. All possibilities of the combinations equal 10,000 different color bins for the histogram. The total number of pixels in the selected region divides these 10,000 numbers, resulting in a highly sparse feature vector. The conjunction of the very high-dimensional feature space and small image patches allows for the image patches to be efficiently represented as sparse vectors. And the use of a sparse histogram representation for the color space makes the calculations largely independent of the feature space dimension while taking advantage of the high-dimensional properties.

First introduced by Gabor [37], complex Gabor filters are complex exponentials with a Gaussian envelope and provide a mathematical approximation of the spatial receptive field of a simple cell in primate's visual area V1 as,

$$G(x, y, \theta, \lambda, \varphi, \sigma, \gamma) = e^{-\frac{\tilde{x}^2 + \gamma^2 \tilde{y}^2}{2\sigma^2}} e^{j(2\pi \frac{\tilde{x}}{\lambda} + \varphi)} \qquad (3.2)$$

where $\tilde{x} = x\cos\theta + y\sin\theta$, $\tilde{y} = y\cos\theta - x\sin\theta$, θ specifies the orientation of the filter, λ specifies the wavelength of the sine and cosine wave and determines the spacing of light and dark bars that produces the maximum response, φ specifies the phase of the sine and cosine wave and determines where the ON-OFF boundaries fall within the receptive field, ó specifies the radius of the Gaussian, and finally, ã specifies the aspect ratio of the Gaussian. In this work, the following set of parameters are used, è∈{0, π/8, 2π/8, 3π/8, 4π/8, 5π/8, 6π/8, 7π/8}; ë∈{ $\sqrt{2}$, 2$\sqrt{2}$, 3$\sqrt{2}$, 4$\sqrt{2}$, 5$\sqrt{2}$ }, ö ∈{0, π/2}, ó = ë, and, finally, ã = 1, resulting in totally 80 sine and cosine modulated Gaussian filters. Given each image, the convolutions of itself with each pair of Gabor filters are calculated as preparation for the generation of the orientation feature. Then the magnitude of each complex Gabor filter is calculated and the convolution results within each $N \times N$ image patch, as used in color histogram generation, are averaged to obtain 40 texture measures, which are appended to the color histogram to complete the formation of the feature vector. The use of local color histogram and Gabor filters as descriptors is quite standard and the values of some of the parameters used here have been justified [38].

3.3.2. Similarity Measure

With the set of features being defined, the next step is to construct a multidimensional space of perceptual similarities and to ascribe to each physical stimulus a particular position within this feature space. In this process, two major issues are to set up an appropriate similarity measure and to design an algorithm to cluster the input patterns based on the adopted similarity measure. It is desired that the similarity measure is given in numerical form. In a K-dimensional space, the most widely used mathematical models for similarities are L_1-norm and L_2-norm, which are obtained by setting $r = 1$ or 2, respectively, in the following Minkowski power formula,

$$d_{ij} = \left(\sum_{k=1}^{K} \left| x_{ik} - x_{jk} \right|^r \right)^{1/r} \tag{3.3}$$

In the multidimensional scaling literature, these two similarity measures are also called the City-block distance and the Euclidean distance, and are believed to represent two types of processing. As indicated by [39-41], for unitary or holistic stimuli, such as the hue, saturation and brightness of colors, the closest approximation to an invariant relation between data and distances has uniformly been achieved in a space endowed with the familiar Euclidean distance; on the other hand, for analyzable or separable stimuli, such as the size and brightness difference (i.e., orientation), the closes approach to invariance between data and distances has generally been achieved with the City-block distance. To implement these suggestions, the Euclidean distance is used for the n dimensional HSV color histogram component of the feature vector and the City-block distance is used for the m dimensional texture measure component of the feature vector. That is,

$$d_{ij} = \sqrt{\sum_{c=1}^{n} \left(x_{ic} - x_{jc}\right)^2} + \sum_{t=1}^{m} | x_{it} - x_{jt} | \qquad (3.4)$$

3.3.3. Percept Formation

Each feature vector appears as a point in the feature space and patterns pertaining to different classes will fall into different regions of the feature space. The learning process is to partition the feature space into sensible clusters based on some properties in common. In the proposed research, a minimum spanning tree (MST) based clustering method is used. Being a graphical analysis of the feature vectors, a minimum spanning tree is a connected and weighted graph with no closed loops. If the distance associated with each edge of an MST denotes the weight, every edge in the MST is the shortest distance between two subtrees connected by that edge. Therefore, removal of the largest edges produces separate clusters. To begin this method, a minimum spanning tree is constructed. Next the edges of the tree are sorted in a non-increasing manner and removed until the desired number of clusters are formed. Each cluster is assigned a name describing its visual property, the so-called visual word, and a closest color for displaying purposes. A detailed description of the MST-based clustering algorithm used here can be found in [42], and will not be dwelt on here further.

3.3.4. Fast Search by Database Tree

With the set of feature vectors being labelled, the next step is to classify new observations. Real-time processing of the information coming from video demands for efficient search algorithms. Design of searching methods that scale well with the size of the database and the dimensionality of the data is a challenging task. Trees present an efficient way to index local image regions, making the nearest neighbor search more efficient by pruning. Scalable Vocabulary Tree (SVT) is a two-stage recognition scheme that provides a more efficient training of the tree which is built by hierarchically k-means clustering on a large set of representative descriptor vectors [43]. Instead of k defining the final number of clusters or quantization cells, k defines the branch factor (i.e., the number of children of each node) of the tree. More specifically, an initial k-means process is run on the training data, defining k cluster centers. The training data is then partitioned into k groups, where each group consists of the descriptor vectors closest to a particular cluster center. The same process is then applied to each newly generated group of descriptor vectors, recursively defining quantization cells by splitting each quantization cell into k new parts. The tree is determined level by level, up to some maximum number of levels, and each division into k parts is only defined by the distribution of the descriptor vectors that belong to its parent quantization cell.

For very large databases consisting of high dimensional highly sparse feature vectors, we choose to follow the notion of Vocabulary tree but with important variations to generate a k-way approximate nearest neighbor search tree. As shown in Fig. 3.2, given the obtained database, at the first level, there is only one node, the tree root. At the second level, a set of k (which is 3 in Fig. 3.2) representative patterns are randomly selected from the whole database as the cluster centers. Then the whole database is clustered into k subsets by assigning each feature vector to its closest center according to the chosen similarity measure. At the third level, for each of the k clusters obtained at the second level, k feature vectors are selected randomly from its pool of feature vectors as its next level cluster centers, resulting in k^2 cluster tree nodes at this level. This procedure continues until either all the feature vectors in a leaf node (the tree node that has no child nodes) belong to the same object class (a pure node) or the number of the feature vectors in a leaf node is below some limit, e.g., 50. Every feature vector has a class label associated with it.

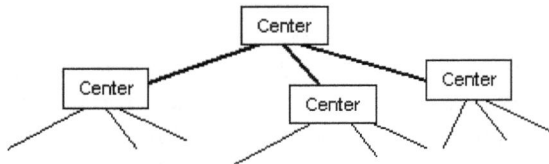

Fig. 3.2. 3-way approximate nearest neighbor search tree.

Given a new feature vector, to search through the tree, its distances to k cluster centers at each level along a certain branch are calculated and the winner is the center among k ones that the new feature vector is nearest to. When a leaf node is reached, if it is pure, assign the associated label and then stop. Otherwise, do a nearest neighbor search over the feature vectors in the leaf node, assign the label associated with the feature vector which gives minimum distance according to the chosen metric, and then stop.

3.4. An On-line Novelty Detection Method

The visual words (i.e., the labelled feature vectors in the database) obtained by image segmentation correspond well to natural features in the image. It is generally agreed that data points which have natural association are contained in the same regions of the feature space. The distance to the closest region is actually an estimated confidence interval of an output given the input. The data points inside such regions significantly indicate the local fitting accuracy. Since the evaluation of data points only involves a relatively small portion of the tree, the detection of novelty becomes computationally efficient. As a result, the proposed on-line novelty detection technique will first find a threshold to detect potential outliers, then apply an 8-connected structure element to erode the obtained binary image in order to find the largest connected outlier groups, and finally, when those groups have sufficient number of novel feature vectors, insert them into the tree as an on-line augmentation. In the following, the methods for threshold selection, erosion operation and tree insertion operation are detailed.

3.4.1. Threshold Selection

Thresholding is one of the most popular methods for novelty detection. Manually tuning threshold values can be tedious. Because it is often impossible to anticipate possible abnormal situations, the main challenges of novelty detection aim to develop fully automatic systems that do not require task-specific thresholds and tuning. Being the most prominent probability distribution in statistics to describe real-valued random variables that cluster around a single mean value, a normal distribution is often used as a first approximation to complex phenomena. To detect those feature vectors with a potential possibility of coming from a novel object, a precise criterion should be obtained from empirically estimated statistical parameters (e.g., mean/median and standard deviation). Drawn from a normal distribution, about 68 % of values are within one standard deviation σ away from the mean; about 95 % of the values lie within two standard deviations; and about 99.7 % are within three standard deviations. Therefore, for novelty detection based on a distance measure with respective to the database, the sum of the unbiased mean and one standard deviation can be a good choice for the threshold value.

To find such a suitable threshold, in the proposed approach, a sufficient number of normal pictures (i.e., with no novel objects in) are collected and distances between feature vectors extracted from each picture and those in the database are calculated using the approximate nearest neighbor search tree. If the size of the pictures taken is Width-by-Height and a moving window of size $N \times N$ is shifted by M pixels evenly in the row and column directions, for each picture, the number of feature vectors (FV) extracted, #FV, are

$$
\begin{aligned}
width &= \left(Width - N\right)/M + 1 \\
height &= \left(Height - N\right)/M + 1 \quad , \\
\#FV &= width \times height
\end{aligned}
\tag{3.5}
$$

where *width* and *height* denote the size of the resulting segmented image. Since the approximate nearest neighbor search through the database tree is used, in order to reduce the sensitivity to noisy outliers, the median value, instead of the mean value, of these #FV distances is calculated as d_{median}. Then an L_1-norm based statistic, std_{L1}, that loosely resembles the standard deviation is calculated as,

$$
std_{L1} = \frac{\sum_{i=1}^{\#FV} \left| d_{FV,i} - d_{median} \right|}{\#FV},
\tag{3.6}
$$

where $d_{FV,i}$ denotes the distance between the ith feature vector (*FV*) and its approximate nearest neighbor in the database tree, $d_{median} = median(d_{FV,1}, \cdots, d_{FV,\#FV})$, and #FV is the total number of feature vectors extracted per image. This results in multiple medians and multiple std_{L1}'s, each for a single training image. If l frames of normal images are used, the sum of the median of these l medians and the median of these l std_{L1}'s is used as the threshold T_{median}. The formal definition of the threshold T_{median} is the following sum,

$$
T_{median} = median(d_{median1}, \cdots, d_{medianl}) + median(std_{L1,1}, \cdots, std_{L1,l}), \tag{3.7}
$$

where the l in the last subscript in each term is the total number of normal images used to obtain the threshold. For comparison purpose, we use T_{mean} in the following to denote the threshold using the sum of the traditional mean and one standard deviation. That is,

$$T_{mean} = \frac{\sum_{j=1}^{\#FV \times l} d_{FV,j}}{\# FV \times l} + \sqrt{\frac{\sum_{j=1}^{\#FV \times l} \left(d_{FV,j} - d_{mean} \right)^2}{\# FV \times l}}, \tag{3.8}$$

3.4.2. Eight-Connected Structure Element Filter

After thresholding each of the width-by-height gray-scale distance images, a binary image is obtained. To find those outliers that have high potentiality of coming from the novel objects, we are more interested in those outliers that are retained in a few largest connected groups. To locate such regions, an eight-connected structure element is applied to find such regions. In other words, in the binary image obtained through thresholding, if an element in the image has a value of 1 and all its 8 neighbours are 1, this element retains a value of 1. Otherwise, it retains a value of 0. Only when the largest connected groups contain more than a certain number of feature vectors, say 50, are these feature vectors collected and stored temporarily for the next step, insertion.

3.4.3. Tree Insertion Operation

With the potential novel feature vectors being identified and collected, the next step is to insert them into the current database search tree for fast process. First, each novel feature vector (together with its new label) is appended to the end of the array of feature vectors holding the database. Its approximate nearest neighbor in a leaf node of the tree is next located. Finally, the index of the novel feature vector in the database list is added to the leaf node's member list and the flag for that leaf node is changed to impure if it is pure before this operation. Using this augmented approximate nearest neighbor search tree, the next image containing the novel object can be segmented.

The whole proposed on-line novelty detection method, as summarized in Fig. 3.3, can be separated into the training phase, which models the normal data, and the testing phase, which identifies the novel data and adds them to the database tree.

Training Phase   Testing Phase

```
┌─────────────────────┐          ┌─────────────────────┐
│   Training Images    │          │  Test Images (Video) │
└─────────────────────┘          └─────────────────────┘
           │                                │
           ▼                                ▼
┌─────────────────────┐          ┌─────────────────────┐
│  Feature Extraction  │          │    Frame Splitting   │
└─────────────────────┘          └─────────────────────┘
           │                                │
           ▼                                ▼
┌─────────────────────┐          ┌─────────────────────┐
│  Classifier Training │          │  Feature Extraction  │
└─────────────────────┘          └─────────────────────┘
           │                                │
           ▼                                ▼
┌─────────────────────┐          ┌──────────────────────────┐
│ Tree Structured Classifier │──▶│ Classifier Distance Output│
└─────────────────────┘          │ Classifier Label Output   │
           ▲                      └──────────────────────────┘
```

Yes

```
        ◇ Enough          No
        Novel FVs ───────────┐
```

┌─────────────────────┐   ┌─────────────────┐   ┌─────────┐
│ Threshold   │   │   Display   │
│ Segmentation   │   └─────────────────┘
└─────────────────────┘

┌─────────────────────┐   ┌─────────────────┐
│ Temporary Novel FV │   │ Binary Image   │
│ Storage   │   │ Erosion   │
└─────────────────────┘   └─────────────────┘

                                  ┌─────────────────┐
                                  │ Isolate New Class│
                                  │ Features   │
                                  └─────────────────┘

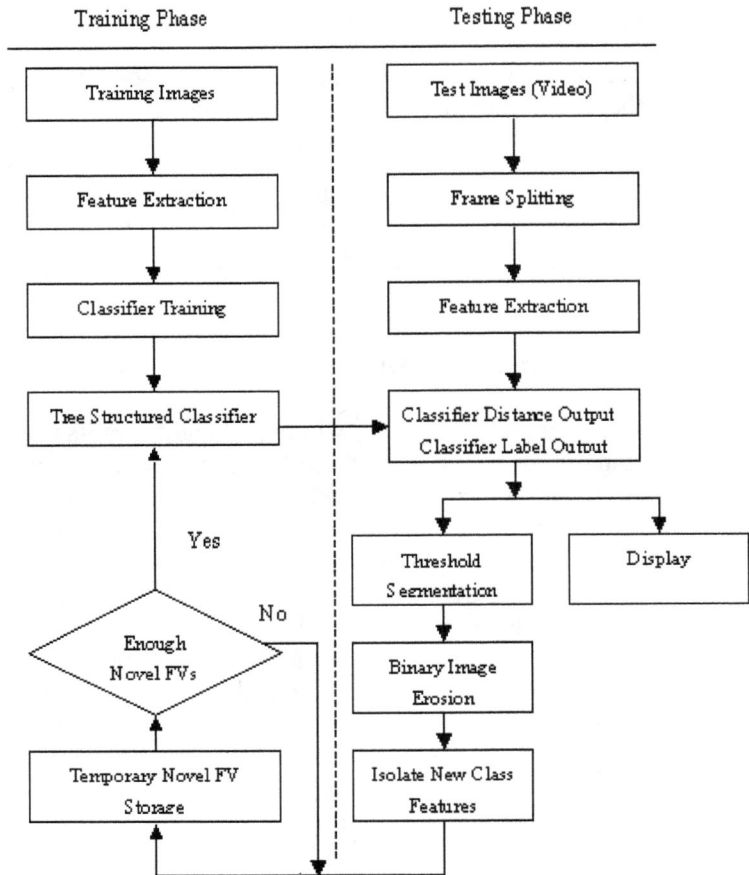

Fig. 3.3. A flow chart of the on-line novelty detection model.

3.5. Experiments and Results

In this section, we empirically investigate the performance of the proposed novel object detection mechanism by conducting a series of experiments in two different environments (an indoor one and an outdoor one). Each experiment consists of two consecutive phases: an exploration phase followed by an inspection phase. In the exploration phase, the robot collects images around the environment and builds a visual model of normality for it. In the inspection phase, the robot uses the learnt model to highlight any abnormal visual object that may appear in the environment.

3.5.1. Experiment I: An Indoor Environment

3.5.1.1. Setup

As shown in the left of Fig. 3.4, the first working environment is an indoor hallway. Typical percepts along the hallway include Wood Panel, White Floor, Yellow Floor, Blue Wall, Blue Railing, Black Stripe Tile, Light Window. As shown in the right of Fig. 3.4, a Pioneer 2 AT four-wheel-drive mobile robot, called Skeeter, is used in the experiment. The robot vision system is a video camera (SONY Digital Handycam DCR VX2000 digital video camera recorder) that sends the data to an on-board laptop computer through IEEE 1394 interface. The intelligent information processing system is an on-board laptop that runs the Fodera core 4 operating system.

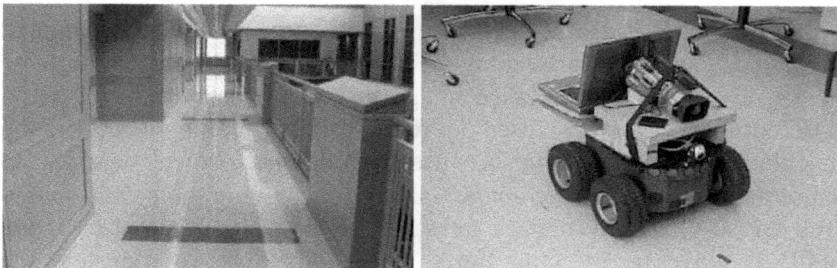

Fig. 3.4. (left) The hallway; (right) mobile robot Skeeter with a camera and a laptop.

3.5.1.2. Exploration Phase

In this phase, a model of the environment is generated so that the robot vision system can identify the major percepts existing in this indoor environment. To build a percept database, 20 pictures of size 720-by-480 along the hallway are taken by the robot, from which 66,740 unlabelled feature vectors are extracted (with N=15 and M=10). According to a minimum spanning tree (MST)-based clustering algorithm given in [42], 11 perceptual groups are obtained. A detailed list of the major percepts present in the video sequences (not all objects appear on the same frames simultaneously but with some objects replacing others as the training sequence progresses) is provided in Table 3.1 together with their semantic meanings and a denoting color.

69

Table 3.1. Major Percepts and Their Semantic Meanings.

Percept Name	Percept Symbol	Denoting Color
BlackStripe	Object1	Black
WoodPanel	Object2	Light Red
LightWondow	Object3	Light Grey
WoodPanel	Object4	Light Red
WhiteFloor	Object5	Dark White
WoodPanel	Object6	Dark Red
WoodPanel	Object7	Red
WoodPanel	Object8	Light Red
WhiteFloor	Object9	White
YellowFloor	Object10	Yellow
YellowFloor	Object11	Blue

From Table 3.1, it can be seen that several perceptual groups obtained by the unsupervised learning algorithm actually correspond to the same object. To make the processed image visually pleasant, several finely differentiated colors are used to refer these different objects in the machine vision perspective. To use the obtained percept database to subsequently segment any image (taken in the environment) reasonably well and fast, 10 Vocabulary trees and 10 proposed random trees are built. The average running times in seconds over 10 runs and the corresponding standard deviations are summarized in Table 3.2. It can be seen that the proposed random trees take much less time to construct.

Table 3.2. Running time for Trees' Construction.

	Vocabulary tree (s)	Random tree (s)
Mean	24228	49
Std	41	2

To show the quality of these trees, segmentation results using Vocabulary tree and the proposed random tree for a sample image are presented in Fig. 3.5, where the left one is the original image, the middle one is the segmented image using the Vocabulary tree, and the right one is the segmented image using the proposed random tree.

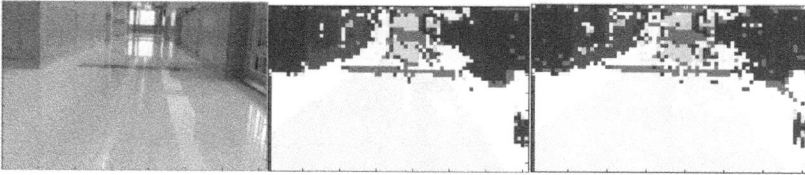

Fig. 3.5. (left) Original image; (middle) VT tree search result;
(right) random tree search result.

From the figure, it can be seen both search trees work very well. In a qualitative visual evaluation, it can be seen that the metric used does a reasonably good job on the whole and clearly identifies the yellow floor, the white floor, and the wood panel wall, although it misclassifies the railing as the wood panel, which is not too surprising since the railing is a hard percept to model. To further quantify the search quality and the technical robustness of both trees, 10 more images along the hallway are taken. Then for each of them, 3337 feature vectors are extracted which are classified first by an exact nearest neighbor search with respect to the whole database and then by an approximate nearest neighbor search through 10 Vocabulary trees (denoted by Vtree) and 10 fast random search trees (denoted by Otree). The average percentages of correct classification (with respect to the exact search) and the corresponding standard deviations are plotted in Fig. 3.6. The horizontal axis denotes the image number. The vertical axis denotes the average percentages of correct classification over 10 runs and the corresponding standard deviation. From the figure, it can be seen both Vocabulary tree and the proposed random tree do a good job and are close in performance.

3.5.1.3. Parameter Extraction

The next step in this experiment is to extract parameters suitable for novelty detection. Threshold calculation is a crucial step. Using different thresholds may result in different detection performance. For this purpose, 80 normal images along the hallway are collected.

Two different sets of threshold values are extracted from them and shown in Table 3.3. In the table, OT-I, OT-II, VT-I, VT-II listed in the first column are used to denote the cases for which threshold values are calculated from Eq. 3.7 and Eq. 3.8 using distances obtained by the proposed tree and Vocabulary tree, respectively. The values in the second and third columns are the estimated values for unbiased mean and standard deviation, respectively. Those in the last column are the final thresholds to use.

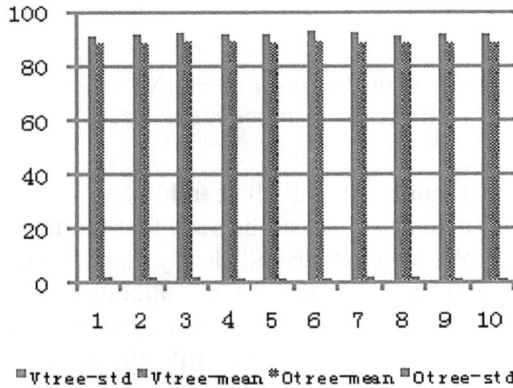

■Vtree-std ■Vtree-mean ＊Otree-mean ■Otree-std

Fig. 3.6. Average percentages of correct classification over ten runs and the corresponding standard deviations.

Table 3.3. Threshold Values for Novelty Detection.

Items	Mean	std	Threshold
OT-I	0.4950	0.2039	0.6989
OT-II	0.5810	0.4914	1.2820
VT-I	0.4475	0.1893	0.6368
VT-II	0.5177	0.3867	0.9044

Additionally, the performance of a neural network, Grow-When-Required (GWR), is also tested. As shown in Fig. 3.7, the GWR network consists of a clustering layer of nodes and a single output node. The connecting synapses are subject to habituation, which reduces in response to repeatedly presented stimuli. The inputs to GWR include some predefined parameters and the input feature vectors from the database.

The training of GWR starts with two topologically unconnected nodes (e.g., any two input vectors), whose weights are randomly initialized. During training, following an unsupervised winner-take-all approach, the Euclidean distance from the input to each node currently in the clustering layer is first computed to find the best matching node, s, (called the winner) and the second best matching node, t. The weights and the output synapses of the winner node and its topological neighbours are updated. If a connection exists between the two nodes, its age is set to 0; otherwise a new connection is created. When the activation and habituation values of the winner node are below predefined thresholds a_T and h_T, the input is decided to be novel and a node can be added to the clustering layer. After inserting a new node, r, to the clustering layer, topological connections are updated by removing the link between the nodes s and t and inserting links between r and s and between r and t. The GWR training iteration is accomplished by removing nodes that no longer have any neighbours and edges whose age is greater than a predefined threshold age_{max}. The set of parameters used in this work are given in Table 3.4. From empirical evidence, it can be observed that, the larger the activity threshold, the larger the number of nodes, and the less the number of false negatives, however, the more time the novelty detection consumed.

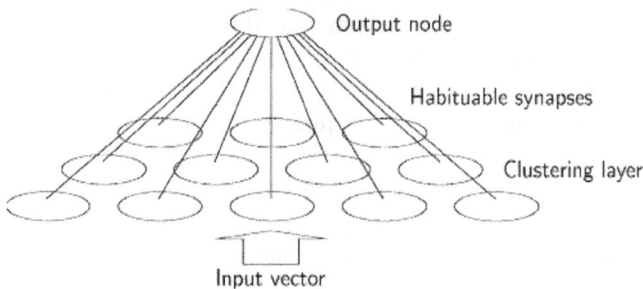

Fig. 3.7. The GWR neural network architecture.

3.5.1.4. On-line Novel Percept Detection

To test the on-line performance of the proposed novelty detection method, a video sequence, consisting of 19 images, is taken as the robot moves along the hallway towards a very conspicuous green ball introduced in it for the robot to approach. Each of the grabbed 19 images is processed on the fly and the calculated approximate nearest neighbor distances are thresholded, giving rise to a binary image of size 71-by-47. Next, the obtained binary image is eroded twice by an 8-connected structure element. Finally, the largest connected group is identified and grows as the robot moves closer and closer to the green ball. If the number of feature vectors it has contained is larger than 50, they are retained and accumulated through temporary storage.

Table 3.4. Parameters Used for GWR in This Work.

Parameter Name	Parameter Values
a_T	{0.3 0.4 0.5 0.6 0.7 0.8 0.9}
ε	{0.1 0.01 0.001}
h_T	0.1
α	1.05
τ_n	0.1
τ_b	0.3
age_{max}	100
$neighbor_{max}$	10

The novelty detection results are presented by 5 of the original 19 images in Fig. 3.8. The ones in the middle column show the results obtained when the robot uses the first 18 images as the training images to collect novel feature vectors, inserts them into the tree at the end of the 18th image, and then uses the updated database to segment the last (i.e., the 19th) image. The ones in the right column show that the robot uses the total set of 19 images as the training images to collect novel feature vectors, inserts them into the tree at the end of the 19th image, and then uses the updated database to segment the whole set (i.e., 19 images) all over again. As found in the experiments, about 600 feature vectors coming from the 19 images with the novel object in are added to the search tree after 2 seconds. This demonstrates that the proposed system is able to continue learning with only a modest processing delay.

3.5.1.5. Off-line Novelty Detection Evaluation

To make further comparisons between both methods, 135 testing images (25 of which contains the novel object) are collected for off-line testing. Every round of the inspection phase consists of ten consecutive loops of these images. The segmented images based on the obtained percept database should yield a set of connected regions in the image. A connected-component labelling algorithm is then applied to find these regions, which can be regarded as the corresponding objects. Since the number of images used in the experiments is not very large, the target and non-target images are manually labelled by human through visual examination.

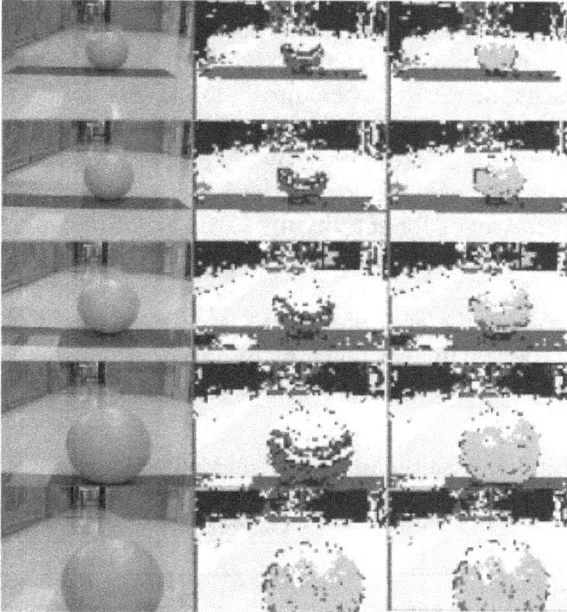

Fig. 3.8. (left) Original images; (middle) processed images; (right) processed images after learning.

For this round of experiments, the performance of each method is evaluated by the accumulated results which are given in the form of a contingency table. A contingency table is a two-dimensional table formed by cross-classifying subjects or events on two categorical variables used to relate the actual presence of novelty and the response of the system. One variable's categories define the rows while the other variable's categories define the columns. The intersection (crosstabulation) of each row and column forms a cell, which displays the count (frequency) of cases classified as being in the applicable category of both variables. Next, the $\chi2$ analysis is performed, which is then followed by the computation of Cramer's V. The Chi-square statistic ($\chi2$) is computed to determine whether or not the row and column categories for the table as a whole are independent of each other. If the computed $\chi2$ statistic exceeds a critical value, then we reject the null hypothesis and conclude that the variable categories are indeed related. Cramer's V is computed to test the strength of such relationships and is the nominal association measure of choice. Large values of Cramer's V's indicate better performance.

The performance evaluation of the proposed system vs. GWR network to detect the simple novel percept is shown in Table 3.5. The first table in Table 3.5 lists the results when threshold OT-I is used. From it, it can be seen that, among 250 test images that contain the novel object, 227 are detected and 23 are missed by the proposed system, and, among 1100 test images that are normal, none is detected. For GWR network approach, seven values for activity threshold and three values for ε have been tried, and the best results (in terms of Cramer's V value) are presented in Table 3.5. From the table, it can be seen that among 250 test images that contain the novel object, all are detected by GWR network, and, among 1100 test images that are normal, 20 are mistakenly detected, yielding statistically significant results.

Table 3.5. Results of the Proposed Method for Green Ball.

OT-I	Novelty Detected	Novelty Not Detected
Novelty Present	227	23
Novelty Not Present	0	1100
Cramer's V = 0.9431		
GWR	Novelty Detected	Novelty Not Detected
Novelty Present	250	0
Novelty Not Present	20	1080
Cramer's V = 0.9535		
$a_T = 0.8$		

3.5.1.6. Other Threshold Values

The calculated values for Cramer's V obtained when other threshold values are considered are given in Fig. 3.9. There are four lines in the figure, each corresponding to one of four cases in Table 3.3. The vertical axis denotes the values of Cramer's V's. The horizontal axis denotes the number of standard deviations used in the threshold calculation (i.e., 1.0, 1.5, 2.0, 2.5 and 3.0). From the figure, it can be seen that the threshold values along those two lines starting with OT-I and VT-I achieve the best results, indicating the proposed threshold selection criterion works well for both trees.

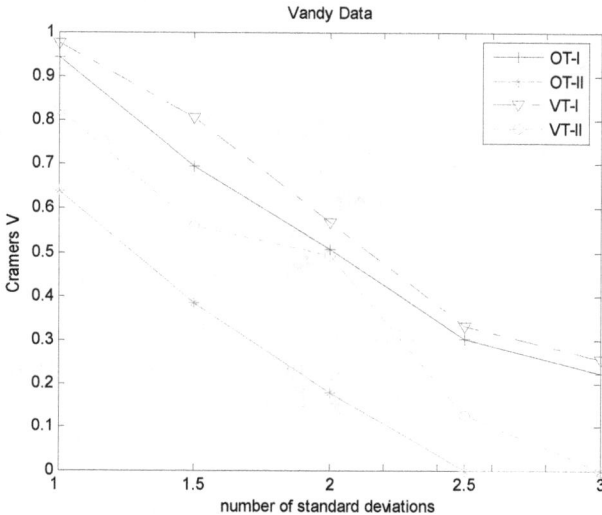

Fig. 3.9. Performance variations with different threshold values.

3.5.1.7. Running Time Issue

To compare the relative time issue of two approaches, for the database of size 66,740, it is observed that the proposed tree has so far at most 20 levels. As a result, at most $3 \times 20 + 50 = 110$ distance calculations are necessary. This is a considerable reduction. It takes no more than 2 seconds to find approximate nearest neighbours for the 3337 feature vectors obtained for each newly grabbed image. For GWR network, the number of nodes generated during training and testing are shown in Table 3.6, respectively. It can tell that GWR method takes several thousands of distance computations for each feature vector extracted from a newly grabbed image. This is to say, GWR method is very computationally expensive in high dimensional feature space.

Table 3.6. The Number of Nodes.

a_T	Training (# of nodes)	Testing (# of nodes)
0.7	2619	2936
0.8	4431	5023

77

3.5.2. Experiment II: An Outdoor Environment

3.5.2.1. Setup

Although the proposed method does a good job in the last experiment, the detection of a single brightly colored green ball, which is very different from all the known objects in the frames, simplifies the novel object detection task significantly and makes the experiment somewhat artificial. For a more convincing demonstration of how the proposed model would perform detecting more realistic objects in a different and less structured environment, and to justify the proposed novelty detection method is realistic and applicable in practice, an outdoor traffic intersection is chosen to be a second working environment, as shown in the left of Fig. 3.10.

Fig. 3.10. (left) Original image; (middle) VT tree search result; (right) random tree search result.

Further the previous experiment is focused on mobile robots and does not use any additional constraints and information that can be derived from a mobile camera. As a result, the extension to a more general environment setting substantially increases the technical soundness of the proposed approach. In this set of experiments, the robot vision system is Canon IXUS 50HS digital video camera recorder, which is somewhat fixed and captures a video with three different vehicles coming in and leaving. The intelligent information processing system is a workstation computer that runs the Fodera core 4 operating system. From the video, 450 images of size 320-by-240 are extracted. The first 20 images are used as the training images to obtain the database. Then 80 images are used for threshold extraction. The last 350 images are the test images for novelty detection.

3.5.2.2. Exploration Phase

To build the percept database, the first 20 images in the video sequence are used, from which 93,220 feature vectors are extracted (with N=6 and M=4). The same MST-based clustering algorithm [42] is applied and 25 perceptual groups are obtained. Typical perceptual groups include those for gray road, white crossing, green grass, dark green trees and blue sky. A list of major percepts present in the video sequences is provided in Table 3.7 together with their semantic meanings and displaying color.

Table 3.7. Major Percepts and Their Semantic Meanings.

Percept Name	Percept Symbol	Denoting Color
Blue Sky	Object1	Blue
Blue Sky	Object4	Light Blue
Blue Sky	Object5	Pale Blue
Blue Sky	Object7	Palest Blue
Tree	Object8	Green
Grass	Object12	Dark Green
Pole	Object16	Dark Gray
Crossing	Object17	White
Ground	Object22	Gray

Based on the obtained database, 10 Vocabulary trees and 10 proposed random trees are constructed, respectively. Their average running times in seconds and the corresponding standard deviations are summarized in Table 3.8. From the table, it can be seen that the construction of the proposed random trees takes much less time.

Table 3.8. Running time for Trees' Construction.

	Vocabulary tree (s)	Random tree (s)
Mean	34961	64
Std	12	4

The segmentation results using Vocabulary tree and the proposed random tree are shown in Fig. 3.10. Like the case in Experiment I, both search trees work very well.

To further quantify the search quality and the technical robustness of both trees, 10 more normal images are chosen. Then for each of the 10 images, 4661 feature vectors are extracted which are then classified first by an exact nearest neighbor search with respect to the whole database and then by an approximate nearest neighbor search through 10 Vocabulary trees and 10 fast random search trees. To demonstrate the performance of these fast search trees, the average percentages of correct classification (with respect to the exact search) and the corresponding standard deviations are plotted in Fig. 3.11. The horizontal axis denotes the image number. The vertical axis denotes the average percentage of correct classification over 10 runs and its standard deviation. From the figure, it can be seen both Vocabulary tree and the proposed random tree do a good job. The performances of Vocabulary tree and the proposed random tree are very close.

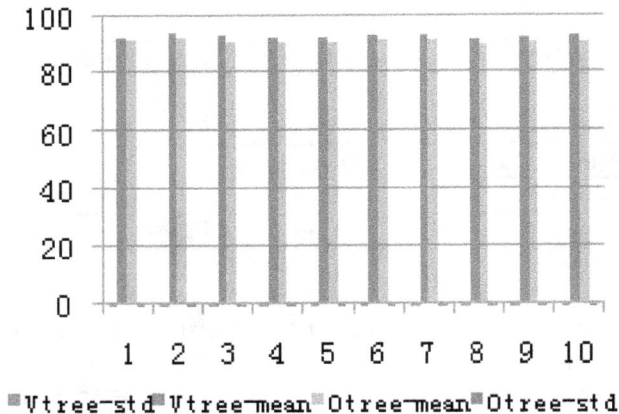

Fig. 3.11. Average percentages of correct classification and the corresponding standard deviations.

3.5.2.3. Parameter Extraction

To extract parameters suitable for novelty detection, similar to Table 3.3, the threshold values for the proposed method are extracted from the collected 80 normal images and shown in Table 3.9. In the table, OT-I, OT-II, VT-I, VT-II listed in the first column are used to denote the cases for which threshold values are calculated from Eq. 3.7 and Eq. 3.8 using distances obtained by the proposed tree and Vocabulary tree, respectively. The values in the second and third columns are the estimated values for unbiased mean and standard deviation, respectively. Those in the last column are the final thresholds to use.

Table 3.9. Threshold Values for Novelty Detection.

Items	Mean	std	Threshold
OT-I	0.5150	0.5454	1.0604
OT-II	0.7176	0.5887	1.3063
VT-I	0.4560	0.4823	0.9383
VT-II	0.7940	0.8170	1.611

It can be clearly seen from the previous experiment that GWR method does not scale well with the size and the dimensionality of large databases. In this experiment, instead of GWR, we evaluate the performance of the proposed method with the more recently proposed E-SVDD. As a one-class classifier, support vector data description method is a very attractive kernel method and has become a standard novelty detection technique. Its run-time complexity is linear in the number of support vectors, which is not very convenient for modern large databases. To improve, Fast-SVDD (F-SVDD) uses only one kernel term in its decision function. However, given the Gaussian kernel, its decision boundary becomes spherical in the input space, which is not suitable for many real cases. As a result, E-SVDD method first partitions the dataset into small clusters. As a result, the decision function of E-SVDD contains some crucial kernel terms, one for each cluster, and its time complexity is linear in the number of the clusters. The clustering algorithm used originally by E-SVDD is partitioning-entropy-based kernel fuzzy c-means algorithm. For the high-dimensional database, *k*-means algorithm is used instead. To more thoroughly cover the data, the number of clusters is set to 100. Then for each cluster, the parameters used in the classifier such as the centroid are calculated. Finally, the decision function is computed according to F-SVDD.

3.5.2.4. On-line Novel Percept Detection

To show how the proposed system would perform in the presence of multiple novel objects, this round of experiments work on a video sequence of 350 images, in which a traffic intersection is inspected when several vehicles come in. In this video sequence, the first novel object, a black car, begins to come in at about the 20th image. Then the second novel object, a purple truck, appears following the black car at about the 120th frame. Finally the last novel object, a white jeep, shows up at about the 180th frame and goes away at about the 300th frame.

Each of the grabbed 350 images is processed on the fly. For the proposed method, the calculated approximate nearest neighbor distances are thresholded, giving rise to a binary image of size 79-by-59. For E-SVDD, the decision function is computed for each extracted feature vector. If the decision value is smaller than 0, the feature vector is normal. Otherwise, it is novel. Next, the obtained binary image is eroded twice by an 8-connected structure element as explained in the previous section. Finally, the largest connected groups are identified. If the number of detected novel feature vectors in the largest connected groups is larger than 100, they are retained and accumulated through temporary storage. To detect multiple objects, spatial information about the position of the largest connected groups in the image space are utilized to separate novel percepts that are apart but appear in the same frame.

The novelty detection results are presented by 6 of the original 350 pictures in Fig. 3.12. The first column shows the original images of some typical novelty detection situations in the video sequence. The second column shows the segmented images using the database just before the novel feature vectors coming from the newly entered vehicles are learned. The third column shows the segmented images after all the novel objects are learned. That is, all 350 images are processed twice. In the first pass, they are used to collect novel feature vectors from three vehicles, which are labelled unsupervisedly. Then in the second pass, the images shown there are segmented ones using the updated database (with the novel feature vectors being added to the database).

More specifically, the images on the first row describe the normal situation when no vehicle comes in. The first image in the second column is a segmented image using the original database (i.e., with no novel feature vectors being added to it). And the first image in the third column is a segmented one using the database augmented with novel feature vectors detected from all three novel objects. The images on the second row describe the situation when the first novel object (i.e., the black car) comes in. The second image in the second column is a segmented one using the original database with no novel feature vectors augmented to it. The small region circled by an ellipse in this image denotes the place where the novel black car should otherwise appear.

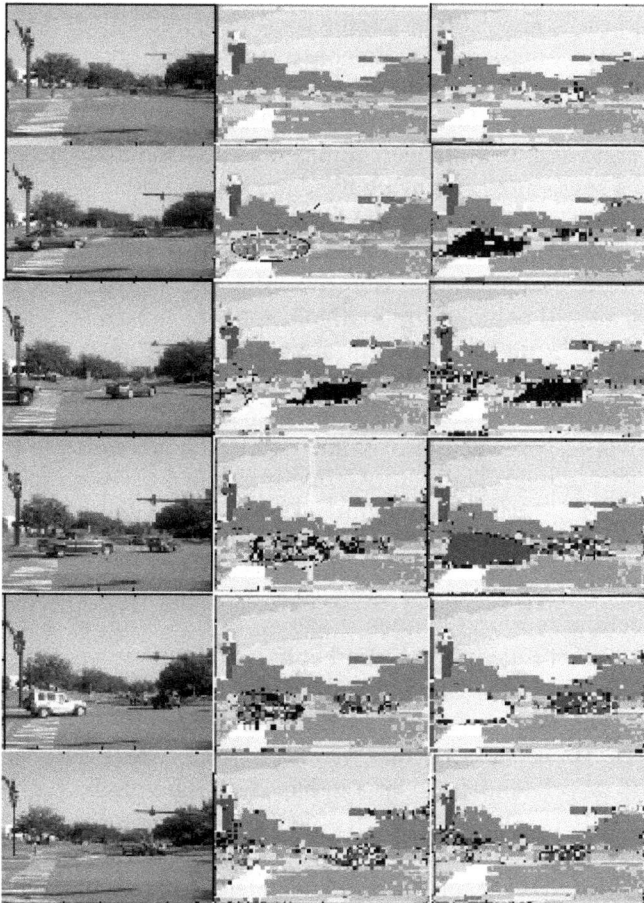

Fig. 3.12. (left) Original images; (middle) processed images; (right) processed images after learning.

When using the updated database with all three novel objects being learned, it can be seen the learned novel black car percept (denoted by black color) shows up clearly in the second image in the third column. The images on the third row manifest the situation when a purple truck enters the scene. The third image in the middle column is a segmented one using the database after the novel feature vectors from the black car have been identified and inserted into the tree. From the segmented one in the third column, the learned novel purple car (denoted by purple color) is partially seen as it shows up into the scene. The images on the fourth row further explore the situation when the purple truck comes deeper into the scene. Therefore, the fourth image in the second column is a segmented one using the database augmented only with the novel feature vectors detected from the black car. From the fourth image in the third column, the purple truck is clearly identified. The images on the fifth row show that a white jeep comes in. The fifth image in the second column is a segmented one using the database augmented with the novel feature vectors detected from the black car and the purple truck. From the fifth image in the third column, the white jeep can be clearly identified. The images on the last row show that the learning process is completed and all vehicles gradually disappear. More than 10,000 novel feature vectors from the black car are added to the search tree after 16 seconds. Then, more than 10,000 novel feature vectors from the purple truck are added to the search tree after 26 seconds. Finally, more than 10,000 novel feature vectors from the white jeep are added to the search tree after 31 seconds. This demonstrates that the proposed system is able to continue learning with only a modest processing delay.

To compare the detecting performances, the numbers of detected novel feature vectors over 350 images is shown in Fig. 3.13. We show in the upper-left plot of Fig. 3.13 the number of novel feature vectors, obtained by the proposed method (using the threshold referred to as OT-I in Table 3.9) and the E-SVDD method, respectively, for each of the 350 images. It can be seen that there are three separated peaks whose number of novel objects is larger than 100. Therefore, both methods successfully detect the three novel objects, consuming similar amount of time, i.e., two seconds for each frame. Shown in the upper-right plot is the number of novel feature vectors detected for each image before and after the collected novel feature vectors from the black car are inserted into the tree. Shown in the lower-left plot is the number of novel feature vectors detected for each image before and after the collected novel feature vectors from the purple truck are inserted into the tree. Shown in the lower-right plot is the number of novel feature vectors detected for each image before and after the collected novel feature vectors from the white jeep are inserted into the tree. As found in the experimental results, the total number of the novel feature vectors added to the database is more than 30,000, 31 % of the original database. For this amount of feature vectors, it can be hard for many clustering algorithms to partition them into three clusters in real time.

3.5.2.5. Off-line Novelty Detection Evaluation

As in Experiment I, to make fair comparisons by using the same dataset for off-line testing, we manually select a set of 254 testing images (154 of which contains the novel objects, i.e., the three vehicles) that we believe could be discriminated relatively well based on their visual feature distribution. Each round of the experiments consists of three consecutive test loops of these images. To evaluate the performance of each method, a contingency table is first used to relate the actual presence of novelty and the response of the system. Next, the $\chi 2$ analysis is performed, which is then followed by the computation of Cramer's V.

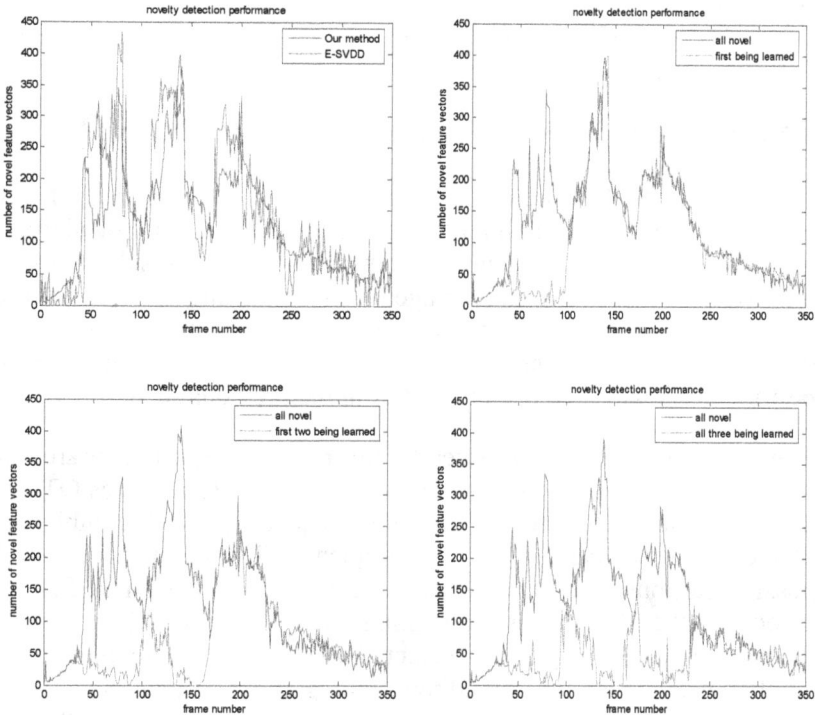

Fig. 3.13. The numbers of detected novel feature vectors over 350 images (upper left) the proposed approach vs. E-SVDD, (upper right) before and after the car is learned, (lower left) before and after the truck is learned, (lower right) before and after the jeep is learned.

85

The accumulated results obtained for this round of experiments are given in Table 3.10. It can be observed from both tables that both systems can detect the novelty percept and, statistically significant results are yielded.

Table 3.10. Results of the Proposed Method vs. E-SVDD.

OT-I	Novelty Detected	Novelty Not Detected
Novelty Present	456	6
Novelty Not Present	0	300
Cramer's V = 0.9837		
GWR	Novelty Detected	Novelty Not Detected
Novelty Present	402	60
Novelty Not Present	3	297
Cramer's V = 0.8421		

3.5.2.6. Off-line Multi-Novelty Detection Evaluation

Unlike the E-SVDD method which can detect the existence of novel object in the environment but requires an additional clustering algorithm to partition the obtained novel feature vectors into subgroups, the proposed method has the advantage of distinguishing between different objects on-line. To evaluate the recognition ability of the updated database (i.e., the augmented database), two more experiments are performed on the same test data set as in the previous section.

The accumulated results obtained for this round of experiments are given in Table 3.11 using two sets of threshold values referred to as OT-I and VT-I in Table 3.9. The rows in the table list all the possible combinations of the novel objects in the video sequence. The columns list all the possible responses of the proposed system. For example, on the first row, among all the 159 images (containing only the black car) presented to the system, 153 images are detected and 6 images are missed. For another example, on the second row, among all the 63 images containing both the black car and the purple truck presented to the system, 48 images are detected to have one novel object, and 15 images are detected to have both novel objects. It can be observed from the first table of it that the proposed system can identify all the learned vehicles. However, when some of them appear at the same time in a test image, it happens that one is not detected in some images.

Table 3.11. Results of the Proposed Method.

OT-I	1 Novelty Detected	2 Novelty Detected	Novelty Not Detected
Only Car Present	153	0	6
Car and Truck Present	48	15	0
Only Truck Present	96	0	0
Truck and Jeep Present	39	30	0
Only Jeep Present	75	0	0
No Novelty Present	0	0	300
Cramer's V = 0.7958			
VT-I	1 Novelty Detected	2 Novelty Detected	Novelty Not Detected
Only Car Present	159	0	0
Car and Truck Present	63	0	0
Only Truck Present	96	0	0
Truck and Jeep Present	39	30	0
Only Jeep Present	75	0	0
No Novelty Present	0	0	300
Cramer's V = 0.8526			

3.5.2.7. Other Threshold Values

The calculated values for Cramer's V obtained when other threshold values are considered are given in Fig. 3.14. There are four lines in the figure, each corresponding to one of four cases in Table 3.9. The vertical axis denotes the values of Cramer's V's. The horizontal axis denotes the number of standard deviations used in the threshold calculation (i.e., 1.0, 1.5, 2.0, 2.5 and 3.0). From the figure, it can be seen that the threshold values along those two lines starting with OT-I and VT-I achieve the best results, indicating the proposed threshold selection criterion works well for both trees.

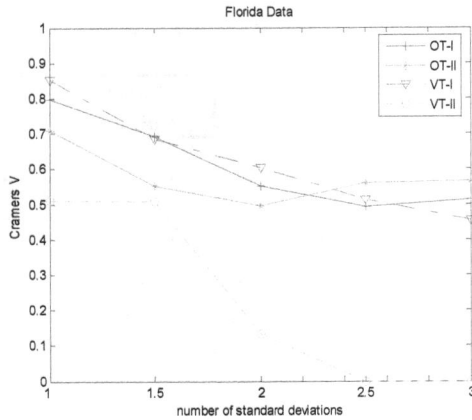

Fig. 3.14. Performance variations with different threshold values.

3.5.2.8. Running Time Issue

To compare the relative time issue of the proposed approach and E-SVDD, the proposed tree has so far at most 18 levels for the database consisting of 93,220 labelled feature vectors from the experiments. As a result, we have had at most $3 \times 18 + 50 = 104$ distance calculations. This is a considerable reduction. It takes no more than 2 seconds to find approximate nearest neighbours for the 4661 feature vectors obtained for each newly grabbed image. For E-SVDD, the number of centroids used in the decision calculation is 100. It takes similar amount of computation time for 4661 feature vectors extracted from a newly grabbed image. Therefore, both methods have similar running time performances.

3.6. Conclusions

In this chapter, an efficient on-line novelty detection method is presented. Its performance is compared with two best known methods, GWR neural network and E-SVDD method. The proposed approach provides a better performance, while offering advantages of near real time ability, which is extremely useful to assess which percepts of the environment are actually learnt by the system. The key innovations of the proposed approach are the use of a fast approximate nearest neighbor search facility (as opposed to the widely used Vocabulary tree), a threshold extraction method for detecting novel feature vectors and the rather standard tree insertion procedure to cluster and unsupervisedly learn the novel feature vectors. In contrast to GWR neural network, the proposed approach is particularly suitable for real-time autonomous applications and has built-in robustness to noise and clutter. In contrast to E-SVDD, the proposed approach is able to use both the spatial information and feature information of novel objects to autonomously identify multiple precepts in a frame in real time. Overall the proposed approach is autonomous and suitable for video-analytical tasks in different environments. Future investigations of the proposed method aim at how the system would perform in the presence of multiple objects of similar features but with different shapes in multiple video sequences.

Acknowledgements

The authors would like to thank the Chinese National Science Foundation for its valuable support of this work under award 61473220.

References

[1]. M. Markou, S. Singh, Novelty detection: a review-part 1: statistical approaches, *Signal Processing*, Vol. 83, Issue 12, 2003, pp. 2481-2497.

[2]. M. Markou, S. Singh, Novelty detection: a review-part 2: neural network based approaches, *Signal Processing*, Vol. 83, Issue 12, 2003, pp. 2499-2521.

[3]. R. O. Duda, P. E. Hart, D. G. Stork, Pattern Classification, *John Wiley & Sons,* NY, USA, 2001.

[4]. C. Bishop, Pattern Recognition and Machine Learning, *Springer,* New York, 2006.

[5]. L. Tarassenko, P. Hayton, N. Cerneaz, M. Brady, Novelty detection for the identification of masses in mammograms, in *Proceedings of the 4th International Conference on Artificial Neural Networks*, July 1995, pp. 442–447.

[6]. J. Quinn, C. Williams, Known unknowns: novelty detection in condition monitoring, in *Proceedings of the Iberian Conference on Pattern Recognition and Image Analysis (IbPRIA'07),* Vol. 4477, 2007, pp. 1–6.

[7]. L. Clifton, D. Clifton, P. Watkinson, L. Tarassenko, Identification of patient deterioration in vital-sign data using one-class support vector machines, in *Proceedings of the Federated Conference on Computer Science and Information Systems (FedCSIS'11)*, Szczecin, Poland, 18-21 September 2011, pp. 125–131.

[8]. L. Tarassenko, D. Clifton, P. Bannister, S. King, D. King, Novelty Detection, *John Wiley & Sons*, 2009, pp. 1–22.

[9]. C. Surace, K. Worden, Novelty detection in a changing environment: a negative selection approach, *Mechanical Systems and Signal Processing*, Vol. 24, Issue 4, 2010, pp. 1114–1128.

[10]. A. Patcha, J. Park, An overview of anomaly detection techniques: existing solutions and latest technological trends, *Computer Networks*, Vol. 51, Issue 12, 2007, pp. 3448–3470.

[11]. V. Jyothsna, V. V. R. Prasad, K. M. Prasad, A review of anomaly based intrusion detection systems, *International Journal of Computer Applications*, Vol. 28, Issue 7, 2011, pp. 26–35.

[12]. C. Diehl, J. Hampshire, Real-time object classification and novelty detection for collaborative video surveillance, in *Proceedings of the International Joint Conference on Neural Networks (IJCNN'02)*, February 2002, Vol. 3, pp. 2620–2625.

[13]. M. Markou, S. Singh, A neural network-based novelty detector for image sequence analysis, *IEEE Transactions on Pattern Analysis and Machine Intelligence*, Vol. 28, Issue 10, 2006, pp. 1664–1677.

[14]. H. Vieira Neto, U. Nehmzow, Real-time automated visual inspection using mobile robots, *Journal of Intelligent and Robotic Systems*, Vol. 49, Issue 3, 2007, pp. 293-307.

[15]. B. Sofman, B. Neuman, A. Stentz, J. Bagnell, Anytime online novelty and change detection for mobile robots, *Journal of Field Robotics*, Vol. 28, Issue 4, 2011, pp. 589–618.

[16]. Y. Zhang, N. Meratnia, P. Havinga, Outlier detection techniques for wireless sensor networks: a survey, *IEEE Communications Surveys & Tutorials*, Vol. 12, Issue 2, 2010, pp. 159–170.

[17]. H. Dutta, C. Giannella, K. Borne, H. Kargupta, Distributed top-k outlier detection from astronomy catalogs using the DEMAC system, in *Proceedings of the 7th SIAM International Conference on Data Mining*, Minneapolis, Minnesota, USA, April 26-28, 2007.

[18]. H. Escalante, A comparison of outlier detection algorithms for machine learning, in *Proceedings of the International Conference on Communications in Computing*, January 2005, pp. 10-15.

[19]. S. Basu, M. Bilenko, R. Mooney, A probabilistic framework for semi-supervised clustering, in *Proceedings of the 10th ACM International Conference on Knowledge Discovery and Data Mining (SIGKDD'04)*, Seattle, WA, USA, 22-25 August 2004, pp. 59–68.

[20]. S. Marsland, U. Nehmzow, J. Shapiro, Vision-based environmental novelty detection on a mobile robot, in *Proceedings of the 8th International Conference on Neural Information Processing (ICONIP'01)*, Shanghai, China, January 2001.

[21]. S. Marsland, U. Nehmzow, J. Shapiro, Environment-specific novelty detection, in *From Animals to Animates: Proceedings of the 7th International Conference on the Simulation of Adaptive Behavior (SAB'02)*, MIT, Edinburgh, UK, August 2002, pp. 36-45.

[22]. H. V. Neto, U. Nehmzow, Visual novelty detection for inspection tasks using mobile robots, in *Proceedings of the 8th Brazilian Symposium on Neural Networks (SBRN'04)*, Sao Luis, Brazil, 2004.

[23]. U. Nehmzow, H. V. Neto, Novelty-based visual inspection using mobile robots, in *Towards Autonomous Robotic Systems: Proceedings of the 5th British Conference on Mobile Robotics (TAROS'04)*, Colchester, UK, January 2004.

[24]. H. V. Neto, U. Nehmzow, Visual novelty detection with automatic scale selection, *Robotics and Autonomous Systems*, Vol. 55, Issue 9, 2007, pp. 693-701.

[25]. D. M. J. Tax, R. P. W. Duin, Support Vector Data Description, *Machine Learning*, Vol. 54, Issue 1, 2004, pp. 45-66.

[26]. Y.-H. Liu, Y.-C. Liu, Y.-J. Chen, Fast support vector data descriptions for novelty detection, *IEEE Transactions on Neural Networks*, Vol. 21, Issue 8, 2010, pp. 1296-1313.

[27]. X. Peng, D. Xu, Efficient support vector data descriptions for novelty detection, *Neural Computing and Applications*, Vol. 21, Issue 8, 2012, pp. 2023-2032.

[28]. D. Nister, H. Stewenius, Scalable recognition with a vocabulary tree, in *Proceedings of the IEEE International Conference on Computer Vision and Pattern Recognition (CVPR'06)*, Vol. 2, pp. 2161-2168.

[29]. E. Meeds, Novelty detection model selection using volume estimation, UTML-TR-2005-004, Technical Report, *University of Toronto,* 2005.

[30]. B. A. Olshausen, D. J. Field, Emergence of simple-cell receptive field properties by learning a sparse code for natural images, *Nature*, Vol. 381, 1996, pp. 607-609.

[31]. R. Goldstone, Perceptual learning, *Annual Review of Psychology*, Vol. 49, 1998, pp. 585-612.

[32]. J. M. Buhmann, T. Lange, U. Ramacher, Image segmentation by networks of spiking neurons, *Neural Computation*, Vol. 17, 2005, pp. 1010-1031.

[33]. M. Tugcu, X. Wang, J. E. Hunter, J. Phillips, D. Noelle, D. M. Wilkes, A computational neuroscience model of working memory with application to robot perceptual learning, in *Proceedings of the Third IASTED International Conference on Computational Intelligence (CI'07)*, Canada, 2-4 July 2007, pp. 120-127.

[34]. X. Wang, M. Tugcu, J. E. Hunter, D. M. Wilkes, Exploration of configural representation in landmark learning using working memory toolkit, *Pattern Recognition Letters*, Vol. 30, Issue 1, 2009, pp. 66-79.

[35]. M. J. Swain, D. Ballard, Color indexing, *International Journal of Computer Vision*, Vol. 7, Issue 1, 1991, pp. 11-32.

[36]. J. E. Hunter, Human motion segmentation and object recognition using Fuzzy rules, in *Proceedings of 14th Annual IEEE International Workshop on Robot and Human Interactive Communication (RO-MAN'05)*, Nashville, TN, 13-15 August 2005, pp. 210-216.

[37]. D. Gabor, Theory of communications, *Journal of Institute of Electrical Engineering*, Vol. 93, Issue 3, 1946, pp. 429-457.

[38]. J. E. Hunter, M. Tugcu, X. Wang, C. Costello, D. M. Wilkes, Exploiting sparse representations in very high-dimensional feature spaces obtained from patch-based processing, *Machine Vision and Applications*, Vol. 22, Issue 3, 2011, pp. 449-460.

[39]. R. N. Shepard, Attention and the metric structure of the stimulus space, *Journal of Mathematical Psychology*, Vol. 1, Issue 1, 1964, pp. 54-87.

[40]. W. R. Garner, The processing of information and structure, *John Wiley & Sons*, New York, USA, 1974.

[41]. R. N. Shepard, Toward a universal law of generalization for psychological science, *Science*, Vol. 237, Issue 4820, 1987, pp. 1317-1323.

[42]. X. Wang, X. L. Wang, D. M. Wilkes, A divide-and-conquer approach for minimum spanning tree-based clustering, *IEEE Transactions on Knowledge and Data Engineering*, Vol. 21, Issue 7, 2009, pp. 945-958.

[43]. D. Lowe, Distinctive image features from scale invariant keypoints, *International Journal of Computer Vision*, Vol. 60, Issue 2, 2004, pp. 91-110.

Chapter 4

Dynamics and Control of a Centrifuge Flight Simulator and a Simulator for Spatial Disorientation

Vladimir Kvrgić

4.1. Introduction

Modern combat aircraft have the capability of developing multi-axis accelerations, especially during the performance of "super-manoeuvres". These "agile" aircraft are capable of unconventional flight with high angles of attack, high agile motions with thrust-vectored propulsion in all 3 aircraft axes, rotations around those axes and accelerations of up to 9 g (g is Earth's acceleration), with acceleration rates (jerk) of up to 9 g/s. These aircraft can fly at unusual orientations. Hence, the devastating effects of the high acceleration forces and the rapid changes of these forces on the pilot's physiology and ability to perform tasks under these flight conditions must be examined.

A human centrifuge is used for the reliable generation of high G onset rates and high levels of sustained G, to test the reactions and the tolerances of the pilots. Here, acceleration force $G=a/g$, $a=(a_n^2+a_t^2+g^2)^{1/2}$ is the magnitude of acceleration acting on the pilot, a_n is normal acceleration, and a_t is the tangential acceleration.

The modern centrifuge flight simulator (CFS) (Fig. 4.1) has the form of a three degree-of-freedom (3DoF) manipulator with rotational axes, where the pilot's head (or chest for some of the training) is considered to be the end-effector. The arm rotation around the vertical (planetary) axis is the main motion that achieves the desired acceleration force. The CFS must achieve velocity, acceleration and jerk of the pilot through suitable rotations of the centrifuge arm about this axis. This chapter presents a centrifuge in which the arm carries a gimballed gondola system, with

Vladimir Kvrgić, Lola Institute, Belgrade, Serbia

two rotational axes providing pitch and roll capabilities. The task of the roll and pitch axes is to direct the acceleration force into the desired direction. It is considered that the pilot's head (chest) is placed in the intersection of the gondola's roll and pitch axes. In this way, the centrifuge produces the orthogonal components G_x, G_y and G_z for the transverse, lateral and longitudinal acceleration force G components, respectively and the orthogonal components $\hat{\omega}_x$, $\hat{\omega}_y$ and $\hat{\omega}_z$ for the roll, pitch, and yaw angular velocity components to simulate the aircraft's acceleration forces and angular velocities.

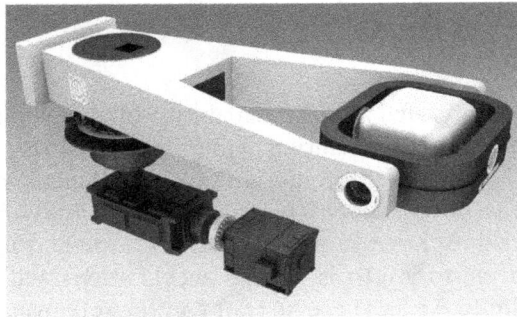

Fig. 4.1. Centrifuge with 3 degrees of freedom.

The presented dynamic environment simulator gimballed centrifuge is aimed not only at improving $+G_z$ tolerance but also at the combined G_y/G_z and G_x/G_z exposure. Multi-axis sustained accelerations can either enhance or reduce the $+G_z$ tolerance of the pilot, depending on the direction of the net gravitoinertial force. The G_y acceleration in conjunction with G_z acceleration can enhance G tolerance. The G_x acceleration in addition to G_z acceleration can reduce the G tolerance. Although the presented centrifuge is capable of generating acceleration forces of up to 15 g for materials testing purposes, forces that are less than or equal to 9 g are used for pilot training.

The spatial disorientation trainer (SDT) examines a pilot's ability to recognise unusual combat aircraft orientations, to adapt to unusual positions and persuade the pilot to believe in the aircraft instruments for orientation and not in his own senses. This trainer offers pilots the situational awareness training. The SDT presented here is a 4DoF manipulator with rotational axes where the pilot's head or chest is considered as the end-effector (Fig. 4.2).

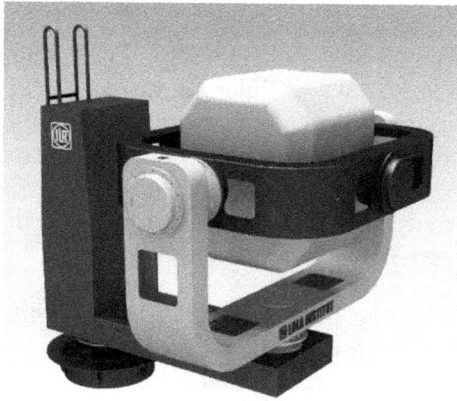

Fig. 4.2. SDT with 4 degrees of freedom.

The control algorithms for the CFS/SDT presented in this chapter calculate their kinematic and dynamic parameters in each interpolation period to predict their dynamic behaviour. This method includes a new algorithms for inverse and forward dynamics of robots that calculates the successive actuator torques and the angular accelerations of the links that are needed for the given motion. The task of forward dynamics is to calculate the joint accelerations taking into account actuator torques and forces and moments acting on the CFS/STD links and pilot. Calculating forward dynamics in a conventional way is extremely complicated. In this chapter, an algorithm which linearizes the nonlinear equations of the robot motion and considerably alleviates the solving of forward dynamics is proposed. Errors due to linearization are negligibly small.

The method determined whether actuators can achieve needed torques and accelerations; if they cannot, it calculates the maximum successive link angular accelerations that the motors can achieve. Instead of sending unachievable commands to actuators, the control unit sends commands that give the maximum possible values for the angular accelerations and speeds. This strategy improves the quality of the motion control and enables a more precise calculation of the forces and moments that act on the CFS/SDT links, which is necessary for the axes bearing and links strength calculations that are performed during its design.

The robot dynamic model can be derived by different methods, such as the Newton–Euler recursive method, the Lagrange equation and the virtual work principle. In that way, the force/torques required for the given motion are calculated – inverse dynamics. The method based on the Lagrange formulation is conceptually simple and systematic. The method based on the Newton–Euler formulation yields a model in a recursive form; it is composed of a forward computation of the velocities and accelerations of each link, followed by a backward computation of the forces and moments in each joint. The Newton-Euler algorithm is computationally more efficient because it exploits the typically open structure of the manipulator kinematic chain.

Another reason why the new approximate algorithm for the inverse dynamics is based on the Newton-Euler equations is because they incorporate all of the forces and moments that act on the robot links. This is essentially for sizing the links and bearings during the design stage.

The presented control algorithms with the given dynamic model can be used for a computer simulation of the CFS/SDT. Through an overview of the behaviour of the model under various operating conditions, it is possible to predict how a real system will behave. Within the design of high-performance machines that imply rapid movements and high values of velocities, such as the motion simulators, the dynamics of the manipulator plays an important role in achieving such high-speed performance. It allows for virtual simulators prototyping and analysis of the interaction between engine control, the simulation of inertial forces acting on the pilot and the structure of the simulators.

A kinematics and dynamics analysis of the centrifuge and SDT, the calculation of the acceleration forces that act on the simulator pilot, and the roll and pitch angles of the gondola for the known forces are given in this chapter. Furthermore, the calculation of the angular acceleration of the centrifuge arm, the smoothing of its profile, the desired and maximal possible values of the CFS/SDT link accelerations, and programming instructions of the centrifuge movement are given. The control algorithm of the centrifuge and SDT movement are also given.

4.2. Kinematics and Dynamics of the Centrifuge

Because of the possibilities of the present actuators, it is very difficult to produce a centrifuge that can realise all of the given changes of the acceleration forces completely accurately. Actuators that have the desired power have very large weights. For this reason, it is necessary to make compromises in the centrifuge design regarding the powers and weights of the chosen motors and the strengths, masses, and mass distribution (mass centre positions and moments of inertia) of the links. The gondola must be large enough to carry the pilot, his seat and the equipment. Easy access for the medical staff in the case of undesired health problems caused by the inertial loads must be provided as well. On the other hand, it has to be as light as possible. The same requirements apply for the roll ring and the centrifuge arm.

For the reason that the forces and moments exerted on link i by link i-1 in link i-1 coordinates do not depend on the position of link 1, the algorithm can calculate them for the angle $q_1=0$. This step reduces the computational cost of the algorithm significantly. For this case, the centrifuge kinematics and dynamics parameters will be as follows.

4.2.1. Forward Geometric Model of the Centrifuge

The roll axis lies in the plane of the arm rotation (around the vertical axis), perpendicular to the main rotational axis, i.e., in the x-axis direction. The pitch (y) axis is perpendicular to the roll axis. Its second axis is perpendicular to the first axis and is along the horizontal line when the centrifuge is in a neutral position (Fig. 4.3). The centrifuge links and their coordinate frames are denoted by using the Denavit-Hartenberg convention (D-H). The base is denoted by 0, the arm by 1, the roll ring by 2 and the gondola by 3. The arm rotation angle is denoted by $q_1=\psi$, the roll ring rotation angle by $q_2=\phi$ and the gondola rotation angle (pitch) by $q_3=\theta$. The centrifuge presented here has the following features: arm length $a_1=8$ m, roll axis rotation range of $\pm180°$ and pitch axis rotation range $\pm360°$. The allowed gondola payload is 250 kg. The centrifuge base coordinates are denoted by $x_0y_0z_0$, the arm coordinates by $x_1y_1z_1$, the roll ring coordinates by $x_2y_2z_2$, the gondola coordinates by $x_3y_3z_3$ and the pilot coordinates by xyz. Here, $x_3=x$, $y_3=y$ and $z_3=z$. The D-H parameters for the 3-axis centrifuge components are given in Table 4.1.

Table 4.1. D-H parameters for the 3-axis centrifuge links.

Link	Variable [°]	a [mm]	d [mm]	α [°]
1	q_1	a_1	0	90
2	q_2+90	0	0	90
3	q_3+90	0	0	-90

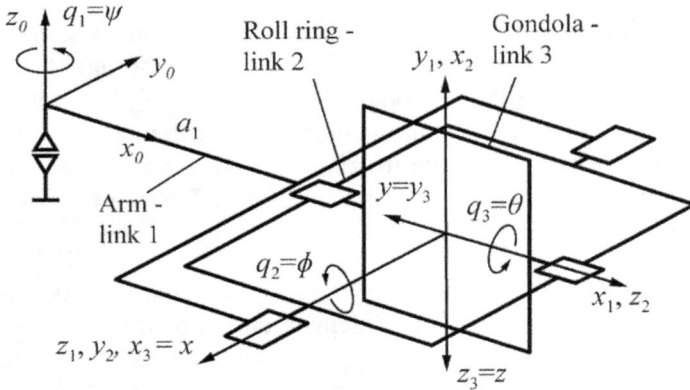

Fig. 4.3. Coordinate frames of the 3-axis centrifuge links.

The 4×4 homogenous matrix that transforms the coordinates of a point from frame $x_n y_n z_n$ to frame $x_m y_m z_m$ is denoted by ${}^n\mathbf{T}_m = \begin{bmatrix} {}^n\mathbf{D}_m & {}^n\mathbf{p}_m \\ 0 \quad 0 \quad 0 & 1 \end{bmatrix}$ and from $x_0 y_0 z_0$ to $x_m y_m z_m$ by $\mathbf{T}_m$. ${}^n\mathbf{D}_m$ is a 3×3 orientation matrix, and ${}^n\mathbf{p}_m$ is a 3×1 position vector. By using the convenient shorthand notation, $\sin(q_i) = s_i$, $\cos(q_i) = c_i$ ($i = 1, 2, 3$) the following homogenous matrices for the relation between the centrifuge link coordinate frames are defined to derive the kinematic equations for the machine, as follows:

$$\mathbf{T}_1 = \mathbf{Rot}(z_0, q_1)\mathbf{Trans}(x_0, a_1)\mathbf{Rot}(x'_0, 90°) = \begin{bmatrix} c_1 & 0 & s_1 & a_1 c_1 \\ s_1 & 0 & -c_1 & a_1 s_1 \\ 0 & 1 & 0 & 0 \\ 0 & 0 & 0 & 1 \end{bmatrix} \quad (4.1)$$

$$^1\mathbf{T}_2 = \mathbf{Rot}(z_1, q_2)\mathbf{Rot}(z_1, 90°)\mathbf{Rot}(x'_1, 90°) = \begin{bmatrix} -s_2 & 0 & c_2 & 0 \\ c_2 & 0 & s_2 & 0 \\ 0 & 1 & 0 & 0 \\ 0 & 0 & 0 & 1 \end{bmatrix} \quad (4.2)$$

$$^2\mathbf{T}_3 = \mathbf{Rot}(z_2, q_3)\mathbf{Rot}(z_2, 90°)\mathbf{Rot}(x'_2, -90°) = \begin{bmatrix} -s_3 & 0 & -c_3 & 0 \\ c_3 & 0 & -s_3 & 0 \\ 0 & -1 & 0 & 0 \\ 0 & 0 & 0 & 1 \end{bmatrix} \quad (4.3)$$

Forward kinematics related to robot geometry is used to calculate the position and orientation of the links and end-effector (i.e., the pilot's head or chest) with respect to the centrifuge variables q_1, q_2 and q_3. These are determined from the following matrices:

$$T_2 = {}^0T_1\, {}^1T_2 = \begin{bmatrix} -c_1 s_2 & s_1 & c_1 c_2 & a_1 c_1 \\ -s_1 s_2 & -c_1 & s_1 c_2 & a_1 s_1 \\ c_2 & 0 & s_2 & 0 \\ 0 & 0 & 0 & 1 \end{bmatrix} \tag{4.4}$$

$$T_3 = {}^0T_1\, {}^1T_2\, {}^2T_3 = \begin{bmatrix} c_1 s_2 s_3 + s_1 c_3 & -c_1 c_2 & c_1 s_2 c_3 - s_1 s_3 & a_1 c_1 \\ s_1 s_2 s_3 - c_1 c_3 & -s_1 c_2 & s_1 s_2 c_3 + c_1 s_3 & a_1 s_1 \\ -c_2 s_3 & -s_2 & -c_2 c_3 & 0 \\ 0 & 0 & 0 & 1 \end{bmatrix} \tag{4.5}$$

4.2.2. Forward Kinematics Related to the Centrifuge Velocities and Accelerations

On the basis of presented transformational matrices, the equations of forward kinematics that relate to the velocities and accelerations of the links and end-effector-pilot's head/chest (for $q_1=0$) are developed.

The centrifuge links angular velocities $\boldsymbol{\omega}_{i+1} = \boldsymbol{\omega}_i + \mathbf{z}_i\, \dot{q}_{i+1}$, $i+1=1, 2, 3$ are:

$$\boldsymbol{\omega}_1 = \begin{bmatrix} 0 & 0 & \dot{q}_1 \end{bmatrix}^T, \quad \boldsymbol{\omega}_2 = \boldsymbol{\omega}_1 + \begin{bmatrix} 0 & -1 & 0 \end{bmatrix}^T \dot{q}_2, \quad \boldsymbol{\omega}_3 = \boldsymbol{\omega}_2 + \begin{bmatrix} c_2 & 0 & s_2 \end{bmatrix}^T \dot{q}_3, \tag{4.6}$$

where $\mathbf{z}_i$ is the unit vector that represents the z_i axis.

The centrifuge links linear velocities $\mathbf{v}_{i+1} = \mathbf{v}_i + \boldsymbol{\omega}_{i+1} \times \mathbf{p}^*_{i+1}$, $i+1=1, 2, 3$ are:

$$\mathbf{v}_1 = \mathbf{v}_2 = \mathbf{v}_3 = a_1 \begin{bmatrix} 0 & 1 & 0 \end{bmatrix}^T \dot{q}_1, \tag{4.7}$$

where $\mathbf{p}^*_{i+1} = \mathbf{p}_{i+1} - \mathbf{p}_i$, $\mathbf{p}^*_1 = \begin{bmatrix} a_1 & 0 & 0 \end{bmatrix}^T$, $\mathbf{p}^*_2 = \mathbf{p}^*_3 = \begin{bmatrix} 0 & 0 & 0 \end{bmatrix}^T$

The centrifuge links angular accelerations $\dot{\boldsymbol{\omega}}_{i+1} = \dot{\boldsymbol{\omega}}_i + \mathbf{z}_i\, \ddot{q}_{i+1} + \boldsymbol{\omega}_i \times \mathbf{z}_i\, \dot{q}_{i+1}$, $i+1=1, 2, 3$ are:

$$\dot{\boldsymbol{\omega}}_1 = \begin{bmatrix} 0 & 0 & \ddot{q}_1 \end{bmatrix}^T \tag{4.8}$$

$$\dot{\boldsymbol{\omega}}_2 = \dot{\boldsymbol{\omega}}_1 + \begin{bmatrix} 0 & -1 & 0 \end{bmatrix}^T \ddot{q}_1 + \begin{bmatrix} 1 & 0 & 0 \end{bmatrix}^T \dot{q}_1 \dot{q}_2 \tag{4.9}$$

$$\dot{\boldsymbol{\omega}}_3 = \dot{\boldsymbol{\omega}}_2 + \begin{bmatrix} c_2 & 0 & s_2 \end{bmatrix}^T \ddot{q}_3 + \begin{bmatrix} 0 & c_2 & 0 \end{bmatrix}^T \dot{q}_1 \dot{q}_3 + \begin{bmatrix} -c_2 & 0 & c_2 \end{bmatrix}^T \dot{q}_2 \dot{q}_3 \tag{4.10}$$

The links linear accelerations $\dot{\mathbf{v}}_{i+1} = \dot{\mathbf{v}}_i + \dot{\boldsymbol{\omega}}_{i+1} \times \mathbf{p}_{i+1}^* + \boldsymbol{\omega}_{i+1} \times (\boldsymbol{\omega}_{i+1} \times \mathbf{p}_{i+1}^*)$, $i+1 = 1, 2, 3$ experienced by the simulator pilot at the intersection point of the centrifuge roll and pitch axes is:

$$\dot{\mathbf{v}}_1 = \dot{\mathbf{v}}_2 = \dot{\mathbf{v}}_3 = \begin{bmatrix} \dot{v}_{1x} & \dot{v}_{1y} & \dot{v}_{1z} \end{bmatrix}^T = a_1 \begin{bmatrix} s_1 \dot{\omega}_1 + c_1 \omega_1^2 & c_1 \dot{\omega}_1 - s_1 \omega_1^2 & 0 \end{bmatrix}^T \tag{4.11}$$

If we assign the position of the link i center of mass (CM) with respect to the link i coordinates expressed in the link i coordinates with $\hat{\mathbf{r}}_i^{cm} = \begin{bmatrix} \hat{r}_{xi} & \hat{r}_{yi} & \hat{r}_{zi} \end{bmatrix}^T$ this vector in the centrifuge base coordinates is $\mathbf{r}_i^{cm} = \begin{bmatrix} r_{xi} & r_{yi} & r_{zi} \end{bmatrix}^T = \mathbf{D}_i \hat{\mathbf{r}}_i^{cm}$ (Fig. 4.4). Here is the external load (weight of the pilot, his seat and equipment) that affects the link 3 marked with 4. Further, we can calculate the linear acceleration of the centrifuge links CM with the following equation:

$$\dot{\mathbf{v}}_i^{cm} = \begin{bmatrix} \dot{v}_{xi}^{cm} & \dot{v}_{yi}^{cm} & \dot{v}_{zi}^{cm} \end{bmatrix}^T = \dot{\mathbf{v}}_i + \dot{\boldsymbol{\omega}}_i \times \mathbf{r}_i^{cm} + \boldsymbol{\omega}_i \times (\boldsymbol{\omega}_i \times \mathbf{r}_i^{cm})$$

Based on this are:

$$\dot{\mathbf{v}}_1^{cm} = \begin{bmatrix} -r_{y1} \\ a_1 + r_{x1} \\ 0 \end{bmatrix} \ddot{q}_1 + \begin{bmatrix} -a_1 - r_{x1} \\ -r_{y1} \\ 0 \end{bmatrix} \dot{q}_1 \dot{q}_1 \tag{4.12}$$

$$\dot{\mathbf{v}}_2^{cm} = \begin{bmatrix} -r_{y2} \\ a_1 + r_{x2} \\ 0 \end{bmatrix} \ddot{q}_1 + \begin{bmatrix} -r_{z2} \\ 0 \\ r_{x2} \end{bmatrix} \ddot{q}_2 + \begin{bmatrix} -a_1 - r_{x2} \\ -r_{y2} \\ 0 \end{bmatrix} \dot{q}_1 \dot{q}_1 + 2 \begin{bmatrix} 0 \\ -r_{z2} \\ 0 \end{bmatrix} \dot{q}_1 \dot{q}_2 + \begin{bmatrix} -r_{x2} \\ 0 \\ -r_{z2} \end{bmatrix} \dot{q}_2 \dot{q}_2 \tag{4.13}$$

$$\dot{\mathbf{v}}_{3,4}^{cm} = \begin{bmatrix} -r_{y3,4} \\ a_1 + r_{x3,4} \\ 0 \end{bmatrix} \ddot{q}_1 + \begin{bmatrix} -r_{z3,4} \\ 0 \\ r_{x3,4} \end{bmatrix} \ddot{q}_2 + \begin{bmatrix} -s_2 r_{y3,4} \\ s_2 r_{x3,4} - c_2 r_{z3,4} \\ c_2 r_{y3,4} \end{bmatrix} \ddot{q}_3 - \begin{bmatrix} a_1 + r_{x3,4} \\ r_{y3,4} \\ 0 \end{bmatrix} \dot{q}_1 \dot{q}_1 + \begin{bmatrix} -r_{x3,4} \\ 0 \\ -r_{z3,4} \end{bmatrix} \dot{q}_2 \dot{q}_2$$

$$+ 2 \begin{bmatrix} 0 \\ -r_{z3,4} \\ 0 \end{bmatrix} \dot{q}_1 \dot{q}_2 + 2 \begin{bmatrix} c_2 r_{z3,4} \\ -s_2 r_{y3,4} \\ 0 \end{bmatrix} \dot{q}_1 \dot{q}_3 + 2 \begin{bmatrix} -c_2 r_{y3,4} \\ 0 \\ -s_2 r_{y3,4} \end{bmatrix} \dot{q}_2 \dot{q}_3 + \begin{bmatrix} s_2 (c_2 r_{z3,4} - s_2 r_{x3,4}) \\ -c_2^2 r_{y3,4} - s_2^2 r_{x3,4} \\ c_2 (s_2 r_{x3,4} - r_{z3,4}) \end{bmatrix} \dot{q}_3 \dot{q}_3$$

$$\tag{4.14}$$

For the centrifuge links shown in this chapter, the mass centre coordinates are: $\hat{\mathbf{r}}_1^{cm} = [-8.344 \; -0.782 \; 0.002]^T$, $\hat{\mathbf{r}}_2^{cm} = [0. \; 0. \; -0.004]^T$, $\hat{\mathbf{r}}_3^{cm} = [0.0 \; 0.0 \; 0.064]^T$ and $\hat{\mathbf{r}}_4^{cm} = [-0.5 \; 0.005 \; 0.55]^T$. Here, we have:

$$\mathbf{r}_1^{cm} = \begin{bmatrix} \hat{r}_{x1} \\ -\hat{r}_{z1} \\ \hat{r}_{y1} \end{bmatrix}, \quad \mathbf{r}_2^{cm} = \begin{bmatrix} c_2\hat{r}_{z2} - s_2\hat{r}_{x2} \\ -\hat{r}_{y2} \\ c_2\hat{r}_{x2} + s_2\hat{r}_{z2} \end{bmatrix}, \quad \mathbf{r}_{3,4} = \begin{bmatrix} s_2\left(s_3\hat{r}_{x3,4} + c_3\hat{r}_{z3,4}\right) - c_2\hat{r}_{y3,4} \\ c_3\hat{r}_{x3,4} + s_3\hat{r}_{z3,4} \\ -c_2\left(s_3\hat{r}_{x3,4} + c_3\hat{r}_{z3,4}\right) - s_2\hat{r}_{y3,4} \end{bmatrix} \quad (4.15)$$

In this chapter's indexing notation, the subscript $3,4$ denotes axis 3 or the external load (weight of the pilot with his seat and equipment).

4.2.3. Centrifuge Dynamics

Inverse dynamics algorithm, which involves the determination of actuator torques required to generate a desired trajectory of the centrifuge and moments and forces in the centrifuge axes, is given in this Section.

The 3x3 matrix on the inertia of link i about the centre of mass of link i (expressed in the base link coordinates) is assigned to $\mathbf{I}_i^{cm}$, and the 3×3 matrix on the inertia of link i about the centre of mass of link i (expressed in link i coordinates) is assigned to $\hat{\mathbf{I}}_i^{cm}$. In matrix $\hat{\mathbf{I}}_i^{cm}$, only the principal moments of inertia $\hat{I}_{xi}$, $\hat{I}_{yi}$ and $\hat{I}_{zi}$ are accounted for; in other words, the cross products of inertia of link i are neglected. Matrix $\mathbf{I}_i^{cm}$ is a function of the position and orientation of link i, whereas $\hat{\mathbf{I}}_i^{cm}$ is constant. Matrix $\mathbf{I}_i^{cm}$ is calculated with:

$$\mathbf{I}_i^{cm} = \mathbf{D}_i\hat{\mathbf{I}}_i^{cm}\mathbf{D}_i^T = \begin{bmatrix} I_{xi} & -I_{xyi} & -I_{xzi} \\ -I_{xyi} & I_{yi} & -I_{yzi} \\ -I_{xzi} & -I_{yzi} & I_{zi} \end{bmatrix}, \text{ where } \hat{\mathbf{I}}_i^{cm} = \begin{bmatrix} \hat{I}_{xi} & 0 & 0 \\ 0 & \hat{I}_{yi} & 0 \\ 0 & 0 & \hat{I}_{zi} \end{bmatrix}, i=1, 2, 3 \quad (4.16)$$

The principal moments of inertia about the link centres of mass in the link coordinate systems and the masses of the centrifuge links are: $\hat{I}_{x1}=91,905$, $\hat{I}_{y1}=219,978$, $\hat{I}_{z1}=243,913$, $\hat{I}_{x2}=3,243$, $\hat{I}_{y2}=1,365$, $\hat{I}_{z2}=2,010$, $\hat{I}_{x3}=666$, $\hat{I}_{y3}=217$, $\hat{I}_{z3}=650$, $\hat{I}_{x4}=47$, $\hat{I}_{y4}=61$, $\hat{I}_{z4}=46$ kgm^2, $m_1=45,5$, $m_2=1,139$, $m_3=566$ and $m_4=250$ kg.

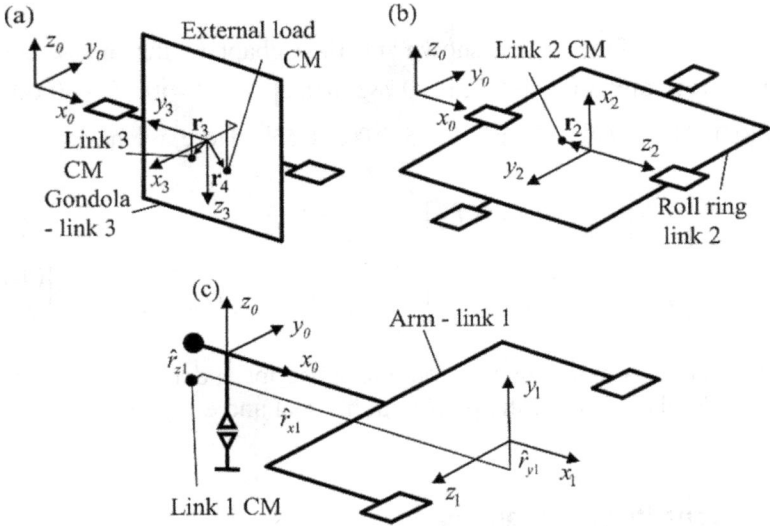

Fig. 4.4. Coordinate frame and the centre of mass of (a) link 3; (b) link 2, and (c) link 1.

In this way,

$$\text{for axis 1: } I_{z1} = \hat{I}_{y1} \text{ (other terms are equal to zero)} \tag{4.17}$$

$$\text{for axis 2: } \mathbf{I}_2^{cm} = \begin{bmatrix} I_{x2} & -I_{xy2} & -I_{xz2} \\ -I_{xy2} & I_{y2} & -I_{yz2} \\ -I_{xz2} & -I_{yz2} & I_{z2} \end{bmatrix} = \begin{bmatrix} \hat{I}_{x2}s_2^2 + \hat{I}_{z2}c_2^2 & 0 & s_2c_2(\hat{I}_{z2} - \hat{I}_{x2}) \\ 0 & \hat{I}_{y2} & 0 \\ s_2c_2(\hat{I}_{z2} - \hat{I}_{x2}) & 0 & \hat{I}_{x2}c_2^2 + \hat{I}_{z2}s_2^2 \end{bmatrix} \tag{4.18}$$

and for axis 3 and external load-4:

$$I_{x3,4} = s_2^2(\hat{I}_{x3,4}s_3^2 + \hat{I}_{z3,4}c_3^2) + \hat{I}_{y3,4}c_2^2, \ I_{y3,4} = -\hat{I}_{x3,4}c_3^2 + \hat{I}_{z3,4}s_3^2$$

$$I_{z3,4} = c_2^2(\hat{I}_{x3,4}s_3^2 + \hat{I}_{z3,4}c_3^2) + \hat{I}_{y3,4}s_2^2, \ I_{xy3,4} = s_2s_3c_3(\hat{I}_{x3,4} - \hat{I}_{z3,4}), \tag{4.19}$$

$$I_{xz3,4} = c_2(s_2(\hat{I}_{x3}s_3s_3 + \hat{I}_{z3,4}c_3c_3) - \hat{I}_{y3}s_2), \ I_{yz3,4} = c_2s_3c_3(-\hat{I}_{x3,4} + \hat{I}_{z3,4})$$

The total force $\mathbf{F}_i$ and the total moment $\mathbf{M}_i$ exerted on link i are obtained from the Newton-Euler equation:

$$\mathbf{F}_i = \begin{bmatrix} F_{xi} & F_{yi} & F_{zi} \end{bmatrix}^T = m_i \begin{bmatrix} \dot{v}_{xi}^{cm} & \dot{v}_{yi}^{cm} & \dot{v}_{zi}^{cm} - g \end{bmatrix}^T$$

$$\mathbf{M}_i = \begin{bmatrix} M_{xi} & M_{yi} & M_{zi} \end{bmatrix}^T = \mathbf{I}_i^{cm}\dot{\boldsymbol{\omega}}_i + \boldsymbol{\omega}_i \times (\mathbf{I}_i^{cm}\boldsymbol{\omega}_i).$$

The total forces $\mathbf{F}_1$, $\mathbf{F}_2$, $\mathbf{F}_3$ and $\mathbf{F}_4$ can be obtained from Eqs. (4.12)-(4.14). The total moments are:

$$\mathbf{M}_1 = \begin{bmatrix} 0 & 0 & I_{z1} \end{bmatrix}^T \ddot{q}_1 \tag{4.20}$$

$$\mathbf{M}_2 = \begin{bmatrix} -I_{xz2} \\ -I_{yz2} \\ I_{z2} \end{bmatrix} \ddot{q}_1 + \begin{bmatrix} -(-I_{xy2}) \\ -I_{y2} \\ -(-I_{yz2}) \end{bmatrix} \ddot{q}_2 + \begin{bmatrix} -(-I_{yz2}) \\ (-I_{xz2}) \\ 0 \end{bmatrix} \dot{q}_1^2 + \begin{bmatrix} I_{x2}+I_{y2}-I_{z2} \\ 0 \\ 2(-I_{xz2}) \end{bmatrix} \dot{q}_1\dot{q}_2 - \begin{bmatrix} (-I_{yz2}) \\ (-I_{yz2}) \\ (-I_{xy2}) \end{bmatrix} \dot{q}_2^2 \tag{4.21}$$

$$\mathbf{M}_{3,4} = \begin{bmatrix} -I_{xz3,4} \\ -I_{yz3,4} \\ I_{z3,4} \end{bmatrix} \ddot{q}_1 + \begin{bmatrix} I_{xy3,4} \\ -I_{y3,4} \\ I_{yz3,4} \end{bmatrix} \ddot{q}_2 + \begin{bmatrix} c_2 I_{x3,4}-s_2 I_{xz3,4} \\ -c_2 I_{xy3,4}-s_2 I_{yz3,4} \\ s_2 I_{z3,4}-c_2 I_{xz3,4} \end{bmatrix} \ddot{q}_3 + \begin{bmatrix} I_{x3,4}+I_{y3,4}-I_{z3,4} \\ 0 \\ -2I_{xz3,4} \end{bmatrix} \dot{q}_1\dot{q}_2$$

$$+ \begin{bmatrix} 2s_2 I_{yz3,4} \\ c_2(I_{x3,4}+I_{y3,4}-I_{z3,4})-2s_2 I_{xz3,4} \\ -2c_2 I_{yz3,4} \end{bmatrix} \dot{q}_1\dot{q}_3 + \begin{bmatrix} I_{yz3,4} \\ -I_{xz3,4} \\ 0 \end{bmatrix} \dot{q}_1^2 + \begin{bmatrix} I_{xz3,4}s_1-I_{yz3,4} \\ 0 \\ I_{xy3,4} \end{bmatrix} \dot{q}_2^2 \tag{4.22}$$

$$+ \begin{bmatrix} -s_2(I_{x3,4}-I_{y3,4}+I_{z3,4}) \\ 2(s_2 I_{xy3,4}-c_2 I_{yz3,4}) \\ -c_2(I_{y3,4}-I_{x3,4}-I_{z3,4}) \end{bmatrix} \dot{q}_2\dot{q}_3 + \begin{bmatrix} s_2 c_2 I_{xy3}-(c_2^2-s_2^2)I_{yz3,4} \\ s_2 c_2[(I_{x3,4}-I_{z3,4})]+(c_2^2-s_2^2)I_{xz3,4} \\ c_2^2 I_{xy3,4}-c_2 s_2 I_{yz3,4} \end{bmatrix} \dot{q}_3^3$$

We will assign the force vector with $\mathbf{f}_i = \begin{bmatrix} f_{xi} & f_{yi} & f_{zi} \end{bmatrix}^T$ and the moment vector with $\mathbf{m}_i = \begin{bmatrix} m_{xi} & m_{yi} & m_{zi} \end{bmatrix}^T$ exerted on link i by link i-1 with respect to the base coordinate frame. From robot dynamics, it is known that the total force and the total moment exerted on link i are: $\mathbf{F}_i = \mathbf{f}_i - \mathbf{f}_{i+1}$ and $\mathbf{M}_i = \mathbf{m}_i - \mathbf{m}_{i+1} - (\mathbf{p}_i^* + \mathbf{r}_i^{cm}) \times \mathbf{f}_i + \mathbf{r}_i^{cm} \times \mathbf{f}_{i+1}$. Solving for $\mathbf{f}_i$ and $\mathbf{m}_i$ gives $\mathbf{f}_i = \mathbf{F}_i + \mathbf{f}_{i+1}$ and $\mathbf{m}_i = \mathbf{M}_i + \mathbf{m}_{i+1} + (\mathbf{p}_i^* + \mathbf{r}_i^{cm}) \times \mathbf{F}_i + \mathbf{p}_i^* \times \mathbf{f}_{i+1}$. The forces and moments exerted on link i by link i-1 in link i-1 coordinates are:

$$\hat{\mathbf{f}}_i = \begin{bmatrix} \hat{f}_{xi} & \hat{f}_{yi} & \hat{f}_{zi} \end{bmatrix}^T = \mathbf{D}_{i-1}^T \mathbf{f}_i \text{ and } \hat{\mathbf{m}}_i = \begin{bmatrix} \hat{m}_{xi} & \hat{m}_{yi} & \hat{m}_{zi} \end{bmatrix}^T = \mathbf{D}_{i-1}^T \mathbf{m}_i \tag{4.23}$$

The effect of the external forces and moments on link 3 will be represented by $\mathbf{f}_4$ and $\mathbf{m}_4$ (Fig. 4.5(a)). In this way, $\mathbf{F}_4 = \mathbf{f}_4$ and $\mathbf{M}_4 = \mathbf{m}_4 - \mathbf{r}_4 \times \mathbf{F}_4$. The forces and the moments of the external load that exert on link 3 with respect to the base coordinate frame are:

$$\mathbf{f}_4 = \mathbf{F}_4, \text{ where } \mathbf{F}_4 = m_4 \begin{bmatrix} \dot{v}_{x4}^{cm} & \dot{v}_{y4}^{cm} & \dot{v}_{z4}^{cm} - g \end{bmatrix}^T \tag{4.24}$$

$$\mathbf{m}_4 = \mathbf{M}_4 + \mathbf{r}_4 \times \mathbf{F}_4 = \mathbf{M}_4 + \left[r_{y4} F_{z4} - r_{z4} F_{y4} \quad r_{z4} F_{x4} - r_{x4} F_{z4} \quad r_{x4} F_{y4} - r_{y4} F_{x4} \right]^T \quad (4.25)$$

The total force and the total moment that exert on link 3 are: $\mathbf{F}_3 = \mathbf{f}_3 - \mathbf{f}_4$ and $\mathbf{M}_3 = \mathbf{m}_3 - \mathbf{m}_4 - \mathbf{r}_3 \times \mathbf{F}_3$. Based on this, the total force and the total moment that exert on link 3 by link 2 with respect to the base coordinates are:

$$\mathbf{f}_3 = \mathbf{F}_3 + \mathbf{f}_4, \text{ where } \mathbf{F}_3 = m_3 \left[\dot{v}_{x3}^{cm} \quad \dot{v}_{y3}^{cm} \quad \dot{v}_{z3}^{cm} - g \right]^T \quad (4.26)$$

$$\mathbf{m}_3 = \mathbf{M}_3 + \mathbf{r}_3 \times \mathbf{F}_3 + \mathbf{m}_4 = \mathbf{M}_3 + \left[r_{y3} F_{z3} - r_{z3} F_{y3} \quad r_{z3} F_{x3} - r_{x3} F_{z3} \quad r_{x3} F_{y3} - r_{y3} F_{x3} \right]^T + \mathbf{m}_4 \quad (4.27)$$

According to Eq. (4.23) this force and moment in link 2 coordinates are:

$$\hat{\mathbf{f}}_3 = \left[-s_2 f_{x3} + c_2 f_{z3} \quad -f_{y3} \quad c_2 f_{x3} + s_2 f_{z3} \right]^T \quad (4.28)$$

$$\hat{\mathbf{m}}_3 = \left[-s_2 m_{x3} + c_2 m_{z3} \quad -m_{y3} \quad c_2 m_{x3} + s_2 m_{z3} \right]^T \quad (4.29)$$

The total force and total moment that exert on link 2 are: $\mathbf{F}_2 = \mathbf{f}_2 - \mathbf{f}_3$, $\mathbf{M}_2 = \mathbf{m}_2 - \mathbf{m}_3 - \mathbf{r}_2 \times \mathbf{F}_2$. Consequently, the force and moment that exert on link 2 by link 1 with respect to the base coordinates are (Fig. 4.5(b)):

$$\mathbf{f}_2 = \mathbf{F}_2 + \mathbf{f}_3 = m_2 \left[\dot{v}_{x2}^{cm} \quad \dot{v}_{y2}^{cm} \quad \dot{v}_{z2}^{cm} - g \right]^T + \mathbf{f}_3 \quad (4.30)$$

$$\mathbf{m}_2 = \mathbf{M}_2 + \mathbf{r}_2 \times \mathbf{F}_2 + \mathbf{m}_3 = \mathbf{M}_2 +$$
$$+ \left[r_{y2} F_{z2} - r_{z2} F_{y2} \quad r_{z2} F_{x2} - r_{x2} F_{z2} \quad r_{x2} F_{y2} - r_{y2} F_{x2} \right]^T + \mathbf{m}_3 \quad (4.31)$$

According to Eq. (4.23), this force and moment in link 1 coordinates are:

$$\hat{\mathbf{f}}_2 = \left[f_{x2} \quad f_{z2} \quad -f_{y2} \right]^T, \quad \hat{\mathbf{m}}_2 = \left[m_{x2} \quad m_{z2} \quad -m_{y2} \right]^T \quad (4.32)$$

The total force and total moment that exert on link 1 are: $\mathbf{F}_1 = \mathbf{f}_1 - \mathbf{f}_2$ and $\mathbf{M}_1 = \mathbf{m}_1 - \mathbf{m}_2 - (\mathbf{p}_1^* + \mathbf{r}_1) \times \mathbf{F}_1 - \mathbf{p}_1^* \times \mathbf{f}_2$. Consequently, the force and moment that exert on link 1 with respect to the base coordinates are (Fig. 4.5(c)):

$$\mathbf{f}_1 = \mathbf{F}_1 + \mathbf{f}_2 = m_1 \left[\dot{v}_{x1}^{cm} \quad \dot{v}_{y1}^{cm} \quad \dot{v}_{z1}^{cm} - g \right]^T + \mathbf{f}_2 \quad (4.33)$$

$$\mathbf{m}_1 = \mathbf{M}_1 + \mathbf{m}_2 + (\mathbf{p}_1^* + \mathbf{r}_1) \times \mathbf{F}_1 + \mathbf{p}_1^* \times \mathbf{f}_2$$

$$= \mathbf{M}_1 + \begin{bmatrix} r_{y1}F_{z1} - r_{z1}F_{y1} \\ -(a_1 + r_{x1})F_{z1} + r_{z1}F_{x1} - a_1 f_{z2} \\ (a_1 + r_{x1})F_{y1} - r_{y1}F_{x1} + a_1 f_{y2} \end{bmatrix} + \mathbf{m}_2 \tag{4.34}$$

(a)   (b)

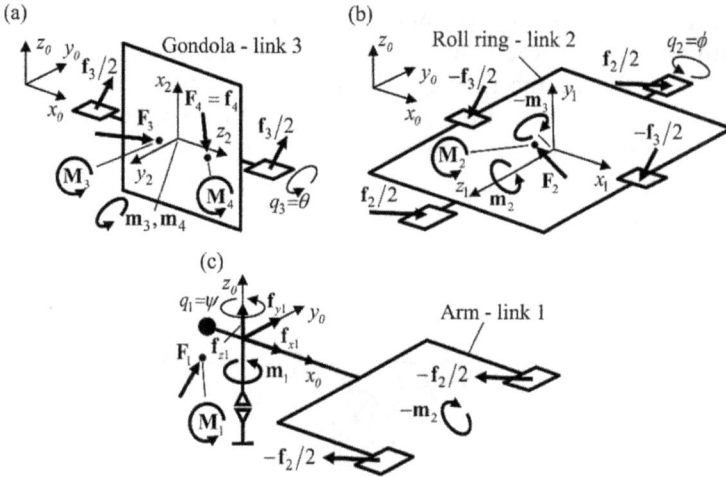

(c)

Fig. 4.5. Forces and moments exerted on (a) link 3; (b) link 2, and (c) link 1.

Based on Eq. (4.23), this force and moment with respect to the base coordinate frame are:

$$\hat{\mathbf{m}}_1 = \begin{bmatrix} \hat{m}_{x1} & \hat{m}_{y1} & \hat{m}_{z1} \end{bmatrix}^T = \begin{bmatrix} m_{x1} & m_{y1} & m_{z1} \end{bmatrix}^T, \quad \hat{\mathbf{f}}_1 = \begin{bmatrix} \hat{f}_{x1} & \hat{f}_{y1} & \hat{f}_{z1} \end{bmatrix}^T = \begin{bmatrix} f_{x1} & f_{y1} & f_{z1} \end{bmatrix}^T \tag{4.35}$$

The torque of the joint i actuator is $\hat{m}_{zi}$, and the moment acting on the bearing i is:

$$\hat{m}_{xyi} = (\hat{m}_{xi}^2 + \hat{m}_{yi}^2)^{1/2} \tag{4.36}$$

The axial force of the bearing i is $\hat{f}_{ai} = \hat{f}_{zi}$, and radial force of the bearing i is:

$$\hat{f}_{ri} = (\hat{f}_{xi}^2 + \hat{f}_{yi}^2)^{1/2} \tag{4.37}$$

105

4.3. Acceleration Forces and Link Angles of the Centrifuge

4.3.1. Calculation of the Simulator Pilot Acceleration Force Components

Fig. 4.6 shows the transverse G_x, lateral G_y and longitudinal G_z acceleration force G components that act on the pilot's head (chest) in the simulator. The three main axes of the coordinate frame attached to the human body are: the x axis, which extends from the face to the back, the y axis, which extends from the pilot's right to the pilot's left side, and the z axis, which extends from the head to the pelvis. Fig. 4.7 shows coordinate frames, angles, angular velocities and acceleration forces of the centrifuge.

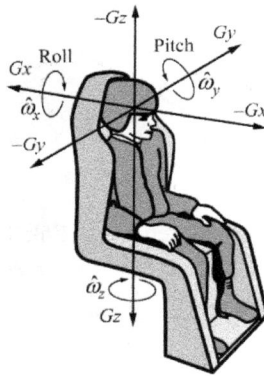

Fig. 4.6. The transverse, lateral and longitudinal acceleration force components G_x, G_y and G_z, acting on the pilot in the simulator.

Based on Eq. (4.11) for $q_1=0$ and adding the gravitational acceleration g, the orthogonal components G_n, G_t and G_v for the normal (radial), tangential and vertical acceleration force G components, respectively, which are experienced by the simulator pilot are the following:

$$\begin{bmatrix} G_{x0} \\ G_{y0} \\ G_{z0} \end{bmatrix} = \frac{1}{g} \begin{bmatrix} -a_n \\ a_t \\ -g \end{bmatrix} = \begin{bmatrix} a_1\omega_1^2/g \\ -a_1\dot{\omega}_1/g \\ -1 \end{bmatrix} = \begin{bmatrix} G_n \\ -G_t \\ -G_v \end{bmatrix} \qquad (4.38)$$

The link angles $q_2=\phi$ and $q_3=\theta$, the angular velocity ω_1, and the angular acceleration $\dot{\omega}_1$ of the arm define the orthogonal components G_x, G_y and G_z of the resultant vector **G** that are experienced by the simulator pilot. Based on Eqs. (4.4), (4.5) and (4.37), the resultant vector **G** is:

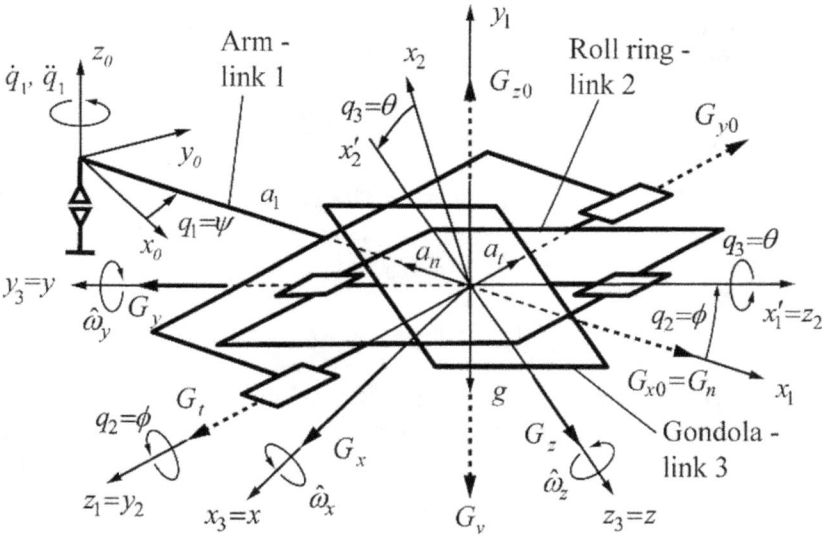

Fig. 4.7. Coordinate frames, angles, angular velocities and acceleration forces of the centrifuge.

$$\mathbf{G}=\begin{bmatrix} G_x & G_y & G_z \end{bmatrix}^T = \mathbf{D}_3^{-1}\begin{bmatrix} G_{x0} & G_{y0} & G_{z0} \end{bmatrix}^T \qquad (4.39)$$

$$G_x = s_3(G_{x0}s_2 + c_2) - G_{y0}c_3 \qquad (4.40)$$

$$G_y = -G_{x0}c_2 + s_2 \qquad (4.41)$$

$$G_z = c_3(G_{x0}s_2 + c_2) + G_{y0}s_3 \qquad (4.42)$$

Angles $q_1=\psi$, $q_2=\phi$ and $q_3=\theta$ and their derivatives define the roll, pitch and yaw angular velocities $\hat{\omega}_x$, $\hat{\omega}_y$ and $\hat{\omega}_z$, which are experienced by the simulator pilot; they are given in the following equations (for $q_1=0$):

$$\hat{\omega} = \begin{bmatrix} \hat{\omega}_x \\ \hat{\omega}_y \\ \hat{\omega}_z \end{bmatrix} = \mathbf{D}_3^{-1}\boldsymbol{\omega}_3 = \begin{bmatrix} \dot{q}_2 c_3 - \dot{q}_1 c_2 s_3 \\ -\dot{q}_3 - \dot{q}_1 s_2 \\ -\dot{q}_2 s_3 - \dot{q}_1 c_2 c_3 \end{bmatrix} \tag{4.43}$$

4.3.2. Calculation of the Centrifuge Roll and Pitch Angles

The centrifuge roll angle is calculated by Eq. (4.11), which uses the given lateral force G_y, in the following way:

$$q_2 = \phi = \text{atan }2(G_{x0} + G_y(1 - G_y^2 + G_{x0}^2)^{1/2}, 1 - G_y^2) \tag{4.44}$$

If $G_y < 0$ and $G_y^2 > 1$, then $q_2 = q_2 + \pi$. The roll angle can be calculated only if $G_{x0}^2 + 1 \geq G_y^2$. Otherwise, it is not possible to achieve the given lateral force G_y. For $G_y = 0$, Eq. (4.44) yields:

$$q_2 = \phi = \text{atan }2(G_{x0}, 1) \tag{4.45}$$

Eqs. (4.40) and (4.42) show that it is not possible to achieve both of the given G_x and G_z forces, even when they are in the allowed ranges. As a result, the centrifuge pitch angle is calculated by Eq. (4.40), using the given transverse force G_x, or by Eq. (4.12), using the given longitudinal G_z force. Eq. (4.40) yields:

$$q_3 = \theta = \text{atan }2(G_{y0}b + G_x(b^2 + G_{y0}^2 - G_x^2)^{1/2}, b^2 - G_x^2), \tag{4.46}$$

where $b = G_{x0}s_2 + c_2$. If $b^2 + G_{y0}^2 < G_x^2$, then it is not possible to achieve the given transverse force G_x. For $G_x = 0$, Eq. (4.45) yields:

$$q_3 = \theta = \text{atan }2(G_{y0}, b) \tag{4.47}$$

Eq. (4.12) yields the following:

$$q_3 = \theta = \text{atan }2(G_{y0}d - G_z(b^2 + G_{y0}^2 - G_z^2)^{1/2}, G_z^2 - G_{y0}^2) \tag{4.48}$$

If $G_z < 0$ and $G_z^2 > G_{y0}^2$, then $q_3 = q_3 - \pi$. If $b^2 + G_{y0}^2 < G_z^2$, then it is not possible to achieve the given longitudinal G_z force. Basic pilot training implies that $G_z = G$ ($G_x = 0$ and $G_y = 0$). Consequently, the roll and pitch angles are given by Eqs. (4.45) and (4.47).

4.4. The Control Algorithm of the Centrifuge Movement

4.4.1. Calculation of the Centrifuge Arm Angular Acceleration $\ddot{q}_1$

Eq. (4.8) gives the resulting force that is experienced by the simulator pilot at the intersection point of the roll and pitch axes (for $q_1=0$) as a function of the angular velocity and acceleration of the centrifuge arm, which is:

$$G=(G_{x0}^2+G_{y0}^2+G_{z0}^2)^{1/2}=\frac{1}{g}(a_n^2+a_t^2+g^2)^{1/2}=\frac{1}{g}[a_1^2(\dot{q}_1^4+\ddot{q}_1^2)+g^2]^{1/2} \quad (4.49)$$

According to the requirement that the increase in the acceleration force G should be constant and equal to n, the following is valid: $dG/dt=n$, which yields $d([a_1^2(\dot{q}_1^4+\ddot{q}_1^2)+g^2]^{1/2})=n\,g\,dt$. If we assign the resulting acceleration with $a=G\,g$, then the previous equation will be:

$$da=d([a_1^2(\dot{q}_1^4+\ddot{q}_1^2)+g^2]^{1/2})=n\,g\,dt \quad (4.50)$$

The previous differential equation does not have a solution in the general case.

In each interpolation period, the robot controller determines the angular velocities of each motor link. An interpolation period of $\Delta t=0.005$ s is adopted here. During this period, the servo system of the controller compares (every 0.001 s) the given and achieved motor rotor positions and corrects rotor angular velocities with the aim of keeping them constant within this period. Based on these observations, an approximated solution from Eq. (4.48) using a discretisation technique is obtained in the following manner. This approach allows us to solve this differential equation for each interpolation period Δt, which simplifies the solution. For the given rate of change of acceleration $\Delta a/\Delta t=ng$, the acceleration a will first be calculated on the basis of this acceleration in the previous interpolation period, a_{prev}, in the following way:

$$a = a_{prev} + \Delta a, \quad \Delta a=n\,g\,\Delta t \quad (4.51)$$

If we assign the angular velocity of the centrifuge arm in the previous interpolation period with $\dot{q}_{1prev}$, we obtain:

$$\dot{q}_1 = \dot{q}_{1prev} + \ddot{q}_1 \, \Delta t \qquad (4.52)$$

If we substitute $\dot{q}_1$, calculated in this manner, into the equation $a^2 = a_1^2 \dot{q}_1^4 + a_1^2 \ddot{q}_1^2 + g^2$ and neglect the terms with Δt^3 and Δt^4, the following equation for calculating the centrifuge arm acceleration is obtained:

$$\ddot{q}_1 = \frac{-2\dot{q}_{1prev}^3 \, \Delta t + [(1 + 6\dot{q}_{1prev}^2 \Delta t^2)(a^2 - g^2)/a_1^2 - 2\dot{q}_{1prev}^6 \Delta t^2 - \dot{q}_{1prev}^4]^{1/2}}{1 + 6\dot{q}_{1prev}^2 \Delta t^2}$$

$$(4.53)$$

The previous equation is valid for the movement that has a positive acceleration onset. For the movement that has a negative acceleration onset, the discriminant $(1 + 6\dot{q}_{1prev}^2 \Delta t^2)(a^2 - g^2)/a_1^2 - 2\dot{q}_{1prev}^6 \Delta t^2 - \dot{q}_{1prev}^4$ is mostly negative, which means that this equation cannot be used directly. In that case, a simple solution is used, in which the values of $\ddot{q}_1$ for the positive acceleration onset n of the same magnitude are reversed. Solving Eq. (4.50) for every interpolation period in the form of Jacobi elliptic integrals gives less precise solution.

4.4.2. Smoothing the Acceleration Force *G* Profile

While a normal acceleration $a_n = a_1 \dot{q}_1^2$ and gravity g act all the time during the centrifuge movement, the tangential acceleration $a_t = a_1 \ddot{q}_1$ does not act during the movement that has a constant load G. The transition from a varying to a constant acceleration causes a sudden increase in $\dot{q}_1$, while the transition from a constant to a varying acceleration causes a sudden decrease in $\dot{q}_1$. The abrupt change in $\dot{q}_1$ causes an abrupt change in q_2 (which can be seen from Eqs. (4.44) and (4.45) and a very large value of $\ddot{q}_1$, which causes an abrupt change in q_3; these relationships can be seen from Eqs. (4.46) and (4.47). To prevent the abrupt change in $\dot{q}_1$, it is adopted that before and after the given linear change in G, i.e., acceleration a, there is a period of smoothing of the acceleration curve. The change in the acceleration has three different stages. Within the first period, smoothing is performed in such a way that the acceleration onset n changes linearly from 0 to a given value (jounce, or snap – the rate of change of the jerk with respect to time). Thereafter, the total acceleration change is denoted by a_e - a_s, where a_s is the initial value of the acceleration, and a_e is the desired acceleration value. The linear acceleration change part of the total acceleration change is denoted by c_{ac}. This part varies depending on the size of the absolute value of a_e - a_s. Here, c_{ac} =0.12 for abs(a_e - a_s) < 1.59 g, and c_{ac} =0.8 for abs(a_e - a_s) > 10.59 g. If 1.59 g < abs(a_e - a_s) < 10.59 g, then c_{ac} changes linearly from 0.12 to 0.8. Within the smoothing algorithm, it is necessary to determine the following parameters: N_a, which is the number of interpolation periods in the starting or ending stage of the G load change; N_c, which is the number of interpolation periods for n=const; n_d, which is the desired acceleration rate of change; and Δn, which is the increase/decrease of n in one interpolation period at the starting or ending stage of the acceleration change. If we assign sign (a_e - a_s)=1 if a increases and sign(a_e - a_s)=-1 if a decreases, then these parameters will be calculated in the following way:

$$N_c = c_{ac}\,(a_e - a_s)/(n_d\ g\ \Delta t) \qquad (4.54)$$

$$N_a = sign(a_e - a_s)\,[(a_e - a_s)/(n_d\ g\ \Delta t) - sign(a_e - a_s)\,N_c] - 1 \quad (4.55)$$

$$\Delta n = n_d/N_a \qquad (4.56)$$

$$n = \frac{sign(a_e - a_s)\,(a_e - a_s) - (N_a + N_c + 1)\,n_d\ g\ \Delta t}{(2N_a + N_c)\ g\ \Delta t} \qquad (4.57)$$

Even though the acceleration profile is smoothed in transitions from a varying to a constant acceleration a (i.e., G and the angular velocity $\dot{q}_1$), in the first interpolation period after the transition, a small tangential acceleration appears. This affects the acceleration load G, which has already been achieved. To nullify these effects, it is necessary to make a small correction in the angular velocity $\dot{q}_1$, which gives a new angular acceleration $\ddot{q}_1 = (\dot{q}_1 - \dot{q}_{1prev})/\Delta t$ of axis 1.

4.4.3. Calculation of the Desired and Maximal Possible Values of $\ddot{q}_1$, $\ddot{q}_2$ and $\ddot{q}_3$

When robot's actuators cannot achieve the desired motion, the control unit sends to the actuators inputs - velocities (positions) which they cannot achieve. Although the control inputs are limited to the maximal possible values, in some interpolation periods the differences between the given and the achieved positions of some links (servo errors) can be too great. Large allowed servo errors reduce the safety during the robot's work. To improve the quality of the robot control, it is very useful to give, in each interpolation period, realistic (feasible) link positions and to allow only acceptably small servo errors. This can increase the safety of the motion and not lessen the smoothing of the link motions.

In addition, the computation of the axes forces and moments based on desired but unachievable velocities and accelerations of the robot links would give bigger values of these forces and moments than existed. This will cause over-dimensioning of the axes bearings and links that will unnecessary increase the robot mass.

For the given reasons, in [11] is given a new algorithm for the calculation of the maximum possible values of the CFS axes accelerations in each interpolation period. It limits maximum given axes velocities as well. In this way, the real forces and moments that act on each link can be calculated. Similar algorithm for the SDT is given in [17].

Because of the possibilities of present actuators, it is very difficult for the motor actuating axis 1 of the centrifuge to achieve complete accuracy given the linear increase in the acceleration force of 9 g/s from the lower baseline level of 1.41 g to the upper baseline level of, for example, 9 g or 15 g, and for the motors of axes 2 and 3 to provide that acceleration components G_x and G_y have zero values during this whole acceleration increase. To accomplish such a demanding request, all of the motors should have been more powerful than the motors that are usually used on the centrifuges. These motors are very heavy and, thus, are unacceptable, because it is very important that the roll ring and the gondola, with their motors, are as light as possible.

The centrifuge arm motion greatly affects the inertial forces that act on the links 2 and 3. The motion of the ink 2 greatly affects the inertial forces and moments that act not only on the link 2 but also on the link 3, and vice versa. Conversely, the motions of the links 2 and 3 have a very small influence on the inertial forces and the moments that act on the link 1 because of its size and mass. For this reason, for each interpolation period, the maximal value of $\ddot{q}_1$ can be examined separately, while the maximal values of $\ddot{q}_2$ and $\ddot{q}_3$ must be examined together.

4.4.3.1. Calculation of the Maximal Possible Value of $\ddot{q}_1$

For axis 1, the DC motor Siemens 1GG5 635-5EV 720V is chosen with a rated torque of $M_{1r}=41200$ Nm, a rated power of $P_{1r}=1610$ kW, a rated number of revolutions of $n_{1r}=374$ min^{-1}, and a maximal number of revolutions through field weakening of $n_{1max}=680$ min^{-1}. If the motor number of revolutions is assigned to be n_{1m}, then for $n_{1m} \leq n_{1r}$, it will be $M_{1r}=const$ and $P_1=P_{1r}n_{1m}/n_{1r}$, and for $n_{1m}>n_{1r}$, it will be $M_{1r}=P_{1r}n_{1r}/n_{1m}$ and $P_1=P_{1r}=const$. If we assign the short time maximal torque (until 5 s) to be M_{1max}, then the overload capability will be $f_{1p}=M_{1max}/M_{1r}=2$. The chosen gearbox has a gear ratio of $k_1=n_{1m}\pi/(30\dot{q}_{1r})=16.5$ and an efficiency of $\eta_1=0.94$. Now, the rated angular speed of link 1 is $\dot{q}_{1r}=2.3736$ s^{-1} and maximal angular speed of link 1 $\dot{q}_{1max}=4.3157$ s^{-1}. The moment of inertia of the rotor of the motor with the gear box elements brought down on that rotor is $I_1=400$ kgm^2.

First, the maximal possible value $\ddot{q}_{1max}$ of $\ddot{q}_1$ will be calculated. If the angular velocity $\dot{q}_1 \in [-\dot{q}_{1r}, \dot{q}_{1r}]$, then we have the following:

$$\text{if } \ddot{q}_1 \, sign(a_e - a_s) \geq 0, \text{ then } \ddot{q}_{1max}=b_1\,\eta_1 \text{ and} \qquad (4.58)$$

$$\text{if } \ddot{q}_1 \, sign(a_e - a_s) < 0, \text{ then } \ddot{q}_{1max}=b_1/\eta_1 \qquad (4.59)$$

If the angular velocity $\dot{q}_1 \in [-\dot{q}_{1max}, -\dot{q}_{1r}]$ or $\dot{q}_1 \in [\dot{q}_{1r}, \dot{q}_{1max}]$, then we have the following:

$$\text{if } \ddot{q}_1 \, sign(a_e - a_s) \geq 0, \text{ then } \ddot{q}_{1max}= b_1\, n_{1r}\, \eta_1\, \pi/(30\,\dot{q}_1) \text{ and} \quad (4.60)$$

$$\text{if } \ddot{q}_1 \, sign(a_e - a_s) < 0, \text{ then } \ddot{q}_{1max}=b_1\, n_{1r}\, \pi/(30\,\dot{q}_1\,\eta_1) \qquad (4.61)$$

If $\ddot{q}_1 > \ddot{q}_{1max}$, then $\ddot{q}_1 = \ddot{q}_{1max}$, and if $\ddot{q}_1 < -\ddot{q}_{1max}$, then $\ddot{q}_1 = -\ddot{q}_{1max}$. Next, we calculate $\dot{q}_1$ by Eq. (4.52). If $\dot{q}_1 > \dot{q}_{1max}$, then $\dot{q}_1 = \dot{q}_{1max}$, and if $\dot{q}_1 < -\dot{q}_{1max}$, then $\dot{q}_1 = -\dot{q}_{1max}$. Now, we have $\ddot{q}_1 = (\dot{q}_1 - \dot{q}_{1prev})/\Delta t$ and $q_1 = q_{1prev} + \dot{q}_1 \Delta t$, with $\dot{q}_{1prev}=\dot{q}_1$, and $q_{1prev}=q_1$. In Eqs. (4.58) – (4.61), we have $b_1=M_{1r}\,f_{1p}\,k_1/I_{t1}$, where the total moment of inertia I_{t1} of all three links and the external load, reduced to the axis z_0 (for $q_1=0$) is as follows:

$$I_{t1}=I_1k_1+\sum_{i=1}^{4}(I_{zi}+[(^{0}r_{xi})^2+(^{0}r_{yi})^2+(^{0}r_{zi})^2]m_i) \qquad (4.62)$$

The values of I_{zi}, $i=1, 2, 3, 4$ are given in Eqs. (4.17), (4.18) and (4.19), respectively; k_1 represents the gear ratio, and I_1 is the moment of inertia of the motor and gear box of axis 1. The components of the vectors $^{0}\mathbf{r}_i^{cm}$ are:

$$^{0}\mathbf{r}_1^{cm}=\begin{bmatrix}^{0}r_{x1} & ^{0}r_{y1} & ^{0}r_{z1}\end{bmatrix}^T =\mathbf{D}_1([a_1 \ 0 \ 0]^T+\hat{\mathbf{r}}_1)=\begin{bmatrix}r_{x1}+a_1 & r_{y1} & r_{z1}\end{bmatrix}^T, \qquad (4.63)$$

$$^{0}\mathbf{r}_i^{cm}=\begin{bmatrix}^{0}r_{xi} & ^{0}r_{yi} & ^{0}r_{zi}\end{bmatrix}^T =\mathbf{D}_1[a_1 \ 0 \ 0]^T+\mathbf{D}_i\hat{\mathbf{r}}_i=\begin{bmatrix}r_{xi}+a_1 & r_{yi} & r_{zi}\end{bmatrix}^T,$$

where $i=2, 3, 4$.

In previous equations, the matrices $\mathbf{D}_1$, $\mathbf{D}_2$ and $\mathbf{D}_{3,4}$ were given in Eqs. (4.1), (4.4) and (4.5), and the vectors $\mathbf{r}_1$, $\mathbf{r}_2$ and $\mathbf{r}_{3,4}$ in Eq. (4.15).

4.4.3.2. Calculation of the Maximal Possible Values of $\ddot{q}_2$ and $\ddot{q}_3$

Axes 2 and 3 are each actuated by two torque motors Siemens 1FW6290-OWB20-OLB2. Each of the four motors has a maximal torque of $M_{imax}=10900$ Nm, a maximal speed for that torque of $\dot{q}_{imax}=3.979$ s^{-1}, a rated torque of $M_{ir}=5760$ Nm and a maximal speed at that torque of $\dot{q}_{ir}=7.1209$ s^{-1}, $i=2, 3$. Their torque decreases linearly from the maximal to the rated value as a function of the speed, what is shown in Fig. 4.8.

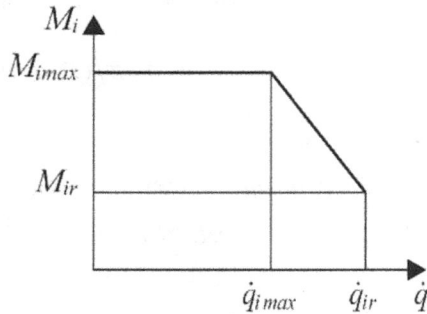

Fig. 4.8. Speed-torque diagram of torque motors.

Calculation of the desired and maximal possible values of $\ddot{q}_2$ and $\ddot{q}_3$ will be performed in the following three steps.

1). The maximum motor speeds are limited in the following way: if $\dot{q}_i > \dot{q}_{ir}$, then $\dot{q}_i = \dot{q}_{ir}$, and if $\dot{q}_i < -\dot{q}_{ir}$, then $\dot{q}_i = -\dot{q}_{ir}$, $i = 2, 3$.

2). The maximum values of the motor torques M_{ia}, $i = 2, 3$, that are related to the motor speed are calculated in the following way (Fig. 4.8):

For $\dot{q}_i \in [-\dot{q}_{imax}, \dot{q}_{imax}]$, we have $M_{ia} = M_{imax}$,

for $\dot{q}_i > \dot{q}_{imax}$, we have $M_{ia} = M_{imax} - (\dot{q}_i - \dot{q}_{imax})(M_{imax} - M_{ir})/(\dot{q}_{ir} - \dot{q}_{imax})$,

for $\dot{q}_i < -\dot{q}_{imax}$, we have $M_{ia} = M_{imax} + (\dot{q}_i + \dot{q}_{imax})(M_{imax} - M_{ir})/(\dot{q}_{ir} - \dot{q}_{imax})$,

$i = 2, 3$

$$(4.64)$$

3). The accelerations $\ddot{q}_i$ for $i = 2, 3$ are limited in the following way. If $\ddot{q}_i > M_{ia}/I_{ti}$ for the accelerated motion in the positive direction and the decelerated motion in the negative direction ($\ddot{q}_i > 0$), then $\ddot{q}_i = M_{ia}/I_{ti}$. If $\ddot{q}_i < -M_{ia}/I_{ti}$ for the decelerated movement in the positive direction and the accelerated movement in the negative direction ($\ddot{q}_i < 0$), then $\ddot{q}_i = -M_{ia}/I_{ti}$. Next, $\dot{q}_i = \dot{q}_{iprev} + \ddot{q}_i \Delta t$ is calculated and restricts $\dot{q}_i$ in the same way as in Step 1. The new value of $\ddot{q}_i$ and q_i will be obtained now from $\ddot{q}_i = (\dot{q}_i - \dot{q}_{iprev})/\Delta t$ and $q_i = q_{iprev} + \dot{q}_i \Delta t$, $i = 2, 3$.

4.4.3.2.1. Total Moments of Inertia for Axes z_1, z_2 and z_3

The total moment of inertia of links 2 and 3 and the external load reduced to axis z_1 is:

$$I_{t2} = {}^1I_{z2} + {}^1I_{z3} + {}^1I_{z4} + [(\hat{r}_{x2})^2 + (\hat{r}_{z2})^2]m_2$$
$$+ [({}^1r_{x3})^2 + ({}^1r_{y3})^2 + ({}^1r_{z3})^2]m_3 + [({}^1r_{x4})^2 + ({}^1r_{y4})^2 + ({}^1r_{z4})^2]m_4$$

$$(4.65)$$

In Eq. (4.65), the value ${}^1I_{z2} = \hat{I}_{y2}$ is obtained from ${}^1\mathbf{I}_2^{cm} = {}^1\mathbf{D}_2 \hat{\mathbf{I}}_2^{cm} {}^1\mathbf{D}_2^{-1}$, the value ${}^1I_{z3} = \hat{I}_{x3}c_3^2 + \hat{I}_{z3}s_3^2$ is obtained from ${}^1\mathbf{I}_3^{cm} = {}^1\mathbf{D}_3 \hat{\mathbf{I}}_3^{cm} {}^1\mathbf{D}_3^{-1}$, and the value ${}^1I_{z4} = \hat{I}_{x4}c_3^2 + \hat{I}_{z4}s_3^2$ is obtained from ${}^1\mathbf{I}_4^{cm} = {}^1\mathbf{D}_3 \hat{\mathbf{I}}_4^{cm} {}^1\mathbf{D}_3^{-1}$. Eq. (4.65) applies to the following:

$$
{}^{1}\mathbf{r}_3 = \begin{bmatrix} {}^{1}r_{x3} \\ {}^{1}r_{y3} \\ {}^{1}r_{z3} \end{bmatrix} = {}^{1}\mathbf{D}_3\ \hat{\mathbf{r}}_3 = \begin{bmatrix} s_2 s_3 \hat{r}_{x3} - c_2 \hat{r}_{y3} + s_2 c_3 \hat{r}_{z3} \\ -s_3 c_3 \hat{r}_{x3} - s_2 \hat{r}_{y3} - c_2 c_3 \hat{r}_{z3} \\ c_3 \hat{r}_{x3} - s_3 \hat{r}_{z3} \end{bmatrix},
\qquad (4.66)
$$

$$
{}^{1}\mathbf{r}_4 = \begin{bmatrix} {}^{1}r_{x4} \\ {}^{1}r_{y4} \\ {}^{1}r_{z4} \end{bmatrix} = {}^{1}\mathbf{D}_3\ \hat{\mathbf{r}}_4 = \begin{bmatrix} s_2 s_3 \hat{r}_{x4} - c_2 \hat{r}_{y4} + s_2 c_3 \hat{r}_{z4} \\ -s_3 c_3 \hat{r}_{x4} - s_2 \hat{r}_{y4} - c_2 c_3 \hat{r}_{z4} \\ c_3 \hat{r}_{x4} - s_3 \hat{r}_{z4} \end{bmatrix}
$$

In previous equations, we have ${}^{1}\mathbf{D}_3 = {}^{1}\mathbf{D}_2\,{}^{2}\mathbf{D}_3$. The matrices ${}^{1}\mathbf{D}_2$ and ${}^{2}\mathbf{D}_3$ are given in Eqs. (4.2) and (4.3), respectively.

The total moment of inertia of link 3 and the external load reduced to the axis z_2 is:

$$
I_{t3} = \hat{I}_{y3} + \hat{I}_{y4} + [(\hat{r}_{x3})^2 + (\hat{r}_{z3})^2]m_3 + [(\hat{r}_{x4})^2 + (\hat{r}_{z4})^2]m_4
\qquad (4.67)
$$

In Eq. (4.67), the value ${}^{2}I_{z3} = \hat{I}_{y3}$ is obtained from ${}^{2}\mathbf{I}_3^{cm} = {}^{2}\mathbf{D}_3 \hat{\mathbf{I}}_3^{cm}\,{}^{2}\mathbf{D}_3^{-1}$, and the value ${}^{2}I_{z4} = \hat{I}_{y4}$ is obtained from ${}^{2}\mathbf{I}_4^{cm} = {}^{2}\mathbf{D}_3 \hat{\mathbf{I}}_4^{cm}\,{}^{2}\mathbf{D}_3^{-1}$.

4.4.3.3. Algorithm for Calculating the Maximum Possible Values of $\ddot{q}_2$ and $\ddot{q}_3$ Based on Approximate Forward Dynamics (Algorithm 4.1)

First, a new algorithm is developed for the approximate inverse dynamics of the robots based on the recursive Newton-Euler algorithm, which calculates the actuator torques $\hat{m}_{z2}$ and $\hat{m}_{z3}$ as functions of the angular accelerations $\ddot{q}_2$ and $\ddot{q}_3$ and some coefficients on which we do not have influence. This algorithm is applied in the following way. If in Eq. (4.12) for $\dot{\mathbf{v}}_2^{cm}$, (4.14) for $\dot{\mathbf{v}}_{3,4}^{cm}$, (4.21) for $\mathbf{M}_2$ and (4.22) for $\mathbf{M}_{3,4}$, the terms $\dot{q}_1$, $\dot{q}_2$ and $\dot{q}_3$ are replaced with $\dot{q}_1 = \dot{q}_{1\,prev} + \ddot{q}_1\,\Delta t$, $\dot{q}_2 = \dot{q}_{2\,prev} + \ddot{q}_2\,\Delta t$ and $\dot{q}_3 = \dot{q}_{3\,prev} + \ddot{q}_3\,\Delta t$, the actuator torques $\hat{m}_{z2}$ and $\hat{m}_{z3}$, given in Eqs. (4.32) and (4.29), will be obtained in the form of the following two equations:

$$
\hat{m}_{z2} = b_2 + (a_{21}\ddot{q}_1 + a_{22}\ddot{q}_2 + a_{23}\ddot{q}_3)\Delta t
$$
$$
+ (a_{211}\ddot{q}_1\ddot{q}_1 + a_{212}\ddot{q}_1\ddot{q}_2 + a_{213}\ddot{q}_1\ddot{q}_3 + a_{222}\ddot{q}_2\ddot{q}_2 + a_{223}\ddot{q}_2\ddot{q}_3 + a_{233}\ddot{q}_3\ddot{q}_3)\Delta t \Delta t
\qquad (4.68)
$$

$$\hat{m}_{z3} = b_3 + (a_{31}\ddot{q}_1 + a_{32}\ddot{q}_2 + a_{33}\ddot{q}_3)\Delta t$$

$$+ (a_{311}\ddot{q}_1\ddot{q}_1 + a_{312}\ddot{q}_1\ddot{q}_2 + a_{313}\ddot{q}_1\ddot{q}_3 + a_{322}\ddot{q}_2\ddot{q}_2 + a_{323}\ddot{q}_2\ddot{q}_3 + a_{333}\ddot{q}_3\ddot{q}_3)\Delta t\Delta t \tag{4.69}$$

In the previous equations, the terms with $\Delta t \Delta t$ can be neglected. Because the motions of links 2 and 3 have a very small influence on the inertial forces and the moments that act on link 1, we could, for each interpolation period, consider that the terms $c_2 = b_2 + a_{21}\ddot{q}_1\Delta t$ and $c_3 = b_3 + a_{31}\ddot{q}_1\Delta t$ are known coefficients. Coefficients $c_{22} = a_{22}\Delta t$, $c_{23} = a_{23}\Delta t$, $c_{32} = a_{32}\Delta t$ and $c_{33} = a_{33}\Delta t$ can also be introduced. In that way, we can obtain the actuator torques m_{z2} and m_{z3} written in the following, much simpler way:

$$\hat{m}_{z2} = c_2 + c_{22}\ddot{q}_2 + c_{23}\ddot{q}_3 \tag{4.70}$$

$$\hat{m}_{z3} = c_3 + c_{32}\ddot{q}_2 + c_{33}\ddot{q}_3 \tag{4.71}$$

The coefficients c_2, c_{22} and c_{23} in Eq. (4.70) are functions of the variables $\ddot{q}_1$, $\dot{q}_{2prev}$, q_1, q_2 and q_3 and the constants Δt, a_1, $\hat{r}_{x2}$, $\hat{r}_{y2}$, $\hat{r}_{z2}$, $\hat{r}_{x3}$, $\hat{r}_{y3}$, $\hat{r}_{z3}$, $\hat{r}_{x4}$, $\hat{r}_{y4}$, $\hat{r}_{z4}$, $\hat{I}_{x2}$, $\hat{I}_{y2}$, $\hat{I}_{z2}$, $\hat{I}_{x3}$, $\hat{I}_{y3}$, $\hat{I}_{z3}$, $\hat{I}_{x4}$, $\hat{I}_{y4}$, $\hat{I}_{z4}$, m_2, m_3 and m_4. The coefficients c_3, c_{32} and c_{33} in Eq. (4.44) are functions of the variables $\ddot{q}_1$, $\dot{q}_{2prev}$, $\dot{q}_{3prev}$, q_1, q_2 and q_3 and the constants Δt, a_1, $\hat{r}_{x3}$, $\hat{r}_{y3}$, $\hat{r}_{z3}$, $\hat{r}_{x4}$, $\hat{r}_{y4}$, $\hat{r}_{z4}$, $\hat{I}_{x3}$, $\hat{I}_{y3}$, $\hat{I}_{z3}$, $\hat{I}_{x4}$, $\hat{I}_{y4}$, $\hat{I}_{z4}$, m_3 and m_4. The performed simulations have shown that the numerical result for $\hat{m}_{z2}$ and $\hat{m}_{z3}$ obtained by Eqs. (4.70) and (4.71) differ by less than 1 % from the results that are obtained from Eqs. (4.32) and (4.29). This difference is negligibly small. Eqs. (4.70) and (4.71) allow a very simple calculation of $\ddot{q}_2$ and $\ddot{q}_3$, while it is almost impossible to perform that calculation with Eqs. (4.32) and (4.29). If necessary, the actuator torque for the first axis could be easily added to the equation $\hat{m}_{z1} = c_1 + c_{11}\ddot{q}_1 + c_{12}\ddot{q}_2 + c_{13}\ddot{q}_3$, and Eqs. (4.43) and (4.44) could be extended with the terms $c_{12}\ddot{q}_1$ and $c_{13}\ddot{q}_1$, respectively. In that way, we would obtain three equations with three unknowns, which we could easily calculate.

After that, check is performed whether the values of required torques of the second and third axis for given angular accelerations are greater than maximum possible. If $\hat{m}_{zi} > M_{ia}$ for the accelerated motion in the positive direction and the decelerated motion in the negative direction ($\ddot{q}_i > 0$), it will be $\hat{m}_{zi} = M_{ia}$. If $\hat{m}_{zi} < -M_{ia}$ for the decelerated motion in the positive direction and the accelerated motion in the negative direction ($\ddot{q}_i < 0$), it will be $\hat{m}_{zi} = -M_{ia}$.

Finally, the maximal possible values of $\ddot{q}_2$ and $\ddot{q}_3$ that motors can achieve will be calculated based on Eqs. (4.70) and (4.71) and the maximal values of the torques $\hat{m}_{z2}$ and $\hat{m}_{z3}$; this calculation of the approximate forward dynamics will be performed in the following way:

$$\ddot{q}_2 = \frac{(\hat{m}_{z2} - c_2)c_{33} - (\hat{m}_{z3} - c_3)c_{23}}{c_{22}c_{33} - c_{23}c_{32}}, \quad \ddot{q}_3 = \frac{(\hat{m}_{z3} - c_3)c_{22} - (\hat{m}_{z2} - c_2)c_{32}}{c_{22}c_{33} - c_{23}c_{32}} \quad (4.72)$$

Next, $\dot{q}_i$ and q_i are calculated again in the following way: $\dot{q}_i = \dot{q}_{iprev} + \ddot{q}_i \Delta t$, $q_i = q_{iprev} + \dot{q}_i \Delta t$, $\dot{q}_{iprev} = \dot{q}_i$, $q_{iprev} = q_i$, $i = 2, 3$.

4.4.4. Centrifuge Control Algorithm (Algorithm 4.2)

The algorithm for the calculation of $\ddot{q}_i$, $\dot{q}_i$ and q_i, $i = 1, 2, 3$ is based on the equations that are presented in Sections 4.4.1 to 4.4.3. First, N_c, N_a, Δn and n are calculated with Eqs. (4.54)-(4.57). Then, the centrifuge kinematic parameters are calculated in three phases. For the positive acceleration onset in the first phase, the onset of the centrifuge inertial force (jerk $= \Delta a/\Delta t$) increases linearly from zero to the programmed value. In the second phase, this force increases linearly ($\Delta a/\Delta t = $const$= n$), and in the third stage, the onset of this force decreases linearly from the programmed value to zero. In the first and third phases, smoothing of the acceleration force profile is performed. For the deceleration motion, the smoothing is performed vice versa. These three phases of the centrifuge control algorithm are shown below.

First phase.

for i=1 to N_a {

$n = n + \Delta n$, $a = a_{prev} + \Delta a = a_{prev} + sign(a_e - a_s) n g \Delta t$;

Step 1. Calculation of $\ddot{q}_1$ by Eq. (4.53) for the positive acceleration onset or the reading of the array $\ddot{q}_1 = -array(\ddot{q}_1[2N_a + N_c - i])$ for the negative acceleration onset;

Step 2. Calculation of the desired and maximal possible value of $\ddot{q}_1$ and the modified values of $\dot{q}_1$ and q_1 (Section 4.4.3.1);

Step 3. Calculation of q_2 by Eq. (4.44) or (4.45) and q_3 by Eq. (4.46) or Eq. (4.47), and $\dot{q}_i = (q_i - q_{iprev})/\Delta t$, $\ddot{q}_i = (\dot{q}_i - \dot{q}_{iprev})/\Delta t$, $i = 2, 3$;

Step 4. Calculation of the desired and maximal possible values of $\ddot{q}_2$ and $\ddot{q}_3$, and the corrected values for $\dot{q}_2$, q_2, $\dot{q}_3$ and q_3 (Algorithm 4.1 - Section 4.4.3.3); }

Second phase.

for $i=1$ to N_c {

$a = a_{prev} + \Delta a = a_{prev} + sign(a_e - a_s) n g \Delta t$;

Calculating $\ddot{q}_1$ by Eq. (4.53) for the positive acceleration onset or reading the array $\ddot{q}_1 = -series(\ddot{q}_1[N_a + N_c - i])$ for the negative acceleration onset;

The remainder of this phase is the same as in the first phase. }

Third phase.

for $i=1$ to N_a {

$n = n - \Delta n$, $a = a_{prev} + \Delta a = a_{prev} + sign(a_e - a_s) n g \Delta t$;

Calculation of $\ddot{q}_1$ by Eq. (4.53) for the positive acceleration onset or reading of the array $\ddot{q}_1 = -series(\ddot{q}_1[N_a - i])$ for the negative acceleration onset;

The remainder of this phase is the same as in the first phase. Steps 2, 3 and 4 are the same as in the first phase.}

4.5. Programming Instruction of the Centrifuge Movement

The proposed control algorithm for the centrifuge motion was tested on the off-line programming system of the robot controller developed at the

Lola Institute. The G_z, G_x and G_y acceleration force profiles are provided by this programming system. Lola-Industrial Robot Language (L-IRL) is here extended with the move instruction (gmove) and additional parameters required for the centrifuge motion programming. The centrifuge movement control algorithm, presented above, is also added. The new additional parameters are the following:

`bl_g` – desired resultant acceleration force G in the gondola centre at the end of a given motion segment - baseline level,

`acc_g` – onset rate of the resultant acceleration force G in the gondola centre (it is zero, if it is not quoted),

time – period in which the desired change of the resultant acceleration force G in the gondola centre should be achieved and

`Gx`, `Gy` and `Gz` - desired G_x, G_y and G_z forces, respectively, in the gondola centre at the end of a given motion segment.

Here, `bl_g`$\in$[1.41, 15], `acc_g`$\in$[-9, 9], `Gx`$\in$[-14, 12], `Gy`$\in$[-14, 10] and `Gz`$\in$[-2, 15]. At least one parameter of `bl_g`, `Gx`, `Gy`, and `Gz` must be quoted. If `bl_g` is not quoted, it is calculated by `bl_g=(Gx`^{2}`+Gy`^{2}`+Gz`^{2}`)`$^{1/2}$. It is not allowed for both `Gx` and `Gz` parameters to be quoted.

4.6. Results: Verification for the Proposed Control Algorithm

An example of the program for the centrifuge motion, written in the extended L-IRL programming language that is used for testing of the human centrifuge motion algorithm is shown in the following code:

```
program test1;
system_specification "centrifuge";
seq
gmove time:= 0.5 bl_g:= 1.41;
gmove acc_g:= 9. bl_g:= 15.0;
gmove time:= 0.7 bl_g:= 15.0;
gmove acc_g:= -9. bl_g:= 1.41;
gmove time:= 0.7 bl_g:= 1.41;
gmove acc_g:= 9. bl_g:= 9.0;
gmove time:= 0.7 bl_g:= 9.0;
gmove acc_g:= -9. bl_g:= 1.41;
```

```
gmove time:= 0.7 bl_g:= 1.41;
gmove acc_g:= 2. bl_g:= 2.0;
gmove time:= 0.7 bl_g:= 2.0;
gmove acc_g:= 5. bl_g:= 5.0;
gmove time:= 0.7 bl_g:= 5.0;
gmove acc_g:= -5. bl_g:= 3.0;
gmove time:= 0.7 bl_g:= 3.0;
gmove acc_g:= -3. bl_g:= 1.41;
gmove time:= 0.7 bl_g:= 1.41;
endseq
endprogram
```

Fig. 4.9 shows kinematics and dynamics parameters of the given example of the CFS motion program. The masses, the mass centre coordinates and the matrices of inertia of the external load and the centrifuge links are given in the Sections 4.2.2 and 4.2.3. Herein, Fig. 4.9 (a_1, a_2, a_3 and a_4) show G, G_x, G_y and G_z. It can be seen that the CFS can achieve the mean growth of the acceleration load of 9 g/s in the period of constant acceleration change (c_{ac}) from the lower baseline level of 1.41 g to the upper baseline level of 15 g. The greatest absolute difference between the programmed and realised G_z change was 0.216 g. During the programmed motion, the deviation of G_x was between 0.38 and -0.5 g. There was no deviation in G_y. Figs. 4.9 (b_1, b_2 and b_3) show $\dot{q}_1$, $\dot{q}_2$ and $\dot{q}_3$. It can be observed that the Algorithm 4.2 limits these velocities: $\dot{q}_1$ is not larger than 4.28 s^{-1}, $\dot{q}_2$ is between 1.47 and -1.35 s^{-1}, and $\dot{q}_3$ is between 3.98 and -3.98s^{-1}. Figs. 4.9 (c_1, c_2 and c_3) show $\ddot{q}_1$, $\ddot{q}_2$ and $\ddot{q}_3$. These values are also limited: $\ddot{q}_1$ is between 3.27 and -3.11 s^{-2}, $\ddot{q}_2$ is between 9.88 and -9.4 s^{-2}, and $\ddot{q}_3$ is between 52.1 and -52.1s^{-2}. Figs. 4.9 (d_1, and d_2) show the angles $q_2=\phi$ and $q_3=\theta$, which are obtained with the Algorithm 4.2. Herein, q_2 is between 0 and 86.2°, and q_3 is between 43.6 and -57.3°. Figs. 4.9 (e_1, e_2 and e_3) show the axial forces $\hat{f}_{a1}$, $\hat{f}_{a2}$ and $\hat{f}_{a3}$ and Fig. 4.9 (f_1, f_2 and f_3) show the radial forces $\hat{f}_{r1}$, $\hat{f}_{r2}$ and $\hat{f}_{r3}$ of the axes bearings. Figs. 4.9 (g_1, g_2 and g_3) show the torques $\hat{m}_{z1}$, $\hat{m}_{z2}$ and $\hat{m}_{z3}$ of the actuators. It can be observed that the Algorithm 4.2 limits these torques: $\hat{m}_{z1}$ is between 1,228,670 and -1,107,000 Nm, $\hat{m}_{z2}$ is between 19,664 and -21777 Nm, and $\hat{m}_{z3}$ is between 21756 and -21696 Nm. Figs. 4.9 (h_1, h_2 and h_3) show the torques $\hat{m}_{xy1}$, $\hat{m}_{xy2}$ and $\hat{m}_{xy3}$ that act on the bearings. Figs. 4.9 (i_1, i_2 and i_3) show the powers P_1, P_2 and P_3 of the motors. It can be observed that the Algorithm 4.2 limits these powers; P_1 is between 2900 and -2559 kW, P_2 is between 10 and -16.417 kW, and P_3 is between 28.646 and -22.115 kW.

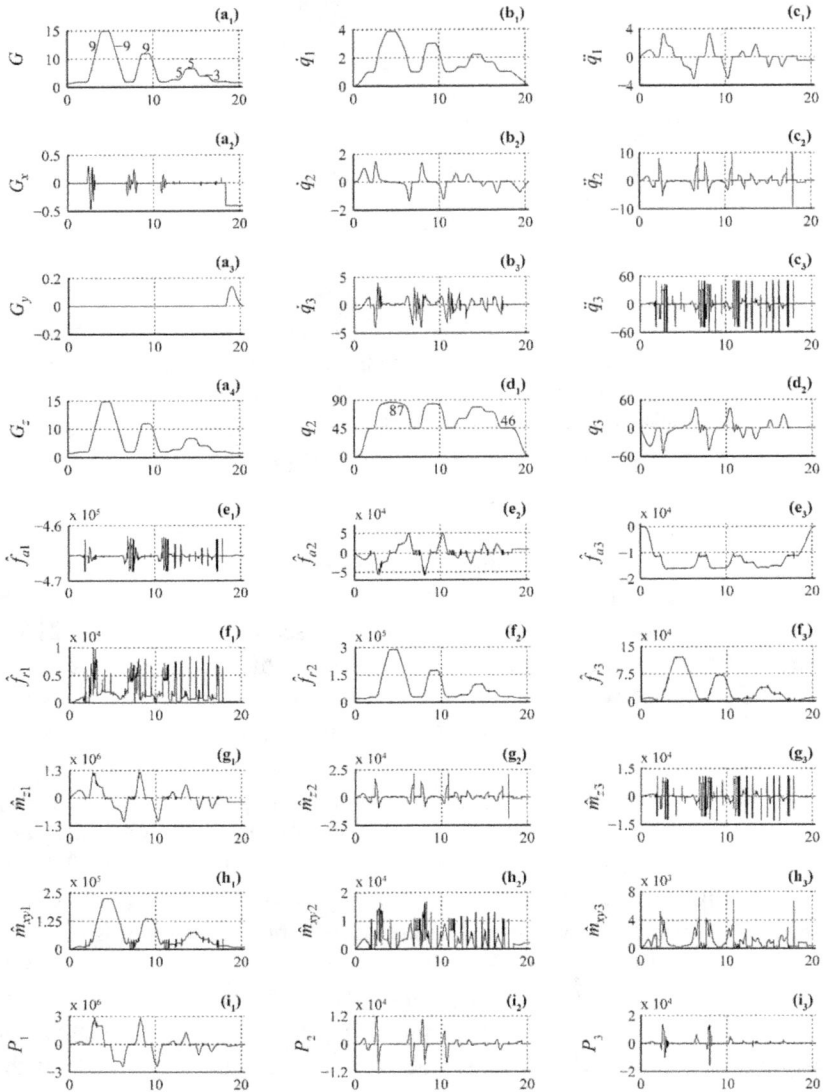

Fig. 4.9. Kinematics and dynamics parameters of the centrifuge motion of the program test 1: (a_1) G, (a_2) G_x, (a_3) G_y, (a_4) G_z, (b_1) $\dot{q}_1$, (b_2) $\dot{q}_2$, (b_3) $\dot{q}_3$ $[s^{-1}]$, (c_1) $\ddot{q}_1$, (c_2) $\ddot{q}_2$, (c_3) $\ddot{q}_3$ $[s^{-2}]$, (d_1) $q_2 = \phi$, (d_2) $q_3 = \theta$ $[°]$, (e_1) $\hat{f}_{a1}$, (e_2) $\hat{f}_{a2}$, (e_3) $\hat{f}_{a3}$ $[N]$, (f_1) $\hat{f}_{r1}$, (f_2) $\hat{f}_{r2}$, (f_3) $\hat{f}_{r3}$ $[N]$, (g_1) $\hat{m}_{z1}$, (g_2) $\hat{m}_{z2}$, (g_3) $\hat{m}_{z3}$ $[Nm]$, (h_1) $\hat{m}_{xy1}$, (h_2) $\hat{m}_{xy2}$, (h_3) $\hat{m}_{xy3}$ $[Nm]$, (i_1) P_1, (i_2) P_2 and (i_3) $P_3[kW]$.

The diagrams for $\hat{m}_{z2}$ and $\hat{m}_{z3}$ (Fig. 4.9 (g_2 and g_3)) are presented for the case when the Algorithm 4.2 has been applied. If this algorithm is not used, the simulation program shows that to completely accurately achieve a given change in G_z, the required torques that the actuators would have to produce are 26.287 Nm for axis 2 and 24.723 Nm for axis 3. Fig. 4.9 shows the kinematic and dynamic parameters that were obtained for the given example, in the case of using roll and pitch motors only in the range of the maximal torques. It can be observed that in that case, the chosen motors fulfil all of the requested demands.

In contrast, Fig. 4.10 (a) shows G_z, Fig. 4.10 (b) shows G_x, and Fig. 4.10 (c) shows q_3 when these motors are used as well in the ranges in which the magnetic fields and torques weaken, until the maximal speeds; in this case, $\dot{q}_i > \dot{q}_{imax}$ or $\dot{q}_i < -\dot{q}_{imax}$, $i = 2, 3$ (in Eq. (4.64)). It can be observed that in that case, the chosen motors do not satisfy the requested characteristics. During the programmed motion in that case, the deviation of G_x is between 5.55 and -8.82 g, while the greatest difference between the programmed and the realised G_z change is 3.529 g. In this way, the implementation of the presented control algorithm shows that using motors for the third axis in the scenario in which the field and torque weakening is included gives unacceptable results.

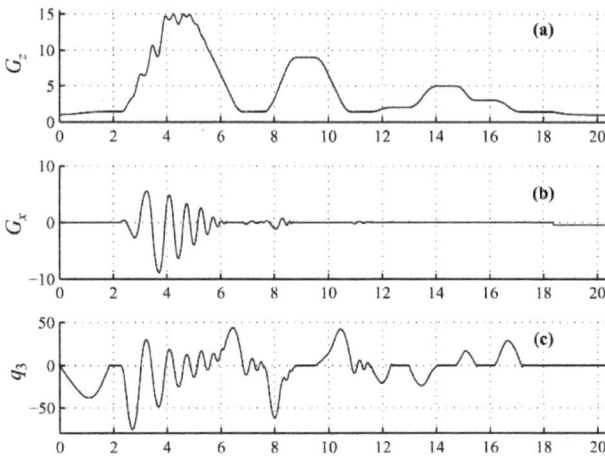

Fig. 4.10. (a) G_z, (b) G_x and (c) $q_3 = \theta$ from the example given by the program test 1, when the roll and pitch motors are used as well as in the scenarios that have the field and torque weakening.

123

Fig. 4.11 shows a case in which a positive or negative G_z is explicitly given. Here, G, G_z, G_x, $\dot{q}_1$, q_2 and q_3 are shown in Fig. 4.11 (a, b, c, d, e and f). Fig. 4.12 shows a case in which a positive or negative G_x is explicitly given. Here, G, G_z, G_x, $\dot{q}_1$, q_2 and q_3 are shown in Fig. 4.12 (a, b, c, d, e and f). Fig. 4.13 shows a case in which a positive or negative G_y is explicitly given. Here, G, G_z, G_y, $\dot{q}_1$, q_2 and q_3 are shown in Fig. 4.13 (a, b, c, d, e and f).

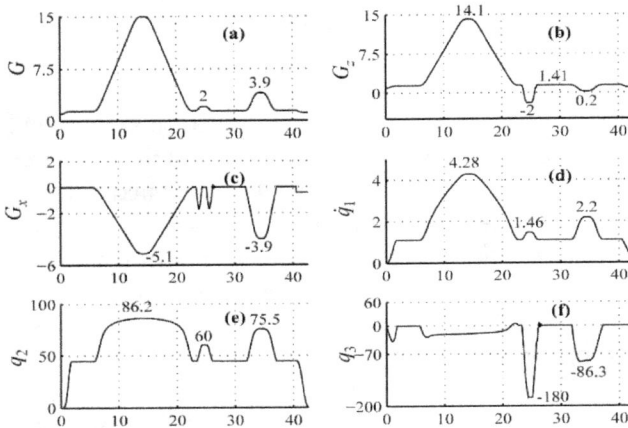

Fig. 4.11. A case in which a positive or negative G_z is explicitly given: (a) G, (b) G_z, (c) G_x, (d) $\dot{q}_1$, (e) q_2 and (f) q_3.

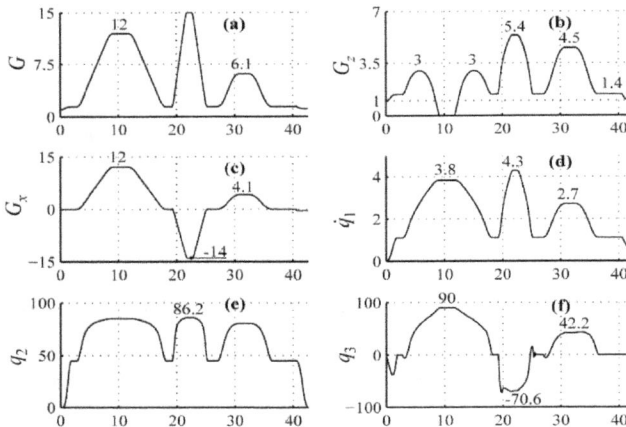

Fig. 4.12. A case in which a positive or negative G_x is explicitly given: (a) G, (b) G_z, (c) G_x, (d) $\dot{q}_1$, (e) q_2 and (f) q_3.

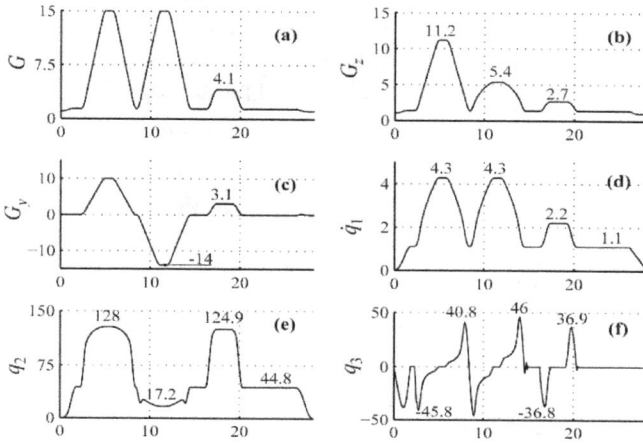

Fig. 4.13. A case in which a positive or negative G_y is explicitly given: (a) G, (b) G_z, (c) G_y, (d) $\dot{q}_1$, (e) q_2 and (f) q_3.

4.7. Kinematics and Dynamics of the SDT

4.7.1. Forward Geometric Model of the SDT

This section defines the coordinate frames for the SDT links (Fig. 4.14) and the matrices that determine their relations. Arm rotation around the vertical (i.e., planetary) axis is the primary motion. The arm carries a gyroscopic gondola system with three rotational axes providing yaw, pitch and roll capabilities. Their task is to achieve any orientation; different acceleration forces acting on the pilot can also be simulated. The yaw axis (z) is parallel with the arm axis. The roll axis lies in the plane of the arm rotation, perpendicular to the main rotational axis (i.e., in the x direction). The pitch (y) axis is perpendicular to the roll axis.

The base is denoted by 0, the arm by 1, the gyroscope frame by 2, the roll ring by 3 and the gondola by 4. The arm rotation angle is denoted by q_1, the gyroscope frame rotation angle by q_2, the roll ring rotation angle by $q_3=\phi$ and the gondola rotation angle (i.e., pitch) by $q_4=\theta$. The yaw angle is $\psi=q_1+q_2$. The SDT presented in this chapter has the following features: arm length $a_1=2.394$ m, gyroscope frame length $d_2=1.957$ m; and q_1, q_2, q_3 and q_4 rotation ranges $\pm360°$. The allowed gondola payload is 180 kg. The SDT base coordinates are denoted by x_0, y_0, z_0; the arm

125

coordinates by x_1, y_1, z_1 (link 1); the gyroscope frame by x_2, y_2, z_2 (link 2); the roll ring coordinates by x_3, y_3, z_3 (link 3); the gondola coordinates by x_4, y_4, z_4 (link 4); and the pilot coordinates by x, y, z. It is assumed that $x_4=x$, $y_4=y$ and $z_4=z$. The D-H parameters for the 4-axis SDT links are given in Table 4.2.

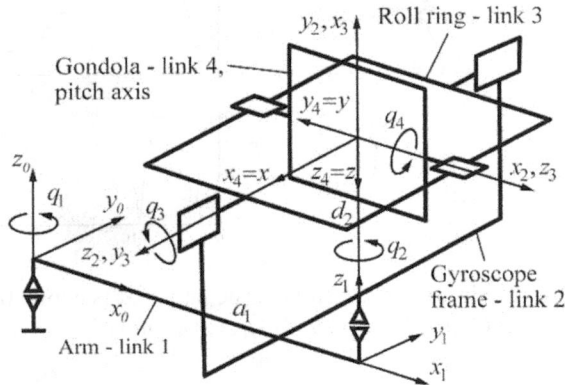

Fig. 4.14. Coordinate frames of the 4-axis SDT.

Table 4.2. D-H parameters for the 4-axis SDT links.

Link	Variable [°]	a [mm]	d [mm]	α [°]
1	q_1	a_1	0	0
2	q_2	0	d_2	90
3	q_2+90	0	0	90
4	q_3+90	0	0	-90

Using shorthand notation, $\sin(q_i)=s_i$, $\cos(q_i)=c_i$ (i=1, 2, 3, 4), $\sin(\psi)=s_\psi$ and $\cos(\psi)=c_\psi$, the following homogenous matrices for the relation between the SDT links coordinate frames are defined to derive the kinematic equations for the machine as follows:

$$\mathbf{T}_1 = \mathbf{Rot}(z_0, q_1)\mathbf{Trans}(x_0, a_1) = \begin{bmatrix} c_1 & -s_1 & 0 & c_1 a_1 \\ s_1 & c_1 & 0 & s_1 a_1 \\ 0 & 0 & 1 & 0 \\ 0 & 0 & 0 & 1 \end{bmatrix} \tag{4.73}$$

$$^1\mathbf{T}_2 = \mathrm{Rot}(z_1, q_2)\mathrm{Trans}(z_1, d_2)\mathrm{Rot}(x_1', 90°) = \begin{bmatrix} c_2 & 0 & s_2 & 0 \\ s_2 & 0 & -c_2 & 0 \\ 0 & 1 & 0 & d_2 \\ 0 & 0 & 0 & 1 \end{bmatrix} \quad (4.74)$$

$$^2\mathbf{T}_3 = \mathrm{Rot}(z_2, q_3)\mathrm{Rot}(z_2, 90°)\mathrm{Rot}(x_2', 90°) = \begin{bmatrix} -s_3 & 0 & c_3 & 0 \\ c_3 & 0 & s_3 & 0 \\ 0 & 1 & 0 & 0 \\ 0 & 0 & 0 & 1 \end{bmatrix} \quad (4.75)$$

$$^3\mathbf{T}_4 = \mathrm{Rot}(z_3, \theta_4)\mathrm{Rot}(z_3, 90°)\mathrm{Rot}(x_3', -90°) = \begin{bmatrix} -s_4 & 0 & -c_4 & 0 \\ c_4 & 0 & -s_4 & 0 \\ 0 & -1 & 0 & 0 \\ 0 & 0 & 0 & 1 \end{bmatrix} \quad (4.76)$$

The position and orientation of the links and end-effector (i.e., the pilot's head or chest) with respect to the SDT variables q_1, q_2, q_3 and q_4. (forward kinematics related to robot geometry) is determined from the following matrix:

$$\mathbf{T}_4 = {}^0\mathbf{T}_1\, {}^1\mathbf{T}_2\, {}^2\mathbf{T}_3\, {}^3\mathbf{T}_4 = \begin{bmatrix} c_\psi s_3 s_4 + s_\psi c_4 & -c_\psi c_3 & c_\psi s_3 c_4 - s_\psi s_4 & a_1 c_1 \\ s_\psi s_3 s_4 - c_\psi c_4 & -s_\psi c_3 & s_\psi s_3 c_4 + c_\psi s_4 & a_1 s_1 \\ -c_3 s_4 & -s_3 & -c_3 c_4 & d_2 \\ 0 & 0 & 0 & 1 \end{bmatrix} \quad (4.77)$$

4.7.2. Forward Kinematics Related to the SDT Velocities and Accelerations

Using the given transformational matrices, equations that describe the forward kinematics relating to velocities and accelerations of the links and end-effector can be developed as with the centrifuge links (Section 4.2.2).

The SDT links angular velocities (Fig. 4.14) are:

$$\boldsymbol{\omega}_1 = [0 \ 0 \ \dot{q}_1]^T, \quad \boldsymbol{\omega}_2 = [0 \ 0 \ \dot{q}_\psi]^T,$$

$$\boldsymbol{\omega}_3 = \boldsymbol{\omega}_2 + [s_\psi \ -c_\psi \ 0]^T \dot{q}_3, \quad \boldsymbol{\omega}_4 = \boldsymbol{\omega}_3 + [c_\psi c_3 \ s_\psi c_3 \ s_3]^T \dot{q}_4 \quad (4.78)$$

The SDT links linear velocities, where $\mathbf{p}_1^* = a_1 \begin{bmatrix} c_1 & s_1 & 0 \end{bmatrix}^T$, $\mathbf{p}_2^* = \begin{bmatrix} 0 & 0 & d_2 \end{bmatrix}^T$, $\mathbf{p}_3^* = \mathbf{p}_4^* = \begin{bmatrix} 0 & 0 & 0 \end{bmatrix}^T$, are:

$$\dot{\mathbf{v}}_1 = \dot{\mathbf{v}}_2 = \dot{\mathbf{v}}_3 = \dot{\mathbf{v}}_4 = \begin{bmatrix} \dot{v}_{1x} & \dot{v}_{1y} & \dot{v}_{1z} \end{bmatrix}^T = a_1 \begin{bmatrix} s_1\dot{\omega}_1 + c_1\omega_1^2 & c_1\dot{\omega}_1 - s_1\omega_1^2 & 0 \end{bmatrix}^T \quad (4.79)$$

The SDT links angular accelerations are:

$$\dot{\boldsymbol{\omega}}_1 = \begin{bmatrix} 0 & 0 & \ddot{q}_1 \end{bmatrix}^T \quad (4.80)$$

$$\dot{\boldsymbol{\omega}}_2 = \begin{bmatrix} 0 & 0 & \ddot{q}_\psi \end{bmatrix}^T \quad (4.81)$$

$$\dot{\boldsymbol{\omega}}_3 = \dot{\boldsymbol{\omega}}_2 + \begin{bmatrix} c_\psi & s_\psi & 0 \end{bmatrix}^T \dot{q}_\psi \dot{q}_3 + \begin{bmatrix} s_\psi & -c_\psi & 0 \end{bmatrix}^T \ddot{q}_3 \quad (4.82)$$

$$\dot{\boldsymbol{\omega}}_4 = \dot{\boldsymbol{\omega}}_3 + \begin{bmatrix} -s_\psi c_3 \dot{q}_\psi - c_\psi s_3 \dot{q}_3 \\ c_\psi c_3 \dot{q}_\psi - s_\psi s_3 \dot{q}_3 \\ c_3 \dot{q}_3 \end{bmatrix} \dot{q}_4 + \begin{bmatrix} c_\psi c_3 \\ s_\psi c_3 \\ s_3 \end{bmatrix} \ddot{q}_4 \quad (4.83)$$

The linear accelerations of the SDT links are:

$$\dot{\mathbf{v}}_1 = \dot{\mathbf{v}}_2 = \dot{\mathbf{v}}_3 = \dot{\mathbf{v}}_4 = \begin{bmatrix} \dot{v}_{1x} & \dot{v}_{1y} & \dot{v}_{1z} \end{bmatrix}^T = a_1 \begin{bmatrix} s_1\dot{\omega}_1 + c_1\omega_1^2 & c_1\dot{\omega}_1 - s_1\omega_1^2 & 0 \end{bmatrix}^T \quad (4.84)$$

The linear acceleration of the SDT links and the external load CM are:

$$\dot{\mathbf{v}}_1^{cm} = \begin{bmatrix} -a_1 s_1 - r_{y1} \\ a_1 c_1 + r_{x1} \\ 0 \end{bmatrix} \ddot{q}_1 - \begin{bmatrix} a_1 c_1 + r_{x1} \\ a_1 s_1 + r_{y1} \\ 0 \end{bmatrix} \dot{q}_1^2 \quad (4.85)$$

$$\dot{\mathbf{v}}_2^{cm} = \begin{bmatrix} -a_1 s_1 \\ a_1 c_1 \\ 0 \end{bmatrix} \ddot{q}_1 + \begin{bmatrix} -r_{y2} \\ r_{x2} \\ 0 \end{bmatrix} \ddot{q}_\psi - a_1 \begin{bmatrix} c_1 \\ s_1 \\ 0 \end{bmatrix} \dot{q}_1^2 - \begin{bmatrix} r_{x2} \\ r_{y2} \\ 0 \end{bmatrix} \dot{q}_\psi^2 \quad (4.86)$$

$$\dot{\mathbf{v}}_3^{cm} = \begin{bmatrix} -a_1 s_1 \\ a_1 c_1 \\ 0 \end{bmatrix} \ddot{q}_1 + \begin{bmatrix} -r_{y3} \\ r_{x3} \\ 0 \end{bmatrix} \ddot{q}_\psi + \begin{bmatrix} -c_\psi r_{z3} \\ -s_\psi r_{z3} \\ s_\psi r_{y3} + c_\psi r_{x3} \end{bmatrix} \ddot{q}_3 - a_1 \begin{bmatrix} c_1 \\ s_1 \\ 0 \end{bmatrix} \dot{q}_1^2$$
$$- \begin{bmatrix} r_{x3} \\ r_{y3} \\ 0 \end{bmatrix} \dot{q}_\psi^2 + 2 \begin{bmatrix} s_\psi r_{z3} \\ -c_\psi r_{z3} \\ 0 \end{bmatrix} \dot{q}_\psi \dot{q}_3 - \begin{bmatrix} c_\psi (s_\psi r_{y3} + c_\psi r_{x3}) \\ s_\psi (s_\psi r_{y3} + c_\psi r_{x3}) \\ r_{z3} \end{bmatrix} \dot{q}_3^2 \quad (4.87)$$

$$\dot{\mathbf{v}}_{4,5}^{cm} = \begin{bmatrix} -a_1 s_\psi \\ a_1 c_\psi \\ 0 \end{bmatrix} \ddot{q}_1 + \begin{bmatrix} -r_{y4,5} \\ r_{x4,5} \\ 0 \end{bmatrix} \ddot{q}_\psi + \begin{bmatrix} -c_\psi r_{z4,5} \\ -s_\psi r_{z4,5} \\ s_\psi r_{y4,5} + c_\psi r_{x4,5} \end{bmatrix} \ddot{q}_3 + \begin{bmatrix} s_\psi c_3 r_{z4,5} - s_3 r_{y4,5} \\ s_3 r_{x4,5} - c_\psi c_3 r_{z4,5} \\ c_3(c_\psi r_{y4,5} - s_\psi r_{x4,5}) \end{bmatrix} \ddot{q}_4 - a_1 \begin{bmatrix} c_1 \\ s_1 \\ 0 \end{bmatrix} \dot{q}_1^2$$

$$- \begin{bmatrix} r_{x4,5} \\ r_{y4,5} \\ 0 \end{bmatrix} \dot{q}_\psi^2 + 2 \begin{bmatrix} s_\psi r_{z4,5} \\ -c_\psi r_{z4,5} \\ 0 \end{bmatrix} \dot{q}_\psi \dot{q}_3 - \begin{bmatrix} c_\psi (s_\psi r_{y4,5} + c_\psi r_{x4,5}) \\ s_\psi (s_\psi r_{y4,5} + c_\psi r_{x4,5}) \\ r_{z4,5} \end{bmatrix} \dot{q}_3^2 + 2 \begin{bmatrix} c_\psi c_3 r_{z4,5} - s_3 r_{x4,5} \\ s_\psi c_3 r_{z4,5} - s_3 r_{y4,5} \\ 0 \end{bmatrix} \dot{q}_\psi \dot{q}_4$$

$$+ 2 \begin{bmatrix} c_\psi c_3 (s_\psi r_{x4,5} - c_\psi r_{y4,5}) \\ s_\psi c_3 (s_\psi r_{x4,5} - c_\psi r_{y4,5}) \\ s_3 (s_\psi r_{x4,5} - c_\psi r_{y4,5}) \end{bmatrix} \dot{q}_3 \dot{q}_4 + \begin{bmatrix} -s_\psi c_3^2 (s_\psi r_{x4,5} - c_\psi r_{y4,5}) + s_3 (c_\psi c_3 r_{z4,5} - s_3 r_{x4,5}) \\ c_\psi c_3^2 (s_\psi r_{x4,5} - c_\psi r_{y4,5}) + s_3 (s_\psi c_3 r_{z4,5} - s_3 r_{x4,5}) \\ c_3 (s_3 (s_\psi r_{y4,5} + c_\psi r_{x4,5}) - c_3 r_{z4,5}) \end{bmatrix} \dot{q}_4^2$$

$$\text{(4.88)}$$

In Eqs. (4.85)-(4.88) the mass centre coordinates $\hat{\mathbf{r}}_i^{cm} = \begin{bmatrix} \hat{r}_{xi} & \hat{r}_{yi} & \hat{r}_{zi} \end{bmatrix}^T$ and $\mathbf{r}_i^{cm} = \begin{bmatrix} r_{xi} & r_{yi} & r_{zi} \end{bmatrix}^T$ of the SDT links (Fig. 4.15) (for $q_1 = 0$) are:

$$\hat{\mathbf{r}}_1^{cm} = \begin{bmatrix} -1.464 & 0.002 & 0.170 \end{bmatrix}^T, \ \hat{\mathbf{r}}_2^{cm} = \begin{bmatrix} -0.004 & -0.904 & 0.237 \end{bmatrix}^T,$$

$$\hat{\mathbf{r}}_3^{cm} = \begin{bmatrix} 0.02 & 0.111 & -0.170 \end{bmatrix}^T, \ \hat{\mathbf{r}}_4^{cm} = \begin{bmatrix} 0. & 0.115 & 0.018 \end{bmatrix}^T, \text{ and}$$

$$\hat{\mathbf{r}}_5^{cm} = \begin{bmatrix} -0.182 & 0.004 & 0.37 \end{bmatrix}^T,$$

$$\mathbf{r}_1^{cm} = \begin{bmatrix} \hat{r}_{x1} & \hat{r}_{y1} & \hat{r}_{z1} \end{bmatrix}^T, \ \mathbf{r}_2^{cm} = \begin{bmatrix} c_\psi \hat{r}_{x2} + s_\psi \hat{r}_{z2} & s_\psi \hat{r}_{x2} - c_\psi \hat{r}_{z2} & \hat{r}_{y2} \end{bmatrix}^T,$$

$$\mathbf{r}_3^{cm} = \begin{bmatrix} c_\psi (c_3 r_{z3} - s_3 \hat{r}_{x3}) + s_\psi \hat{r}_{y3} & s_\psi (c_3 \hat{r}_{z3} - s_3 \hat{r}_{x3}) - c_\psi \hat{r}_{y3} & c_3 \hat{r}_{x3} + s_3 \hat{r}_{z3} \end{bmatrix}^T,$$

$$\mathbf{r}_{4,5}^{cm} = \begin{bmatrix} (c_\psi s_3 s_4 + s_\psi c_4) \hat{r}_{x4,5} - c_\psi c_3 \hat{r}_{y4,5} + (c_\psi s_3 c_4 - s_\psi s_4) \hat{r}_{z4,5} \\ (s_\psi s_3 s_4 - c_\psi c_4) \hat{r}_{x4,5} - s_\psi c_3 \hat{r}_{y4,5} + (s_\psi s_3 c_4 + c_\psi s_4) \hat{r}_{z4,5} \\ -c_3 (s_4 \hat{r}_{x4,5} + c_4 \hat{r}_{z4,5}) - s_3 \hat{r}_{y4,5} \end{bmatrix} \qquad \text{(4.89)}$$

Here is the external load (i.e., weight of the pilot with his seat and equipment) that affects link 4 denoted with 5. The subscript $4,5$ denotes axis 4 or the external load.

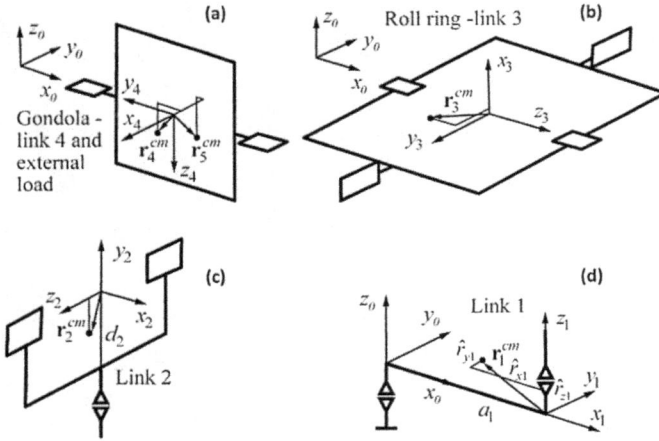

Fig. 4.15. Coordinate frame and the centre of mass of (a) link 4, (b) link 3, (c) link 2 and (d) link 1.

4.7.3. SDT Dynamics

Using these equations of forward kinematics, the inverse dynamics algorithm, which refers to the determination of the actuator torques required to generate a desired movement of the SDT and all forces and moments that act on the SDT axes, is here developed in the same manner with the CFS (Section 4.2.3).

The principal moments of inertia about the link centres of mass in the link coordinate systems and masses of the SDT links are: $\hat{I}_{x1}=1022$, $\hat{I}_{y1}=5048$, $\hat{I}_{z1}=4432$, $\hat{I}_{x2}=2194$, $\hat{I}_{y2}=1238$, $\hat{I}_{z2}=1021$, $\hat{I}_{x3}=1042$, $\hat{I}_{y3}=417$, $\hat{I}_{z3}=670$, $\hat{I}_{x4}=273$, $\hat{I}_{y4}=146$, $\hat{I}_{z4}=260\,\text{kgm}^2$, $\hat{I}_{x5}=57$, $\hat{I}_{y5}=73$, $\hat{I}_{z5}=36\,\text{kgm}^2$, $m_1=1223$, $m_2=669$, $m_3=573$, $m_4=348$ and $m_5=180$ kg. The SDT matrixes of the inertia of link i about the centre of mass of link i expressed in the base link coordinates are:

For axis 1: $I_{z1}=\hat{I}_{z1}$ (other terms are equal to zero) $\hspace{2cm}$ (4.90)

For axis 2: $I_{z2} = \hat{I}_{y2}$ (other terms are equal to zero) $\hspace{2cm}$ (4.91)

For axis 3:

$$\mathbf{I}_3^{cm}=\begin{bmatrix}\hat{I}_{x3}c_\psi^2 s_3^2+\hat{I}_{y3}s_\psi^2+\hat{I}_{z3}c_\psi^2 c_3^2 & s_\psi c_\psi(\hat{I}_{x3}s_3^2-\hat{I}_{y3}+\hat{I}_{z3}c_3^2) & c_\psi s_3 c_3(\hat{I}_{z3}-\hat{I}_{x3})\\ s_\psi c_\psi(\hat{I}_{x3}s_3^2-\hat{I}_{y3}+\hat{I}_{z3}c_3^2) & \hat{I}_{x3}s_\psi^2 s_3^2+\hat{I}_{y3}c_\psi^2+\hat{I}_{z3}s_\psi^2 c_3^2 & s_\psi s_3 c_3(\hat{I}_{z3}-\hat{I}_{x3})\\ c_\psi s_3 c_3(\hat{I}_{z3}-\hat{I}_{x3}) & s_\psi s_3 c_3(\hat{I}_{z3}-\hat{I}_{x3}) & \hat{I}_{x3}c_3^2+\hat{I}_{z3}s_3^2\end{bmatrix}$$

$$(4.92)$$

For axis 4 and the external load:

$$I_{x4,5}=\hat{I}_{x4,5}(c_\psi s_3 s_4+s_\psi c_4)^2+\hat{I}_{y4,5}c_\psi^2 c_3^2+\hat{I}_{z4,5}(c_\psi s_3 c_4-s_\psi s_4)^2$$

$$I_{y4,5}=\hat{I}_{x4,5}(s_\psi s_3 s_4-c_\psi c_4)^2+\hat{I}_{y4,5}s_\psi^2 c_3^2+\hat{I}_{z4,5}(s_\psi s_3 c_4+c_\psi s_4)^2$$

$$I_{z4,5}=c_3^2(\hat{I}_{x4,5}s_4^2+\hat{I}_{z4,5}c_4^2)+\hat{I}_{y4,5}s_3^2$$

$$I_{xy4,5}=-\hat{I}_{x4,5}[c_\psi s_\psi(s_3^2 s_4^2-c_4^2)+s_3 c_4 s_4(s_\psi^2-c_\psi^2)]$$
$$-\hat{I}_{y4,5}c_\psi s_\psi c_3^2-\hat{I}_{z4,5}[c_\psi s_\psi(s_3^2 c_4^2-s_4^2)-s_3 c_4 s_4(s_\psi^2-c_\psi^2)] \qquad (4.93)$$

$$I_{xz4,5}=\hat{I}_{x4,5}(c_\psi s_3 s_4+s_\psi c_4)c_3 s_4-\hat{I}_{y4,5}c_\psi c_3 s_3+\hat{I}_{z4,5}(c_\psi s_3 c_4-s_\psi s_4)c_3 c_4$$

$$I_{yz4,5}=\hat{I}_{x4,5}(s_\psi s_3 s_4-c_\psi c_4)c_3 s_4-\hat{I}_{y4,5}s_\psi c_3 s_3+\hat{I}_{z4,5}(s_\psi s_3 c_4+c_\psi s_4)c_3 c_4$$

The total forces $\mathbf{F}_1$, $\mathbf{F}_2$, $\mathbf{F}_3$, $\mathbf{F}_4$ and $\mathbf{F}_5$ can be obtained from Eqs. (4.85)-(4.88) and the total moments $\mathbf{M}_1$, $\mathbf{M}_2$, $\mathbf{M}_3$, $\mathbf{M}_4$ and $\mathbf{M}_5$ can be obtained from Eqs. (4.77), (4.80)-(4.83) and (4.90)-(4.93) in the following way:

$$\mathbf{M}_1=\begin{bmatrix}0 & 0 & I_{z1}\end{bmatrix}^T\ddot{q}_1 \qquad (4.94)$$

$$\mathbf{M}_2=\begin{bmatrix}0 & 0 & I_{z2}\end{bmatrix}^T\ddot{q}_\psi \qquad (4.95)$$

$$\mathbf{M}_3=\begin{bmatrix}-I_{xz3}\\ -I_{yz3}\\ I_{z3}\end{bmatrix}\ddot{q}_\psi+\begin{bmatrix}I_{x3}s_\psi+I_{xy3}c_\psi\\ -I_{y3}c_\psi-I_{xy3}s_\psi\\ -I_{xz3}s_\psi+I_{yz3}c_\psi\end{bmatrix}\ddot{q}_3+\begin{bmatrix}I_{yz3}\\ -I_{xz3}\\ 0\end{bmatrix}\dot{q}_\psi^2$$
$$+\begin{bmatrix}c_\psi(I_{x3}+I_{y3}-I_{z3})\\ s_\psi(I_{x3}+I_{y3}-I_{z3})\\ -2(I_{xz3}c_\psi+I_{yz3}s_\psi)\end{bmatrix}\dot{q}_\psi\dot{q}_3+\begin{bmatrix}c_\psi(I_{xz3}s_\psi-I_{yz3}c_\psi)\\ s_\psi(I_{xz3}s_\psi-I_{yz3}c_\psi)\\ s_\psi c_\psi(I_{x3}-I_{y3})-I_{xy3}(s_\psi^2-c_\psi^2)\end{bmatrix}\dot{q}_3^2$$

$$(4.96)$$

131

$$\mathbf{M}_{4,5} = \begin{bmatrix} -I_{xz4,5} \\ -I_{yz4,5} \\ I_{z4,5} \end{bmatrix} \ddot{q}_{\psi} + \begin{bmatrix} I_{x4,5}s_{\psi} + I_{xy4,5}c_{\psi} \\ -I_{y4,5}c_{\psi} - I_{xy4,5}s_{\psi} \\ -I_{xz4,5}s_{\psi} + I_{yz4,5}c_{\psi} \end{bmatrix} \ddot{q}_3 + \begin{bmatrix} c_3(I_{x4,5}c_{\psi} - I_{xy4,5}s_{\psi}) - I_{xz4,5}s_3 \\ c_3(I_{y4,5}s_{\psi} - I_{xy4,5}c_{\psi}) - I_{yz4,5}s_3 \\ I_{z4,5}s_3 - c_3(I_{xz4,5}c_{\psi} + I_{yz4,5}s_{\psi}) \end{bmatrix} \ddot{q}_4 + \begin{bmatrix} I_{yz4,5} \\ -I_{xz4,5} \\ 0 \end{bmatrix} \dot{q}_{\psi}^2$$

$$+ \begin{bmatrix} c_{\psi}(I_{x4,5} + I_{y4,5} - I_{z4,5}) \\ s_{\psi}(I_{x4,5} + I_{y4,5} - I_{z4,5}) \\ -2(c_{\psi}I_{xz4,5} + s_{\psi}I_{yz4,5}) \end{bmatrix} \dot{q}_{\psi}\dot{q}_3 + \begin{bmatrix} -s_{\psi}c_3(I_{x4,5} + I_{y4,5} - I_{z4,5}) + 2s_3 I_{yz4,5} \\ c_{\psi}c_3(I_{x4,5} + I_{y4,5} - I_{z4,5}) - 2s_3 I_{xz4,5} \\ -2c_3(c_{\psi}I_{yz4,5} - s_{\psi}I_{xz4,5}) \end{bmatrix} \dot{q}_{\psi}\dot{q}_4$$

$$+ \begin{bmatrix} c_{\psi}(I_{xz4,5}s_{\psi} - I_{yz4,5}c_{\psi}) \\ s_{\psi}(I_{xz4,5}s_{\psi} - I_{yz4,5}c_{\psi}) \\ s_{\psi}c_{\psi}(I_{x4,5} - I_{y4,5}) - (s_{\psi}^2 - c_{\psi}^2)I_{xy4,5} \end{bmatrix} \dot{q}_3^2 \qquad (4.97)$$

$$+ \begin{bmatrix} c_{\psi}s_3(-I_{x4,5} + I_{y4,5} - I_{z4,5}) + 2s_{\psi}(s_3 I_{xy4,5} - s_{\psi}c_3 I_{xz4,5} + c_{\psi}c_3 I_{yz4,5}) \\ s_{\psi}s_3(I_{x4,5} - I_{y4,5} - I_{z4,5}) + 2c_{\psi}(s_3 I_{xy4,5} + s_{\psi}c_3 I_{xz4,5} - c_{\psi}c_3 I_{yz4,5}) \\ c_3[(s_{\psi}^2 - c_{\psi}^2)(I_{y4,5} - I_{x4,5}) + I_{z4,5}] - 4s_{\psi}c_{\psi}I_{xy4,5} \end{bmatrix} \dot{q}_3\dot{q}_4$$

$$+ \begin{bmatrix} c_3 s_3[s_{\psi}(I_{z4,5} - I_{y4,5}) + c_{\psi}I_{xy4,5}] - s_{\psi}c_{\psi}c_3^2 I_{xz4,5} - (s_{\psi}^2 c_3^2 - s_3^2)I_{yz4,5} \\ c_3 s_3[c_{\psi}(I_{x4,5} - I_{z4,5}) - s_{\psi}(I_{xy4,5})] + (c_{\psi}^2 c_3^2 - s_3^2)I_{xz4,5} + s_{\psi}c_{\psi}c_3^2 I_{yz4,5} \\ c_3^2[s_{\psi}c_{\psi}(I_{y4,5} - I_{x4,5}) - (c_{\psi}^2 - s_{\psi}^2)I_{xy4,5}] - c_3 s_3(c_{\psi}I_{yz4,5} - s_{\psi}I_{xz4,5}) \end{bmatrix} \dot{q}_4^2$$

Equivalently to the CFS, Eqs. (4.23)-(4.35), the forces and moments exerted on the SDT links are:

$$\mathbf{f}_5 = \mathbf{F}_5 = m_5 \begin{bmatrix} \dot{v}_{x5}^{cm} & \dot{v}_{y5}^{cm} & \dot{v}_{z5}^{cm} - g \end{bmatrix}^T, \text{ (Fig. 4.16(a))} \qquad (4.98)$$

$$\mathbf{m}_5 = \mathbf{M}_5 + \mathbf{r}_5^{cm} \times \mathbf{F}_5 = \mathbf{M}_5 + \begin{bmatrix} r_{y5}F_{z5} - r_{z5}F_{y5} & r_{z5}F_{x5} - r_{x5}F_{z5} & r_{x5}F_{y5} - r_{y5}F_{x5} \end{bmatrix}^T \qquad (4.99)$$

$$\mathbf{f}_4 = \mathbf{F}_4 + \mathbf{f}_5, \text{ where } \mathbf{F}_4 = m_4 \begin{bmatrix} \dot{v}_{x4}^{cm} & \dot{v}_{y4}^{cm} & \dot{v}_{z4}^{cm} - g \end{bmatrix}^T, \text{ (Fig. 16(a))} \qquad (4.100)$$

$$\mathbf{m}_4 = \mathbf{M}_4 + \mathbf{r}_4^{cm} \times \mathbf{F}_4 + \mathbf{m}_5 = \mathbf{M}_4 + \begin{bmatrix} r_{y4}F_{z4} - r_{z4}F_{y4} & r_{z4}F_{x4} - r_{x4}F_{z4} & r_{x4}F_{y4} - r_{y4}F_{x4} \end{bmatrix}^T + \mathbf{m}_5 \qquad (4.101)$$

$$\hat{\mathbf{f}}_4 = \begin{bmatrix} -s_3(c_{\psi}f_{x4} + s_{\psi}f_{y4}) + c_3 f_{z4} \\ s_{\psi}f_{x4} - c_{\psi}f_{y4} \\ c_3(c_{\psi}f_{x4} + s_{\psi}f_{y4}) + s_3 f_{z4} \end{bmatrix}, \hat{\mathbf{m}}_4 = \begin{bmatrix} -s_3(c_{\psi}m_{x4} + s_{\psi}m_{y4}) + c_3 m_{z4} \\ s_{\psi}m_{x4} - c_{\psi}m_{y4} \\ c_3(c_{\psi}m_{x4} + s_{\psi}m_{y4}) + s_3 m_{z4} \end{bmatrix}$$

$$(4.102)$$

$\mathbf{f}_3 = \mathbf{F}_3 + \mathbf{f}_4$, where $\mathbf{F}_3 = m_3 \begin{bmatrix} \dot{v}_{x3}^{cm} & \dot{v}_{y3}^{cm} & \dot{v}_{z3}^{cm} - g \end{bmatrix}^T$, (Fig. 16(b)) $\qquad$ (4.103)

$$\mathbf{m}_3 = \mathbf{M}_3 + \mathbf{r}_3^{cm} \times \mathbf{F}_3 + \mathbf{m}_4 = \mathbf{M}_3 + \begin{bmatrix} r_{y3}F_{z3} - r_{z3}F_{y3} & r_{z3}F_{x3} - r_{x3}F_{z3} & r_{x3}F_{y3} - r_{y3}F_{x3} \end{bmatrix}^T + \mathbf{m}_4 \tag{4.104}$$

$$\hat{\mathbf{f}}_3 = \begin{bmatrix} c_\psi f_{x3} + s_\psi f_{y3} & f_{z3} & s_\psi f_{x3} - c_\psi f_{y3} \end{bmatrix}^T,$$
$$\hat{\mathbf{m}}_3 = \begin{bmatrix} c_\psi m_{x3} + s_\psi m_{y3} & m_{z3} & s_\psi m_{x3} - c_\psi m_{y3} \end{bmatrix}^T \tag{4.105}$$

$\mathbf{f}_2 = \mathbf{F}_2 + \mathbf{f}_3 = m_2 \begin{bmatrix} \dot{v}_{x2}^{cm} & \dot{v}_{y2}^{cm} & \dot{v}_{z2}^{cm} - g \end{bmatrix}^T + \mathbf{f}_3$, (Fig. 16(c)) $\qquad$ (4.106)

$$\mathbf{m}_2 = \mathbf{M}_2 + (\mathbf{p}_2^* + \mathbf{r}_2^{cm}) \times \mathbf{F}_2 + \mathbf{p}_2^* \times \mathbf{f}_3 + \mathbf{m}_3 = \mathbf{M}_2 + \begin{bmatrix} r_{y2}F_{z2} - (r_{z2}+d_2)F_{y2} - d_2 f_{y3} \\ (r_{z2}+d_2)F_{x2} - r_{x2}F_{z2} + d_2 f_{x3} \\ r_{x2}F_{y2} - r_{y2}F_{x2} \end{bmatrix} + \mathbf{m}_3 \tag{4.107}$$

$$\hat{\mathbf{f}}_2 = \begin{bmatrix} c_1 f_{x2} + s_1 f_{y2} & -s_1 f_{x2} + c_1 f_{y2} & f_{z2} \end{bmatrix}^T,$$
$$\hat{\mathbf{m}}_2 = \begin{bmatrix} c_1 m_{x2} + s_1 m_{y2} & -s_1 m_{x2} + c_1 m_{y2} & m_{z2} \end{bmatrix}^T \tag{4.108}$$

$\mathbf{f}_1 = \mathbf{F}_1 + \mathbf{f}_2 = m_1 \begin{bmatrix} \dot{v}_{x1}^{cm} & \dot{v}_{y1}^{cm} & \dot{v}_{z1}^{cm} - g \end{bmatrix}^T + \mathbf{f}_2$, (Fig. 16(d)) $\qquad$ (4.109)

$$\mathbf{m}_1 = \mathbf{M}_1 + \mathbf{m}_2 + (\mathbf{p}_1^* + \mathbf{r}_1^{cm}) \times \mathbf{F}_1 + \mathbf{p}_1^* \times \mathbf{f}_2$$
$$= \mathbf{M}_1 + \begin{bmatrix} (a_1 s_1 + r_{y1})F_{z1} - r_{z1}F_{y1} + a_1 s_1 f_{z2} \\ -(a_1 c_1 + r_{x1})F_{z1} + r_{z1}F_{x1} - a_1 c_1 f_{z2} \\ (a_1 c_1 + r_{x1})F_{y1} - (a_1 s_1 + r_{y1})F_{x1} + a_1(c_1 f_{y2} - s_1 f_{x2}) \end{bmatrix} + \mathbf{m}_2 \tag{4.110}$$

$$\hat{\mathbf{f}}_1 = \begin{bmatrix} \hat{f}_{x1} & \hat{f}_{y1} & \hat{f}_{z1} \end{bmatrix}^T = \begin{bmatrix} f_{x1} & f_{y1} & f_{z1} \end{bmatrix}^T,$$
$$\hat{\mathbf{m}}_1 = \begin{bmatrix} \hat{m}_{x1} & \hat{m}_{y1} & \hat{m}_{z1} \end{bmatrix}^T = \begin{bmatrix} m_{x1} & m_{y1} & m_{z1} \end{bmatrix}^T \tag{4.111}$$

The torque of the joint i actuator is $\hat{m}_{zi}$, the power of the joint i actuator is $P_i = \hat{m}_{zi} \ddot{q}_i$, the moment acting on the link bearing i is given with Eq. (4.36), the axial force of the bearing i is $\hat{f}_{ai} = \hat{f}_{zi}$ and the radial force of the bearing i is given with Eq. (4.37).

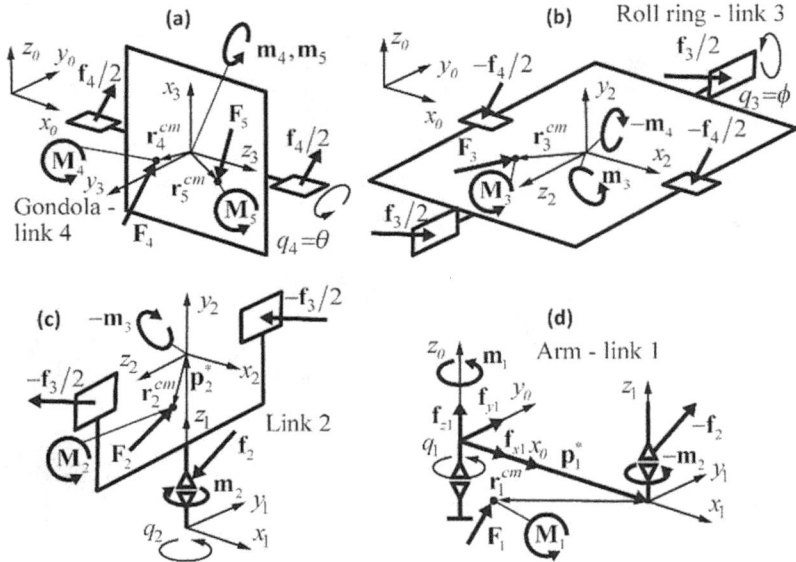

Fig. 4.16. Forces and moments exerted on (a) link 4, (b) link 3, (c) link 2 and (d) link 1.

4.8. Acceleration Forces and Link Angles of the SDT

4.8.1. Calculation of the SDT Simulator Pilot Acceleration Force Components

The transverse G_x, lateral G_y and longitudinal G_z acceleration force components acting on the pilot's head or chest in the SDT are the same as in the CFS (Fig. 4.5). Fig. 4.17 shows the coordinate frames, angles, angular velocities and acceleration forces of the SDT. The linear acceleration experienced by the simulator pilot at the intersection point of the roll and pitch axes is calculated by Eq. (4.79). Based on this equation, for $q_1=q_2=0$ and a gravitational acceleration g, the orthogonal components G_n, G_t and G_v for normal (i.e., radial), tangential and vertical acceleration forces, respectively, experienced by the simulator pilot are the same as in CFS, which means that they are given with Eq. (4.38). As with the CFS, Eq. (4.38) gives the resulting force experienced by the simulator pilot at the intersection point of the roll and pitch axes (i.e., for $q_1=q_2=0$), as a function of angular velocity and the acceleration of the SDT arm, Eq. (4.49).

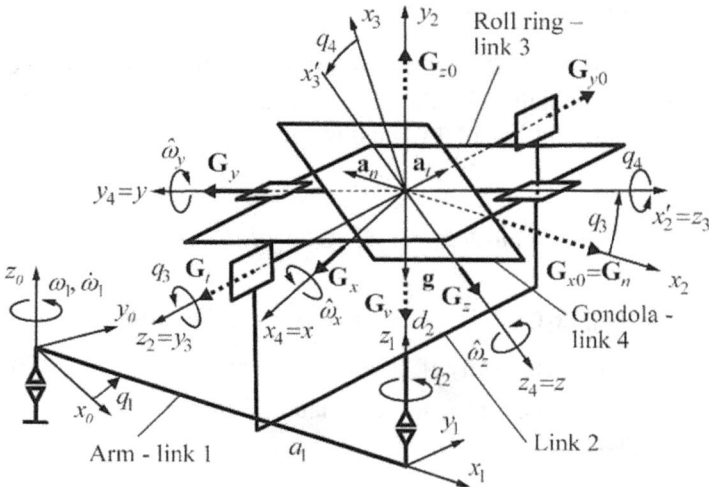

Fig. 4.17. Coordinate frames, angles, angular velocities and acceleration forces of the 4-axis SDT.

Angles q_2, q_3 and q_4, the angular velocity ω_1, and the angular acceleration $\dot{\omega}_1$ of the arm define the orthogonal components G_x, G_y and G_z of the resultant vector **G** experienced by the simulator pilot. Based on Eqs. (4.49) and (4.77), the resultant vector **G** is:

$$\mathbf{G}=\begin{bmatrix} G_x & G_y & G_z \end{bmatrix}^T = \mathbf{D}_4^{-1}\begin{bmatrix} G_{x0} & G_{y0} & G_{z0} \end{bmatrix}^T \tag{4.112}$$

$$G_x=(c_2 s_3 s_4 + s_2 c_4)G_{x0}+(s_2 s_3 s_4 - c_2 c_4)G_{y0}+c_3 s_4 \tag{4.113}$$

$$G_y=-c_3(c_2 G_{x0}+s_2 G_{y0})+s_3 \tag{4.114}$$

$$G_z=(c_2 s_3 c_4 - s_2 s_4)G_{x0}+(s_2 s_3 c_4 + c_2 s_4)G_{y0}+c_3 c_4 \tag{4.115}$$

The link angles $q_3=\phi$ and $q_4=\theta$ and their derivatives, and the derivative of the angle q_2 define the roll, pitch, and yaw angular velocities $\hat{\omega}_x$, $\hat{\omega}_y$ and $\hat{\omega}_z$ experienced by the simulator pilot. These are given in the following equations for $q_1= 0$:

$$\hat{\boldsymbol{\omega}}=\begin{bmatrix} \hat{\omega}_x \\ \hat{\omega}_y \\ \hat{\omega}_z \end{bmatrix}=\mathbf{D}_4^{-1}\boldsymbol{\omega}_4=\begin{bmatrix} c_4\dot{q}_3-c_3 s_4\dot{q}_\psi \\ -\dot{q}_4-s_3\dot{q}_\psi \\ -s_4\dot{q}_3-c_3 c_4\dot{q}_\psi \end{bmatrix} \tag{4.116}$$

135

4.8.2. Calculation of the Roll and Pitch Angles of the SDT

The roll and pitch angles of the gondola for the known acceleration forces can be calculated for the known angle q_2. The roll angle of the SDT is calculated by Eq. (4.114) using the given lateral force G_y, in the following way:

$$q_3 = \phi = \operatorname{atan} 2(p_1 + G_y(1 - G_y^2 + p_1^2)^{1/2}, 1 - G_y^2), \qquad (4.117)$$

where $p_1 = c_2 G_{x0} + s_2 G_{y0}$. If $G_y < 0$ and $G_y^2 > 1$, then $q_3 = q_3 + \pi$. The roll angle can be calculated only if $p_1^2 + 1 \geq G_y^2$ is satisfied; otherwise it is not possible to achieve the given lateral force G_y. For $G_y = 0$, Eq. (4.117) yields:

$$q_3 = \phi = \operatorname{atan} 2(p_1, 1) \qquad (4.118)$$

Eqs. (4.113) and (4.114) show that it is not possible to achieve both of the given G_x and G_z forces, even when they are in the allowed ranges. As a result, the pitch angle of the SDT is calculated by Eq. (4.113), using the given transverse force G_x, or by Eq. (4.115), using the given longitudinal G_z force. Eq. (4.113) yields:

$$q_4 = \theta = \operatorname{atan} 2(-p_2 p_3 + G_x(p_2^2 + p_3^2 - G_x^2)^{1/2}, p_2^2 - G_x^2), \qquad (4.119)$$

where $p_2 = c_2 s_3 G_{x0} + s_2 s_3 G_{y0} + c_3$, $p_3 = s_2 G_{x0} - c_2 G_{y0}$. If $p_2^2 + p_3^2 < G_x^2$, then it is not possible to achieve the given transverse force G_x. For $G_x = 0$, Eq. (4.120) yields:

$$q_4 = \theta = \operatorname{atan} 2(-p_3, p_2) \qquad (4.120)$$

Eq. (4.116) yields the following:

$$q_4 = \theta = \operatorname{atan} 2(p_2 p_3 - G_z(p_2^2 + p_3^2 - G_z^2)^{1/2}, p_3^2 - G_z^2) \qquad (4.121)$$

If $G_z < 0$ and $G_z^2 < p_3^2$, then is $q_4 = q_4 - \pi$. If $p_2^2 + p_3^2 < G_z^2$, then it is not possible to achieve the given longitudinal G_z force.

4.9. The Control Algorithm of the SDT Movement

The SDT can be programmed to achieve the quoted G_z, G_x and G_y acceleration forces. In this case, the roll and pitch angles are calculated in the way as described in Section 4.8.2. But, the basic aim of the SDT is to perform point-to-point movements in order to achieve given orientation. During these movements, the axes which have to move start and finish movements simultaneously. The axis that has the most rotation (i.e., the leading axis) moves at the programmed speed, while the other axes rotate to reach their programmed positions at the same time as leading axis (i.e., they move proportionally more slowly based on how far they must rotate). The actuator torques required for the desired motions depend on the links positions of the SDT. Thus, for each link, the total moment of inertia about the axes of rotation must be known, which consists of the moments of inertia of that link, of the links to the end of the manipulator and of the external load, reduced to the axis of rotation of that link.

To prevent that the control unit sends the velocities to actuators that cannot be achieved and to prevent the over-dimensioning of the axes bearings and links during the design stage of the SDT, the algorithm which is similar to the algorithm given in the Section 4.3 for the CFS is given here. This algorithm calculates the maximum possible values of the SDT axes acceleration based on the positions using the total moment of inertia about the axis of rotation of that link, and accelerations of the other SDT axes. The real forces and moments that act on each SDT link are thus calculated.

4.9.1. Calculation of the Maximum Possible Value of $\ddot{q}_1$

For axis 1, an AC motor Siemens 1FT7 108-5SF71-ZJ13 is chosen with a rated torque of $M_{1r}=37$ Nm, a rated power of $P_{1r}=12$ kW, a rated number of revolutions of $n_{1r}=3000$ min^{-1}, and a maximum number of revolutions through field weakening of $n_{1max}=5600$ min^{-1}. If the motor's number of revolutions is assigned to be n_{1m}, then for $n_{1m} \leq n_{1r}$, $M_{1r}=const$ and $P_1=P_{1r}n_{1m}/n_{1r}$; for $n_{1m}>n_{1r}$, $M_{1r}=P_{1r}n_{1r}/n_{1m}$ and $P_1=P_{1r}=const$. If a short duration (i.e., up to 5 s) is assigned for the maximum torque M_{1max}, then the overload capability will be $f_{1p}=M_{1max}/M_{1r}=5.9$. The chosen gearbox has a gear ratio of $k_1=67.2$ and an efficiency of $\eta_1=0.93$. The rated angular speed of link 1 is $\dot{q}_{1r}=n_{1r}\,\pi/(30k_1)=4.675$ s^{-1} and the maximum angular speed of link 1 $\dot{q}_{1max}$=8.7266 s⁻¹. The moment of inertia of the rotor of the motor with the gearbox elements on the rotor is I_1=0.039 kgm². For axis 1, the maximum torque for the accelerated motion in the positive direction and the decelerated motion in the negative direction $(\ddot{q}_i>0)$ is $\hat{m}_{z1}=M_{1r}\,f_{1p}\,k_1\,\eta_1$=13643 Nm, and for the decelerated movement in the positive direction and the accelerated movement in the negative direction $(\ddot{q}_i<0)$ is $\hat{m}_{z1}=M_{1r}\,f_{1p}\,k_1/\eta_1$=15774 Nm.

First, the maximum possible value $\ddot{q}_{1max}$ of $\ddot{q}_1$ will be calculated in the same way as for the CFS, using Eqs. (4.58)-(4.61) in which a_e is replaced with $\dot{q}_{1e}$ - the desired angular speed of link 1 - and a_s is replaced with $\dot{q}_{1s}$ - the initial angular speed of link 1.

The total moment of inertia I_{t1} of all four links and the external load, reduced to the axis z_0 (for q_1=0), is given with Eq. (4.62) in which i=1-5 and the values of I_{zi} are given in Eqs. (4.90)-(4.93). The components of the vectors $^0\mathbf{r}_i^{cm}$ are:

$$^0\mathbf{r}_1^{cm}=\begin{bmatrix}^0r_{x1} & ^0r_{y1} & ^0r_{z1}\end{bmatrix}^T=\mathbf{D}_1\left(\begin{bmatrix}a_1 & 0 & 0\end{bmatrix}^T+\hat{\mathbf{r}}_1\right)=\begin{bmatrix}r_{x1}+a_1 & r_{y1} & r_{z1}\end{bmatrix}^T \quad (4.122)$$

$$^0\mathbf{r}_i^{cm}=\begin{bmatrix}^0r_{xi} & ^0r_{yi} & ^0r_{zi}\end{bmatrix}^T=\mathbf{D}_1\begin{bmatrix}a_1 & 0 & d_2\end{bmatrix}^T+\mathbf{D}_i\hat{\mathbf{r}}_i=\begin{bmatrix}r_{xi}+a_1 & r_{yi} & r_{zi}+d_2\end{bmatrix}^T, \quad (4.123)$$

where i=2, 3, 4, 5.

In previous equations, the matrices $\mathbf{D}_1$, $\mathbf{D}_2$, $\mathbf{D}_3$ and $\mathbf{D}_{4,5}$ were obtained from Eqs. (4.73)-(4.77), and the vectors $\mathbf{r}_1^{cm}$, $\mathbf{r}_2^{cm}$, $\mathbf{r}_3^{cm}$ and $\mathbf{r}_{4,5}^{cm}$ were given in Eq. (4.89).

4.9.2. Calculation of the Maximum Possible Values of $\ddot{q}_2$, $\ddot{q}_3$ and $\ddot{q}_4$

Axes 2, 3 and 4 are actuated by torque motors. The motor for axis 2 – Siemens 1FW6230-OWB15-8FB2.1 - has a maximum torque of M_{2max}=3950 Nm, a maximum speed at that torque of $\dot{q}_{2max}$=4.6077 s⁻¹, a rated torque of M_{2r}=2380 Nm and a maximum speed at that torque of $\dot{q}_{2r}$=8.3776 s⁻¹. The motors for axes 3 and 4 – Siemens 1FW6160-OWB15-21xx1 - have: M_{imax}=2150 Nm, $\dot{q}_{imax}$=3.5605 s⁻¹, M_{ir}=1350 Nm and $\dot{q}_{ir}$=6.9115 s⁻¹, where i=3, 4. These torques decrease linearly from the maximum to the rated values as functions of their speeds (Fig. 4.8). Calculation of the desired and maximum possible values of $\ddot{q}_i$, $i = 2, 3, 4$ can be done in the three steps given in Section 4.4.3.2.

4.9.2.1. Total Moments of Inertia for SDT Axes z_1, z_2 and z_3

The total moment of inertia of links 2, 3 and 4 and the external load reduced to axis z_1 is:

$$I_{t2} = \sum_{i=2}^{5} (^1I_{zi} + [(^1r_{xi})^2 + (^1r_{yi})^2 + (^1r_{zi})^2] m_i) \qquad (4.124)$$

In Eq. (4.124), the value $^1I_{z2} = \hat{I}_{y2}$ is obtained from $^1\mathbf{I}_2^{cm} = {}^1\mathbf{D}_2 \hat{\mathbf{I}}_2^{cm} {}^1\mathbf{D}_2^{-1}$, the value $^1I_{z3} = c_3^2 \hat{I}_{x3} + s_3^2 \hat{I}_{z3}$ is obtained from $^1\mathbf{I}_3^{cm} = {}^1\mathbf{D}_3 \hat{\mathbf{I}}_3^{cm} {}^1\mathbf{D}_3^{-1}$, the value $^1I_{z4} = c_3^2 (s_4^2 \hat{I}_{x4} + c_4^2 \hat{I}_{z4}) + s_3^2 \hat{I}_{y4}$ is obtained from $^1\mathbf{I}_4^{cm} = {}^1\mathbf{D}_4 \hat{\mathbf{I}}_4^{cm} {}^1\mathbf{D}_4^{-1}$ and the value $^1I_{z5} = c_3^2 (s_4^2 \hat{I}_{x5} + c_4^2 \hat{I}_{z5}) + s_3^2 \hat{I}_{y5}$ is obtained from $^1\mathbf{I}_5^{cm} = {}^1\mathbf{D}_4 \hat{\mathbf{I}}_5^{cm} {}^1\mathbf{D}_4^{-1}$. Eq. (4.76) applies to the following:

$$^1\mathbf{r}_2^{cm} = \begin{bmatrix} ^1r_{x2} & ^1r_{y2} & ^1r_{z2} \end{bmatrix}^T = {}^1\mathbf{D}_2 \, \hat{\mathbf{r}}_2^{cm} = \begin{bmatrix} c_2\hat{r}_{x2} + s_2\hat{r}_{z2} & s_2\hat{r}_{x2} - c_2\hat{r}_{z2} & \hat{r}_{y2} \end{bmatrix}^T \quad (4.125)$$

$$^1\mathbf{r}_3^{cm} = \begin{bmatrix} ^1r_{x3} & ^1r_{y3} & ^1r_{z3} \end{bmatrix}^T = {}^1\mathbf{D}_3 \, \hat{\mathbf{r}}_3 = \begin{bmatrix} c_2(c_3\hat{r}_{z3} - s_3\hat{r}_{x3}) + s_2\hat{r}_{y3} \\ s_2(c_3\hat{r}_{z3} - s_3\hat{r}_{x3}) - c_2\hat{r}_{y3} \\ c_3\hat{r}_{x3} + s_3\hat{r}_{z3} \end{bmatrix} \qquad (4.126)$$

$$^1\mathbf{r}_{4,5}^{cm} = \begin{bmatrix} ^1r_{x4,5} & ^1r_{y4,5} & ^1r_{z4,5} \end{bmatrix}^T = {}^1\mathbf{D}_4 \, \hat{\mathbf{r}}_{4,5} = \begin{bmatrix} c_2[s_3(s_4\hat{r}_{x4,5} + c_4\hat{r}_{z4,5}) - c_3\hat{r}_{y4,5}] + s_2(c_4\hat{r}_{x4,5} - s_4\hat{r}_{z4,5}) \\ s_2[s_3(c_4\hat{r}_{z4,5} - s_4\hat{r}_{x4,5}) - c_3\hat{r}_{y4,5}] - c_2(c_4\hat{r}_{x4,5} + s_4\hat{r}_{z4,5}) \\ -c_3(s_4\hat{r}_{x4,5} + c_4\hat{r}_{z4,5}) - s_3\hat{r}_{y4,5} \end{bmatrix}$$

$$(4.127)$$

In previous equations, $^1\mathbf{D}_3 = {}^1\mathbf{D}_2 \, {}^2\mathbf{D}_3$ and $^1\mathbf{D}_4 = {}^1\mathbf{D}_2 \, {}^2\mathbf{D}_3 \, {}^3\mathbf{D}_4$. The matrices $^1\mathbf{D}_2$, $^2\mathbf{D}_3$ and $^3\mathbf{D}_4$ are given in Eqs. (4.74), (4.75) and (4.76).

The total moment of inertia of link 3, 4 and the external load reduced to the axis z_2 is:

$$I_{t3} = \sum_{i=3}^{5} (^2I_{zi} + [(^2r_{xi})^2 + (^2r_{yi})^2 + (^2r_{zi})^2] m_i) \qquad (4.128)$$

In previous equation,

the value $^2I_{z3}=\hat{I}_{y3}$ is obtained from $^2\mathbf{I}_3^{cm}=\,^2\mathbf{D}_3\hat{\mathbf{I}}_3^{cm}\,^2\mathbf{D}_3^{-1}$,

the value $^2I_{z4}=c_4^2\hat{I}_{x4}+s_4^2\hat{I}_{z4}$ is obtained from $^2\mathbf{I}_4^{cm}=\,^2\mathbf{D}_4\hat{\mathbf{I}}_4^{cm}\,^1\mathbf{D}_4^{-1}$

and the value $^2I_{z5}=c_4^2\hat{I}_{x5}+s_4^2\hat{I}_{z5}$ is obtained from $^2\mathbf{I}_5^{cm}=\,^2\mathbf{D}_4\hat{\mathbf{I}}_5^{cm}\,^2\mathbf{D}_4^{-1}$.

Eq. (4.128) applies to the following:

$$^2\mathbf{r}_3^{cm}=\begin{bmatrix}^2r_{x3} & ^2r_{y3} & ^2r_{z3}\end{bmatrix}^T=\,^2\mathbf{D}_3\,\hat{\mathbf{r}}_3^{cm}=\begin{bmatrix}-s_3\hat{r}_{x3}+c_3\hat{r}_{z3} & c_3\hat{r}_{x3}+s_3\hat{r}_{z3} & \hat{r}_{y3}\end{bmatrix}^T \quad (4.129)$$

$$^2\mathbf{r}_{4,5}^{cm}=\begin{bmatrix}^2r_{x4,5} & ^2r_{y4,5} & ^2r_{z4,5}\end{bmatrix}^T=\,^2\mathbf{D}_4\,\hat{\mathbf{r}}_{4,5}^{cm}=\begin{bmatrix}s_3(s_4\hat{r}_{x4,5}+c_4\hat{r}_{z4,5})-c_3\hat{r}_{y4,5}\\-c_3(s_4\hat{r}_{x4,5}+c_4\hat{r}_{z4,5})-s_3\hat{r}_{y4,5}\\c_4\hat{r}_{x4,5}-s_4\hat{r}_{z4,5}\end{bmatrix} \quad (4.130)$$

In previous equations, $^2\mathbf{D}_4=\,^2\mathbf{D}_3\,^3\mathbf{D}_4$.

The total moment of inertia of link 4 and the external load reduced to the axis z_3 is:

$$I_{t4}=\,^3I_{z4}+\,^3I_{z5}+[(^3r_{x4})^2+(^3r_{y4})^2+(^3r_{z4})^2]m_4+[(^3r_{x5})^2+(^3r_{y5})^2+(^3r_{z5})^2]m_5 \quad (4.131)$$

In previous equation, the value $^3I_{z4,5}=\hat{I}_{y4,5}$ is obtained from $^3\mathbf{I}_{4,5}^{cm}=\,^3\mathbf{D}_4\hat{\mathbf{I}}_{4,5}^{cm}\,^3\mathbf{D}_4^{-1}$. Eq. (4.131) applies the following:

$$^3\mathbf{r}_{4,5}^{cm}=\begin{bmatrix}^3r_{x4,5} & ^3r_{y4,5} & ^3r_{z4,5}\end{bmatrix}^T=\,^3\mathbf{D}_4\,\hat{\mathbf{r}}_{4,5}^{cm}=\begin{bmatrix}-s_4\hat{r}_{x4,5}-c_4\hat{r}_{z4,5} & c_4\hat{r}_{x4,5}-s_4\hat{r}_{z4,5} & -\hat{r}_{y4,5}\end{bmatrix}^T \quad (4.132)$$

Eqs. (4.131) and (4.132) yield the next equation, which shows that I_{t4} is always constant: $I_{t4}=\hat{I}_{y4}+\hat{I}_{y5}+(\hat{r}_{x4}^2+\hat{r}_{y4}^2+\hat{r}_{z4}^2)m_4+(\hat{r}_{x5}^2+\hat{r}_{y5}^2+\hat{r}_{z5}^2)m_5$.

4.9.3. Algorithm for Calculating the Maximum Possible Values of $\ddot{q}_1$, $\ddot{q}_2$, $\ddot{q}_3$ and $\ddot{q}_4$ Based on Approximate Forward Dynamics (Algorithm 4.3)

In Section 4.4.3.3, the new control algorithm of the centrifuge is shown; it contains a new algorithm for the approximate inverse and forward dynamics of the robot based on the recursive Newton-Euler algorithm and accounts for the possible motor actions. In this chapter, similar algorithm to calculate the maximum possible values of $\ddot{q}_i$, i=1, 2, 3, 4 of all SDT links is shown. This algorithm is applied in the following way.

In the first step, it calculates the SDT actuator torques $\hat{m}_{zi}$, i=1, 2, 3, 4 as functions of the angular accelerations $\ddot{q}_i$ of all SDT links; some coefficients do not influence the results. In Eqs. (4.85)-(4.88) for $\dot{\mathbf{v}}_i^{cm}$, i=1, 2, 3, 4 and (4.94)-(4.97) for $\mathbf{M}_1$, $\mathbf{M}_2$, $\mathbf{M}_3$ and $\mathbf{M}_{4,5}$, the terms $\dot{q}_i$ are replaced with $\dot{q}_i = \dot{q}_{iprev} + \ddot{q}_i \Delta t$. The terms with Δt^2 in Eqs. (4.85)-(4.88) and (4.94)-(4.97) are neglected as in Section 4.4.3.3. Thus, the following equations, without the terms with $\ddot{q}_1\ddot{q}_1$, $\ddot{q}_1\ddot{q}_\psi$, $\ddot{q}_1\ddot{q}_3$, $\ddot{q}_1\ddot{q}_4$, $\ddot{q}_\psi\ddot{q}_\psi$, $\ddot{q}_\psi\ddot{q}_3$, $\ddot{q}_\psi\ddot{q}_4$, $\ddot{q}_3\ddot{q}_3$, $\ddot{q}_3\ddot{q}_4$ and $\ddot{q}_4\ddot{q}_4$, are obtained:

$$\dot{\mathbf{v}}_1^{cm} = \mathbf{c}_1 + \mathbf{c}_{11}\ddot{q}_1 \tag{4.133}$$

$$\dot{\mathbf{v}}_2^{cm} = \mathbf{c}_2 + \mathbf{c}_{21}\ddot{q}_1 + \mathbf{c}_{22}\ddot{q}_\psi \tag{4.134}$$

$$\dot{\mathbf{v}}_3^{cm} = \mathbf{c}_3 + \mathbf{c}_{31}\ddot{q}_1 + \mathbf{c}_{32}\ddot{q}_\psi + \mathbf{c}_{33}\ddot{q}_3 \tag{4.135}$$

$$\dot{\mathbf{v}}_4^{cm} = \mathbf{c}_4 + \mathbf{c}_{41}\ddot{q}_1 + \mathbf{c}_{42}\ddot{q}_\psi + \mathbf{c}_{43}\ddot{q}_3 + \mathbf{c}_{44}\ddot{q}_4 \tag{4.136}$$

$$\mathbf{M}_1 = \mathbf{d}_{11}\ddot{q}_1 \tag{4.137}$$

$$\mathbf{M}_2 = \mathbf{d}_{22}\ddot{q}_\psi \tag{4.138}$$

$$\mathbf{M}_3 = \mathbf{d}_3 + \mathbf{d}_{32}\ddot{q}_\psi + \mathbf{d}_{33}\ddot{q}_3 \tag{4.139}$$

$$\mathbf{M}_{4,5} = \mathbf{d}_4 + \mathbf{d}_{41}\ddot{q}_1 + \mathbf{d}_{42}\ddot{q}_\psi + \mathbf{d}_{43}\ddot{q}_3 + \mathbf{d}_{44}\ddot{q}_4 \tag{4.140}$$

The coefficients in Eqs. (4.133)-(4.140) are as follows:

$$\mathbf{c}_{11} = -\dot{q}_{1prev}^2 \mathbf{v}_{12}, \quad \mathbf{c}_{12} = \mathbf{v}_{11} - 2\dot{q}_{1prev}\mathbf{v}_{12}\Delta t \tag{4.141}$$

$$\mathbf{v}_{11} = \begin{bmatrix} -a_1 s_1 - r_{y1} & a_1 c_1 + r_{x1} & 0 \end{bmatrix}^T, \mathbf{v}_{12} = \begin{bmatrix} a_1 c_1 + r_{x1} & a_1 s_1 + r_{y1} & 0 \end{bmatrix}^T$$

$$\mathbf{c}_{21} = -a_1\dot{q}_{1prev}^2 \mathbf{v}_{21} - \dot{q}_{\psi prev}^2 \mathbf{v}_{22},$$

$$\mathbf{c}_{22} = a_1(\mathbf{v}_{23} - 2\mathbf{v}_{21}\dot{q}_{1prev}\Delta t), \tag{4.142}$$

$$\mathbf{c}_{23} = \mathbf{v}_{24} - 2\mathbf{v}_{22}\dot{q}_{\psi prev}\Delta t$$

$$\mathbf{v}_{21} = \begin{bmatrix} c_1 & s_1 & 0 \end{bmatrix}^T, \mathbf{v}_{22} = \begin{bmatrix} r_{x2} & r_{y2} & 0 \end{bmatrix}^T, \mathbf{v}_{23} = \begin{bmatrix} -s_1 & c_1 & 0 \end{bmatrix}^T, \mathbf{v}_{24} = \begin{bmatrix} -r_{y2} & r_{x2} & 0 \end{bmatrix}^T$$

$$\mathbf{c}_{31} = -a_1 \dot{q}_{1\,prev}^2 \mathbf{v}_{21} - \dot{q}_{\psi\,prev}^2 \mathbf{v}_{31} + 2\dot{q}_{\psi\,prev}\dot{q}_{3\,prev}\mathbf{v}_{32} - \dot{q}_{3\,prev}^2 \mathbf{v}_{33} \qquad (4.143)$$

$$\mathbf{c}_{32} = a_1 \mathbf{v}_{34} - 2a_1 \dot{q}_{1\,prev}\mathbf{v}_{21}\Delta t \,, \quad \mathbf{c}_{33} = \mathbf{v}_{35} + 2(\dot{q}_{3\,prev}\mathbf{v}_{32} - \dot{q}_{\psi\,prev}\mathbf{v}_{31})\Delta t \qquad (4.144)$$

$$\mathbf{c}_{34} = \mathbf{v}_{36} + 2(\dot{q}_{\psi\,prev}\mathbf{v}_{32} - \dot{q}_{3\,prev}\mathbf{v}_{33})\Delta t \qquad (4.145)$$

$$\mathbf{v}_{31} = \begin{bmatrix} r_{x3} & r_{y3} & 0 \end{bmatrix}^T, \quad \mathbf{v}_{32} = \begin{bmatrix} s_\psi r_{z3} & -c_\psi r_{z3} & 0 \end{bmatrix}^T,$$

$$\mathbf{v}_{33} = \begin{bmatrix} c_\psi (s_\psi r_{y3} + c_\psi r_{x3}) & s_\psi (s_\psi r_{y3} + c_\psi r_{x3}) & r_{z3} \end{bmatrix}^T$$

$$\mathbf{v}_{34} = \mathbf{v}_{23}, \quad \mathbf{v}_{35} = \begin{bmatrix} -r_{y3} & r_{x3} & 0 \end{bmatrix}^T, \quad \mathbf{v}_{36} = \begin{bmatrix} -c_\psi r_{z3} & -s_\psi r_{z3} & s_\psi r_{y3} + c_\psi r_{x3} \end{bmatrix}^T$$

$$\mathbf{c}_{41} = -a_1 \dot{q}_{1\,prev}^2 \mathbf{v}_{21} + \dot{q}_{\psi\,prev}^2 \mathbf{v}_{41} - \dot{q}_{3\,prev}^2 \mathbf{v}_{43} + \dot{q}_{4\,prev}^2 \mathbf{v}_{46}$$
$$+ 2(\dot{q}_{\psi\,prev}\dot{q}_{3\,prev}\mathbf{v}_{42} + \dot{q}_{\psi\,prev}\dot{q}_{4\,prev}\mathbf{v}_{44} + \dot{q}_{3\,prev}\dot{q}_{4\,prev}\mathbf{v}_{45}) \qquad (4.146)$$

$$\mathbf{c}_{42} = \mathbf{v}_{47} - 2a_1 \dot{q}_{1\,prev}\mathbf{v}_{21}\Delta t \qquad (4.147)$$

$$\mathbf{c}_{43} = \mathbf{v}_{48} + 2(\dot{q}_{\psi\,prev}\mathbf{v}_{41} + \dot{q}_{3\,prev}\mathbf{v}_{42} + \dot{q}_{4\,prev}\mathbf{v}_{43})\Delta t \qquad (4.148)$$

$$\mathbf{c}_{44} = \mathbf{v}_{49} + 2(\dot{q}_{\psi\,prev}\mathbf{v}_{42} - \dot{q}_{3\,prev}\mathbf{v}_{43} + \dot{q}_{4\,prev}\mathbf{v}_{45})\Delta t \qquad (4.149)$$

$$\mathbf{c}_{45} = \mathbf{v}_{410} + 2(\dot{q}_{\psi\,prev}\mathbf{v}_{44} + \dot{q}_{3\,prev}\mathbf{v}_{411} + \dot{q}_{4\,prev}\mathbf{v}_{46})\Delta t \qquad (4.150)$$

$$\mathbf{v}_{41} = \begin{bmatrix} -r_{x4,5} \\ -r_{y4,5} \\ 0 \end{bmatrix}, \quad \mathbf{v}_{42} = \begin{bmatrix} s_\psi r_{z4,5} \\ -c_\psi r_{z4,5} \\ 0 \end{bmatrix}, \quad \mathbf{v}_{43} = \begin{bmatrix} c_\psi (s_\psi r_{y4,5} + c_\psi r_{x4,5}) \\ s_\psi (s_\psi r_{y4,5} + c_\psi r_{x4,5}) \\ r_{z4,5} \end{bmatrix},$$

$$\mathbf{v}_{44} = \begin{bmatrix} c_\psi c_3 r_{z4,5} - s_3 r_{x4,5} \\ s_\psi c_3 r_{z4,5} - s_3 r_{y4,5} \\ 0 \end{bmatrix}, \quad \mathbf{v}_{45} = \begin{bmatrix} c_\psi c_3 (s_\psi r_{x4,5} - c_\psi r_{y4,5}) \\ s_\psi c_3 (s_\psi r_{x4,5} - c_\psi r_{y4,5}) \\ s_3 (s_\psi r_{x4,5} - c_\psi r_{y4,5}) \end{bmatrix},$$

$$\mathbf{v}_{46} = \begin{bmatrix} -s_\psi c_3^2 (s_\psi r_{x4,5} - c_\psi r_{y4,5}) + s_3 (c_\psi c_3 r_{z4,5} - s_3 r_{x4,5}) \\ c_\psi c_3^2 (s_\psi r_{x4,5} - c_\psi r_{y4,5}) + s_3 (s_\psi c_3 r_{z4,5} - s_3 r_{x4,5}) \\ c_3 (s_3 (s_\psi r_{y4,5} + c_\psi r_{x4,5}) - c_3 r_{z4,5}) \end{bmatrix},$$

$$\mathbf{v}_{47} = \begin{bmatrix} -a_1 s_\psi \\ a_1 c_\psi \\ 0 \end{bmatrix}, \quad \mathbf{v}_{48} = \begin{bmatrix} -r_{y4,5} \\ r_{x4,5} \\ 0 \end{bmatrix}, \quad \mathbf{v}_{49} = \begin{bmatrix} -c_\psi r_{z4,5} \\ -s_\psi r_{z4,5} \\ s_\psi r_{y4,5} + c_\psi r_{x4,5} \end{bmatrix},$$

$$\mathbf{v}_{410} = \begin{bmatrix} s_\psi c_3 r_{z4,5} - s_3 r_{y4,5} \\ s_3 r_{x4,5} - c_\psi c_3 r_{z4,5} \\ c_3 (c_\psi r_{y4,5} - s_\psi r_{x4,5}) \end{bmatrix}, \quad \mathbf{v}_{411} = \begin{bmatrix} c_\psi c_3 (s_\psi r_{x4,5} - c_\psi r_{y4,5}) \\ s_\psi c_3 (s_\psi r_{x4,5} - c_\psi r_{y4,5}) \\ s_3 (s_\psi r_{x4,5} - c_\psi r_{y4,5}) \end{bmatrix}$$

$$\mathbf{d}_{12} = \begin{bmatrix} 0 & 0 & I_{z1} \end{bmatrix}^T, \quad \mathbf{d}_{22} = \begin{bmatrix} 0 & 0 & I_{z2} \end{bmatrix}^T \tag{4.151}$$

$$\mathbf{d}_{31} = \dot{q}^2_{\psi\,prev}\,\mathbf{i}_{31} + \dot{q}_{\psi\,prev}\dot{q}_{3\,prev}\,\mathbf{i}_{33} + \dot{q}^2_{3\,prev}\,\mathbf{i}_{34} \tag{4.152}$$

$$\mathbf{d}_{33} = \mathbf{i}_{32} + (2\dot{q}_{\psi\,prev}\mathbf{i}_{31} + \dot{q}_{3\,prev}\mathbf{i}_{33})\Delta t \tag{4.153}$$

$$\mathbf{d}_{34} = \mathbf{i}_{35} + (\dot{q}_{\psi\,prev}\mathbf{i}_{33} + 2\dot{q}_{3\,prev}\mathbf{i}_{34})\Delta t \tag{4.154}$$

$$\mathbf{i}_{31} = \begin{bmatrix} I_{yz3} \\ -I_{xz3} \\ 0 \end{bmatrix}, \quad \mathbf{i}_{32} = \begin{bmatrix} -I_{xz3} \\ -I_{yz3} \\ I_{z3} \end{bmatrix}, \quad \mathbf{i}_{33} = \begin{bmatrix} c_\psi (I_{x3} + I_{y3} - I_{z3}) \\ s_\psi (I_{x3} + I_{y3} - I_{z3}) \\ -2(I_{xz3}c_\psi + I_{yz3}s_\psi) \end{bmatrix},$$

$$\mathbf{i}_{34} = \begin{bmatrix} c_\psi (I_{xz3}s_\psi - I_{yz3}c_\psi) \\ s_\psi (I_{xz3}s_\psi - I_{yz3}c_\psi) \\ s_\psi c_\psi (I_{x3} - I_{y3}) - I_{xy3}(s_\psi^2 - c_\psi^2) \end{bmatrix}, \quad \mathbf{i}_{35} = \begin{bmatrix} I_{x3}s_\psi + I_{xy3}c_\psi \\ -I_{y3}c_\psi - I_{xy3}s_\psi \\ -I_{xz3}s_\psi + I_{yz3}c_\psi \end{bmatrix}$$

$$\mathbf{d}_{41} = \dot{q}^2_{1\,prev}\mathbf{i}_{41} + \dot{q}_{\psi\,prev}\dot{q}_{3\,prev}\mathbf{i}_{42} + \dot{q}_{\psi\,prev}\dot{q}_{4\,prev}\mathbf{i}_{43} + \dot{q}^2_{3\,prev}\mathbf{i}_{44} + \dot{q}_{3\,prev}\dot{q}_{4\,prev}\mathbf{i}_{45} + \dot{q}^2_{4\,prev}\mathbf{i}_{46}$$
$$\tag{4.155}$$

$$\mathbf{d}_{42} = 2\dot{q}_{1\,prev}\,\mathbf{i}_{41}\,\Delta t \tag{4.156}$$

$$\mathbf{d}_{43} = \mathbf{i}_{47} + (\dot{q}_{3\,prev}\mathbf{i}_{42} + \dot{q}_{4\,prev}\mathbf{i}_{43})\Delta t \tag{4.157}$$

$$\mathbf{d}_{44} = \mathbf{i}_{48} + (\dot{q}_{\psi\,prev}\mathbf{i}_{42} + 2\dot{q}_{3\,prev}\mathbf{i}_{44} + \dot{q}_{4\,prev}\mathbf{i}_{45})\Delta t \tag{4.158}$$

$$\mathbf{d}_{45} = \mathbf{i}_{49} + (\dot{q}_{\psi\,prev}\mathbf{i}_{43} + \dot{q}_{3\,prev}\mathbf{i}_{45} + 2\dot{q}_{4\,prev}\mathbf{i}_{46})\Delta t \tag{4.159}$$

$$\mathbf{i}_{41} = \begin{bmatrix} I_{yz4,5} \\ -I_{xz4,5} \\ 0 \end{bmatrix}, \quad \mathbf{i}_{42} = \begin{bmatrix} c_\psi (I_{x4,5} + I_{y4,5} - I_{z4,5}) \\ s_\psi (I_{x4,5} + I_{y4,5} - I_{z4,5}) \\ -2(c_\psi I_{xz4,5} + s_\psi I_{yz4,5}) \end{bmatrix},$$

$$\mathbf{i}_{43} = \begin{bmatrix} -s_\psi c_3 (I_{x4,5} + I_{y4,5} - I_{z4,5}) + 2s_3 I_{yz4,5} \\ c_\psi c_3 (I_{x4,5} + I_{y4,5} - I_{z4,5}) - 2s_3 I_{xz4,5} \\ -2c_3 (c_\psi I_{yz4,5} - s_\psi I_{xz4,5}) \end{bmatrix},$$

$$\mathbf{i}_{44} = \begin{bmatrix} c_\psi (I_{xz4,5}s_\psi - I_{yz4,5}c_\psi) \\ s_\psi (I_{xz4,5}s_\psi - I_{yz4,5}c_\psi) \\ s_\psi c_\psi (I_{x4,5} - I_{y4,5}) - (s_\psi^2 - c_\psi^2)I_{xy4,5} \end{bmatrix},$$

$$\mathbf{i}_{45} = \begin{bmatrix} c_\psi s_3 (-I_{x4,5} + I_{y4,5} - I_{z4,5}) + 2s_\psi (s_3 I_{xy4,5} - s_\psi c_3 I_{xz4,5} + c_\psi c_3 I_{yz4,5}) \\ s_\psi s_3 (I_{x4,5} - I_{y4,5} - I_{z4,5}) + 2c_\psi (s_3 I_{xy4,5} + s_\psi c_3 I_{xz4,5} - c_\psi c_3 I_{yz4,5}) \\ c_3 [(s_\psi^2 - c_\psi^2)(I_{y4,5} - I_{x4,5}) + I_{z4,5}] - 4s_\psi c_\psi I_{xy4,5} \end{bmatrix},$$

$$\mathbf{i}_{46} = \begin{bmatrix} c_3 s_3 [s_\psi (I_{z4,5} - I_{y4,5}) + c_\psi I_{xy4,5}] - s_\psi c_\psi c_3^2 I_{xz4,5} - (s_\psi^2 c_3^2 - s_3^2)I_{yz4,5} \\ c_3 s_3 [c_\psi (I_{x4,5} - I_{z4,5}) - s_\psi (I_{xy4,5})] + (c_\psi^2 c_3^2 - s_3^2)I_{xz4,5} + s_\psi c_\psi c_3^2 I_{yz4,5} \\ c_3^2 [s_\psi c_\psi (I_{y4,5} - I_{x4,5}) - (c_\psi^2 - s_\psi^2)I_{xy4,5}] - c_3 s_3 (c_\psi I_{yz4,5} - s_\psi I_{xz4,5}) \end{bmatrix},$$

$$\mathbf{i}_{47} = \begin{bmatrix} -I_{xz4,5} \\ -I_{yz4,5} \\ I_{z4,5} \end{bmatrix}, \quad \mathbf{i}_{48} = \begin{bmatrix} I_{x4,5}s_\psi + I_{xy4,5}c_\psi \\ -I_{y4,5}c_\psi - I_{xy4,5}s_\psi \\ -I_{xz4,5}s_\psi + I_{yz4,5}c_\psi \end{bmatrix},$$

$$\mathbf{i}_{49} = \begin{bmatrix} c_3 (I_{x4,5}c_\psi - I_{xy4,5}s_\psi) - I_{xz4,5}s_3 \\ c_3 (I_{y4,5}s_\psi - I_{xy4,5}c_\psi) - I_{yz4,5}s_3 \\ I_{z4,5}s_3 - c_3 (I_{xz4,5}c_\psi + I_{yz4,5}s_\psi) \end{bmatrix}$$

Using Eqs. (4.133)-(4.140) and (4.98)-(4.111), the actuator torques $\hat{m}_{zi}$ can be calculated in the form of the following system of four linear equations with four unknowns:

$$\hat{m}_{z1} = b_1 + b_{11}\ddot{q}_1 + b_{12}\ddot{q}_2 + b_{13}\ddot{q}_3 + b_{14}\ddot{q}_4 \tag{4.160}$$

$$\hat{m}_{z2} = b_2 + b_{21}\ddot{q}_1 + b_{22}\ddot{q}_2 + b_{23}\ddot{q}_3 + b_{24}\ddot{q}_4 \tag{4.161}$$

$$\hat{m}_{z3} = b_3 + b_{31}\ddot{q}_1 + b_{32}\ddot{q}_2 + b_{33}\ddot{q}_3 + b_{34}\ddot{q}_4 \tag{4.162}$$

$$\hat{m}_{z4} = b_4 + b_{41}\ddot{q}_1 + b_{42}\ddot{q}_2 + b_{43}\ddot{q}_3 + b_{44}\ddot{q}_4 \tag{4.163}$$

The system of Eqs. (4.160)-(4.163) can be relatively easily solved – approximate forward dynamics - while it is almost impossible to perform an analytical calculation of $\ddot{q}_i$ with Eqs. (4.85)-(4.88), (4.94)-(4.97) and (4.98)-(4.111). Similar simulation as with CFS in which Δt was 0.005 s has shown that the numerical result for $\hat{m}_{zi}$ differs by less than 1 % from the results obtained with the terms with Δt^2. This difference is negligible.

As with the CFS (Section 4.4.3.3), a check is performed as to whether the values of the required torques for the given angular accelerations are greater than the maximum allowed. If $\hat{m}_{zi} > M_{ia}$ when $\ddot{q}_i > 0$, it will be $\hat{m}_{zi} = M_{ia}$. If $\hat{m}_{zi} < -M_{ia}$ when $\ddot{q}_i < 0$, it will be $\hat{m}_{zi} = -M_{ia}$.

Finally, the maximum possible values of $\ddot{q}_i$ that the motors can achieve are calculated based on Eqs. (4.160)-(4.163); the maximum values of the torques $\hat{m}_{zi}$ were determined in previous step using the well-known Cramer's Rule. Next, $\dot{q}_i$ and q_i are calculated again in the following way: $\dot{q}_i = \dot{q}_{iprev} + \ddot{q}_i \Delta t$, $q_i = q_{iprev} + \dot{q}_i \Delta t$, $\dot{q}_{iprev} = \dot{q}_i$, $q_{iprev} = q_i$, $i = 1, 2, 3, 4$.

4.10. Results: Verification for the Proposed Control Algorithm

The proposed control algorithm for the SDT motion was also tested on the off-line programming system of the robot controller developed at the Lola Institute. For the SDT motion only point-to-point move instructions were used. Figs. 4.18 - 4.26 show the kinematic and dynamic parameters of an example of a SDT motion program.

Figs. 4.18 (a, b, c and d) show the angles q_1, q_2, $q_3 = \phi$ and $q_4 = \theta$, and Figs. 4.19 (a, b, c and d) show the angular velocities $\dot{q}_1$, $\dot{q}_2$, $\dot{q}_3$ and $\dot{q}_4$ when Algorithm 4.3 has been applied. These angles and angular velocities are nearly independent of algorithm.

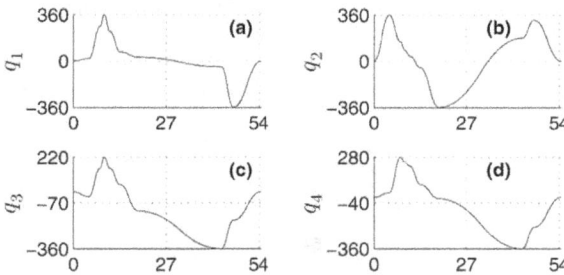

Fig. 4.18. Angles of the links in an example of the SDT motion when Algorithm 4.3 has been applied: (a) q_1, (b) q_2, (c) $q_3 = \phi$ and (d) $q_4 = \theta$ [°].

Figs. 4.20 (a_1, b_1, c_1 and d_1) show the angular accelerations $\ddot{q}_1$, $\ddot{q}_2$, $\ddot{q}_3$ and $\ddot{q}_4$ when Algorithm 4.3 has not been applied, and Figs. 4.20 (a_2, b_2, c_2 and d_2) show these accelerations when this algorithm has been applied. Figs. 4.21 (a, b, c and d) show the acceleration forces G, G_x, G_y and G_z, and Figs. 4.22 (a, b, c and d) show the axial forces of the link bearings

$\hat{f}_{a1}$, $\hat{f}_{a2}$, $\hat{f}_{a3}$ and $\hat{f}_{a4}$ when Algorithm 4.3 has been applied. The values of these forces obtained without this algorithm are very similar.

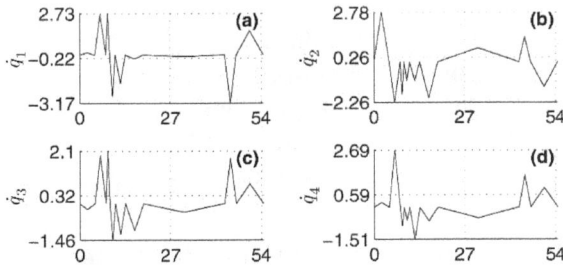

Fig. 4.19. Angular velocities of the links when Algorithm 4.3 has been applied: (a) $\dot{q}_1$, (b) $\dot{q}_2$, (c) $\dot{q}_3$ and (d) $\dot{q}_4$ [°/s].

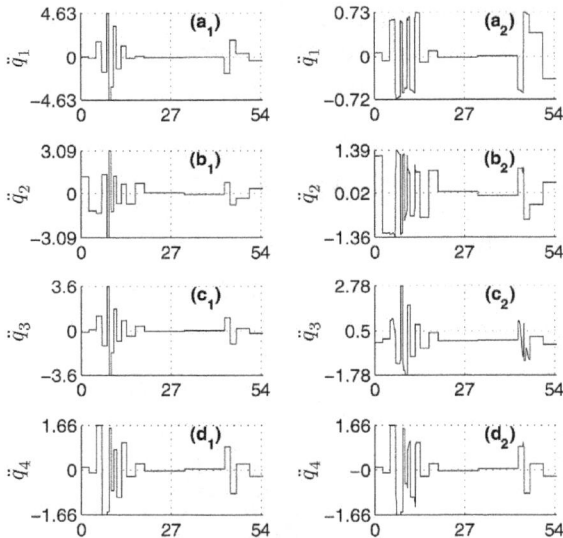

Fig. 4.20. Angular accelerations of the links when Algorithm 4.3 has not been applied: (a₁) $\ddot{q}_1$, (b₁) $\ddot{q}_2$, (c₁) $\ddot{q}_3$ and (d₁) $\ddot{q}_4$, and when this algorithm has been applied: (a₂) $\ddot{q}_1$, (b₂) $\ddot{q}_2$, (c₂) $\ddot{q}_3$ and (d₂) $\ddot{q}_4$ [°/s].

Figs. 4.23 (a₁, b₁, c₁ and d₁) show the radial forces $\hat{f}_{r1}$, $\hat{f}_{r2}$, $\hat{f}_{r3}$ and $\hat{f}_{r4}$ of the axes bearings when Algorithm 4.3 has not been applied, and Figs. 4.23 (a₂, b₂, c₂ and d₂) show these forces when this algorithm has been applied. The radial forces obtained without this algorithm in this example are approximately 15 % larger than the values shown.

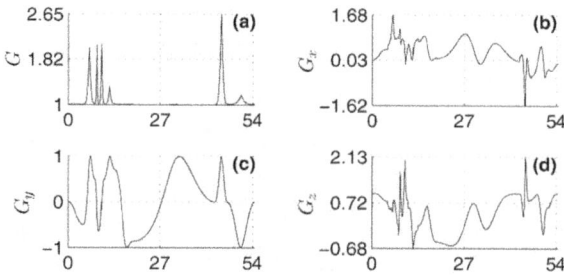

Fig. 4.21. Simulator pilot's acceleration forces when Algorithm 4.3 has been applied: (a) G, (b) G_x, (c) G_y and (d) G_z.

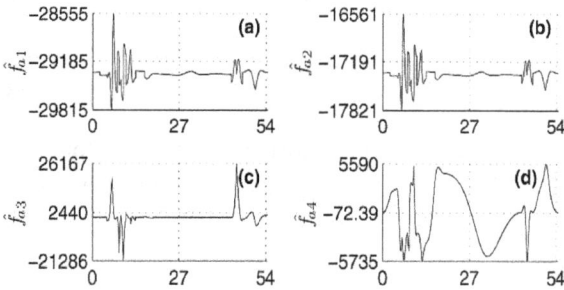

Fig. 4.22. Axial forces of the link bearings when Algorithm 4.3 has been applied: (a) $\hat{f}_{a1}$, (b) $\hat{f}_{a2}$, (c) $\hat{f}_{a3}$ and (d) $\hat{f}_{a4}$ [N].

Figs. 4.24 (a_1, b_1, c_1 and d_1) show the moments $\hat{m}_{xy1}$, $\hat{m}_{xy2}$, $\hat{m}_{xy3}$ and $\hat{m}_{xy4}$ when Algorithm 4.3 has not been applied, and Figs. 4.24 (a_2, b_2, c_2 and d_2) show these moments when this algorithm has been applied. The values of these moments obtained without this algorithm are larger by 21 % in this example. Figs. 4.25 (a_1, b_1, c_1 and d_1) show the torques $\hat{m}_{z1}$, $\hat{m}_{z2}$, $\hat{m}_{z3}$ and $\hat{m}_{z4}$ when Algorithm 4.3 has not been applied, and Figs. 4.25 (a_2, b_2, c_2 and d_2) show these torques when this algorithm has been applied. The values of these torques obtained without this algorithm are larger by 340 % in this example.

Figs. 4.26 (a_1, b_1, c_1 and d_1) show the motor powers P_1, P_2, P_3 and P_4 when the algorithm has not been applied, and Figs. 4.26 (a_2, b_2, c_2 and d_2) show these powers when this algorithm has been applied. The values of these powers obtained without this algorithm are larger by 325 %.

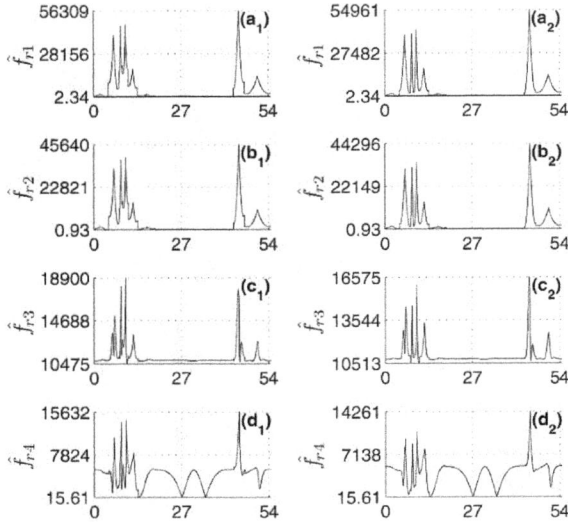

Fig. 4.23. Radial forces of the link bearings when Algorithm 4.3 has not been applied: (a₁) $\hat{f}_{r1}$, (b₁) $\hat{f}_{r2}$, (c₁) $\hat{f}_{r3}$ and (d₁) $\hat{f}_{r4}$, and when this algorithm has been applied: (a₂) $\hat{f}_{r1}$, (b₂) $\hat{f}_{r2}$, (c₂) $\hat{f}_{r3}$ and (d₂) $\hat{f}_{r4}$ [N].

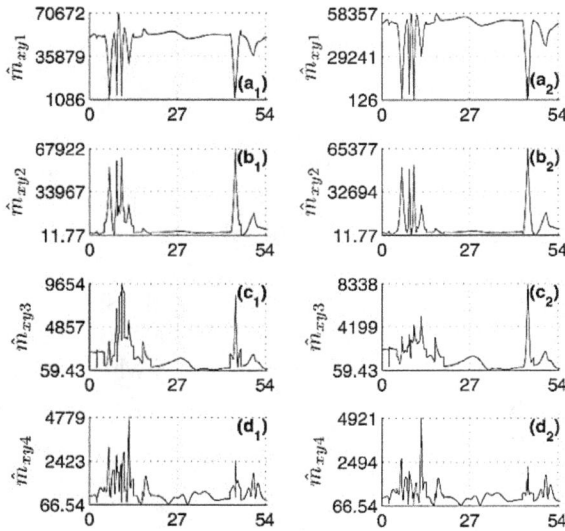

Fig. 4.24. Moments acting on the link bearings when Algorithm 4.3 has not been applied: (a₁) $\hat{m}_{xy1}$, (b₁) $\hat{m}_{xy2}$, (c₁) $\hat{m}_{xy3}$ and (d₁) $\hat{m}_{xy4}$, and when this algorithm has been applied: (a₂) $\hat{m}_{xy1}$, (b₂) $\hat{m}_{xy2}$, (c₂) $\hat{m}_{xy3}$ and (d₂) $\hat{m}_{xy4}$ [Nm].

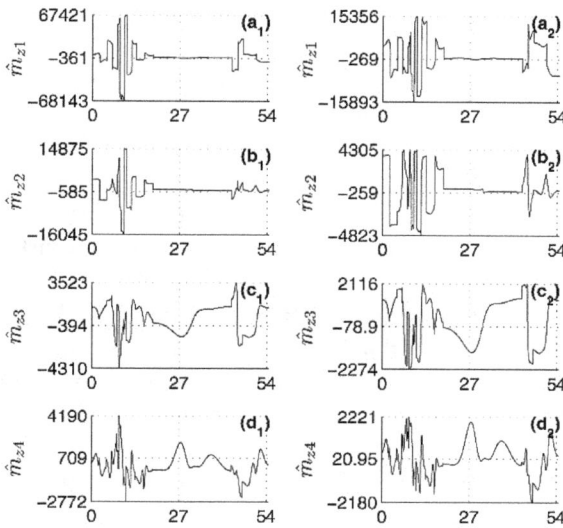

Fig. 4.25. Torques of the link actuators when Algorithm 4.3 has not been applied: (a$_1$) $\hat{m}_{z1}$, (b$_1$) $\hat{m}_{z2}$, (c$_1$) $\hat{m}_{z3}$ and (d$_1$) $\hat{m}_{z4}$, and when this algorithm has been applied: (a$_2$) $\hat{m}_{z1}$, (b$_2$) $\hat{m}_{z2}$, (c$_2$) $\hat{m}_{z3}$ and (d$_2$) $\hat{m}_{z4}$ [Nm].

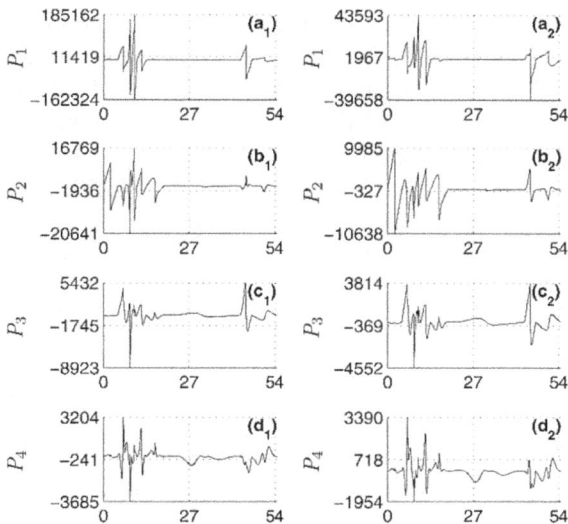

Fig. 4.26. Powers of the link actuators when Algorithm 4.3 has not been applied: (a$_1$) P_1, (b$_1$) P_2, (c$_1$) P_3 and (d$_1$) P_4, and when this algorithm has been applied: (a$_2$) P_1, (b$_2$) P_2, (c$_2$) P_3 and (d$_2$) P_4 [W].

4.11. Conclusions

In this Chapter a CFS for pilot training has been presented. This device is modelled as a manipulator that has three rotational axes. The centrifuge produces the desired transverse G_x, the lateral G_y and the longitudinal G_z acceleration forces and the roll, $\hat{\omega}_x$, pitch, $\hat{\omega}_y$ and yaw, $\hat{\omega}_z$ angular velocities, which simulate the acceleration forces and angular velocities that are accomplished by state-of-the-art aircraft.

A SDT for pilot training has been presented as well. This device is modelled as a manipulator that has four rotational axes and can thus bring the simulator pilot into any orientation. Additionally, some acceleration forces acting on an aircraft pilot can be simulated.

A new control algorithm for a CFS/SDT that contains an algorithm using inverse dynamics of robots based on the recursive Newton-Euler algorithm is presented. This algorithm accounts for motor characteristics. The method first calculates the successive actuator torques of the links that are required for the given motion in each interpolation period. Next, the algorithm checks whether the actuators can achieve these torques in practice; if they cannot, it calculates the maximum successive link angular accelerations that the motors can achieve. These maximum CFS/SDT link accelerations are calculated based on the link positions, which requires the total moment of inertia about the axes of rotation of these links, the acceleration of the other CFS/SDT axes, and the motors' possibilities. The algorithm does not allow motion commands to be sent to the CFS/SDT motors that the motors cannot realise. This approach has improved the quality of the motion control and enabled the precise calculation of the motor torques and the forces and moments that act on the CFS/SDT links and bearings, which was necessary for sizing the links and bearings during the design stage of the SCF/SDT.

The calculation of the acceleration forces and angular velocities experienced by the simulator pilot in the gondola and the roll and pitch angles of the gondola for the known acceleration forces is also given in this Chapter.

Acknowledgements

This research is supported by the Ministry of Education, Science and Technological Development of Serbia under the project "Development of devices for pilot training and dynamic simulation of modern fighter planes flights: 3DoF centrifuge and 4DoF spatial disorientation trainer" (2011-2015), No. 35023.

References

The capability of the modern combat to develop multi-axis accelerations can be found in [1-5].

For the human consequences of combat agile flight environment, see [3]. It also shows how the G_x acceleration in addition to the G_z acceleration can reduce the G tolerance. The connection between the jerk and the movements of the human body is shown in [6–10].

The form of the 3DoF centrifuge with rotational axes is presented in [11-13]. [11] presents a centrifuge in which the arm carries a gimballed gondola system, with two rotational axes providing pitch and roll capabilities. A similar centrifuge, driven by three hydraulic actuators, is described in [14]. In [15, 16], another realisation of the centrifuge is described.

The SDT with four rotational axes is presented in [17].

The different methods for deriving the robot dynamic model are presented in [18-25].

3D simulations for a human centrifuge that can be used for testing and verification of the presented control algorithms are described in [26].

In [27], solution to the Eq. (4.45) is obtained in the form of Jacobi elliptic integrals.

Data about motors chosen for the CFS and SDT are given in [28] and [29].

[1]. W. B. Albery, Human Centrifuges: The Old and the New, in *Proceedings of the 36ᵗʰ Annual Symposium of the SAFE Association*, Phoenix AZ, 14-16 September 1998.
[2]. W. B. Albery, Current and Future Trends in Human Centrifuge Development, *SAFE Journal*, Vol. 29, Issue 2, September 1999, pp. 107-111.

[3]. W. B. Albery, Acceleration in Other Axes Affects +G$_z$ Tolerance: Dynamic Centrifuge Simulation of Agile Flight, *Aviat Space & Environmental Medicine*, Vol. 75, Issue 1, Jan 2004, pp. 1-6.

[4]. S. H. Schot, Jerk: the time rate of change acceleration, *American Journal of Physics*, Vol. 46, Issue 11, 1978, pp. 1090–1094.

[5]. J. Gallardo-Alvarado, Jerk analysis of a six-degrees-of-freedom three-legged parallel manipulator, *Robotics and Computer-Integrated Manufacturing*, Vol. 28, 2012, pp. 220-226.

[6]. P. Morasso, Spatial control arm movements, *Experimental Brain Research*, Vol. 42, 1981, pp. 223-227.

[7]. T. Flash, N. Hogan, The coordination of arm movements: an experimentally confirmed mathematical model, *Journal of Neuroscience*, Vol. 5, Issue 7, 1985, pp. 1688-1703.

[8]. Y. Uno, M. Kawato, R. Suzuki, Formation and control of optimal trajectory in human multi joint arm movements, *Biological Cybernetics*, Vol. 61, Issue 2, 1989, pp. 89-101.

[9]. P. Viviani, R. Schneider, A developmental study of the relationship between geometry and kinematics in drawing movements, *Journal of Experimental Psychology: Human Perception and Performance*, Vol. 17, Issue 1, 1991, pp. 198-218.

[10]. P. Viviani, T. Flash, Minimum-jerk, two-thirds power, law and isochrony: converging approaches to movement planning, *Journal of Experimental Psychology: Human Perception and Performance*, Vol. 21, Issue 1, 1995, pp. 32-53.

[11]. V. Kvrgic, J. Vidakovic, M. Lutovac, G. Ferenc, V. Cvijanovic, A control algorithm for a centrifuge motion simulator, *Robotics and Computer-Integrated Manufacturing*, Vol. 30, Issue 4, 2014, pp. 399-412.

[12]. R. J. Crosbie, Dynamic flight simulator control system, *United States Patent*, No. 4751662, 14 Jun. 1988.

[13]. R. J. Crosbie, D. A. Kiefer, Controlling the Human Centrifuge as a Force and Motion Platform for the Dynamic Flight Simulator Technologies, in *Proceedings of the AIAA Flight Simulation Conference*, St Louis, MO, AIAA, 1985, pp. 37-45.

[14]. M.-H. Tsai, M.-C. Shih, *G*-load tracking control of a centrifuge driven by servo hydraulic systems, in *Proceedings of the Institution of Mechanical Engineers, Part G: J. Aerospace Engineering*, Vol. 223, 2009, pp. 669-682.

[15]. Y. C. Chen, D. W. Reperger, A study of the kinematics dynamics and control algorithms for a centrifuge motion simulator, *Mechatronics*, Vol. 6, No. 7, 1996, pp. 829-852.

[16]. Y. C. Chen, D. W. Reperger, Roberts R., Study of the kinematics, dynamics and control algorithms for a centrifuge motion simulator, in *Proceedings of the American Control Conference*, Vol. 3, 1995, pp. 1901-1905.

[17]. V. Kvrgic, Z. Visnjic, V. Cvijanovic, D. Divnic, S. Mitrovic, Dynamics and control of a spatial disorientation trainer, *Robotics and Computer-Integrated Manufacturing*, Vol. 35, Issue C, 2015, pp. 104-125.

[18] J. Wu, J. S. Wang, Z. You, An overview of dynamic parameter identification of robots, *Robotics and Computer-Integrated Manufacturing*, Vol. 26, Issue 5, 2010, pp. 414-419.

[19]. J. Wu, J. Wang, L. Wang, T. Li, Dynamics and control of a planar 3-DOF parallel manipulator with actuation redundancy, *Mechanism and Machine Theory*, Vol. 44, Issue 4, 2009, pp. 835-849.

[20]. J. Wu, X. Chen, T. Li, L. Wang, Optimal design of a 2-DOF parallel manipulator with actuation redundancy considering kinematics and natural frequency, Robotics and Computer Integrated Manufacturing, Vol. 29, Issue 1, 2013, pp. 80-85.

[21]. B. Siciliano, L. Sciavicco, L. Villani, G. Oriolo, Robotics: Modelling, Planning and Control, *Springer-Verlag London Limited*, 2009.

[22]. L.-W. Tsai, Robot analysis: the mechanics of serial and parallel manipulators, *John Wiley and Sons, Inc.*, 1999.

[23]. S. Y. Nof, Handbook of industrial robotics, *John Wiley and Sons, Inc.*, 1985.

[24]. S. Staicu, X. J. Liu, J. Li, Explicit dynamics equations of the constrained robotic systems, *Nonlinear Dynamics*, Vol. 58, Issue 1-2, 2009, pp. 217-235.

[25]. B. Gherman, D. Pisla, C. Vaida, N. Plitea, Development of inverse dynamic model for a surgical hybrid parallel robot with equivalent lumped masses, *Robotics and Computer-Integrated Manufacturing*, Vol. 28, Issue 3, 2012, pp. 402-415.

[26]. M. Lutovac, V. Kvrgic, G. Ferenc, Z. Dimic, J. Vidakovic, 3D simulator for human centrifuge motion testing and verification, *in Proceedings of 2nd Mediterranean Conference on Embedded Computing MECO 2013*, pp. 160-163.

[27]. J. Vidakovic, G. Ferenc, M. Lutovac, V. Kvrgic, Development and implementation of an algorithm for calculating angular velocity of main arm of human centrifuge Source, in *Proceedings of 15th International Power Electronics and Motion Conference, EPE-PEMC'12 Europe Congress*, 2012, pp. 452-457.

[28]. Catalog Siemens DA.

[29]. Catalog Sinamics S120 1FWG built-in-motors, Configuration 05/2009.

Chapter 5

PCBN Tool Wear Modes and Mechanisms in Finish Hard Turning

Seamus Gordon, Cora Lahiff and Pat Phelan

5.1. Introduction

Since polycrystalline cubic boron nitride (PCBN) was introduced as a cutting tool material nearly forty years ago, there has been a significant amount of research dedicated to understanding the cutting mechanisms and the tool wear. It enabled the hard turning of alloy steels (45–65HRC), an application area which continues to experience rapid growth and expansion. The current market trends are towards higher cutting speeds and increased material removal rates. In response, the number of commercially available grades is increasing with many being tailored for very specific applications. A number of factors are known to determine the tool life and performance of PCBN materials. The composition of the PCBN material is an important consideration but the workpiece, tool geometry, cutting conditions and machine setup also have significant influence on the cutting process. The definition of cutting tool performance is also important as there are usually several different criteria to be achieved by a single process. This is particularly true for finish hard turning operations and Fig. 5.1 [1] illustrates the various parameters that are used to measure cutting tool performance. Other considerations include machine downtime, material removal rates and the number of parts produced. This chapter presents an overview of the published literature in the area of continuous hard turning. The various wear mechanisms of PCBN tool materials are discussed with a view to identifying the critical factors that determine their behaviour in application.

Gordon S.
University of Limerick, Plassey, Limerick, Ireland

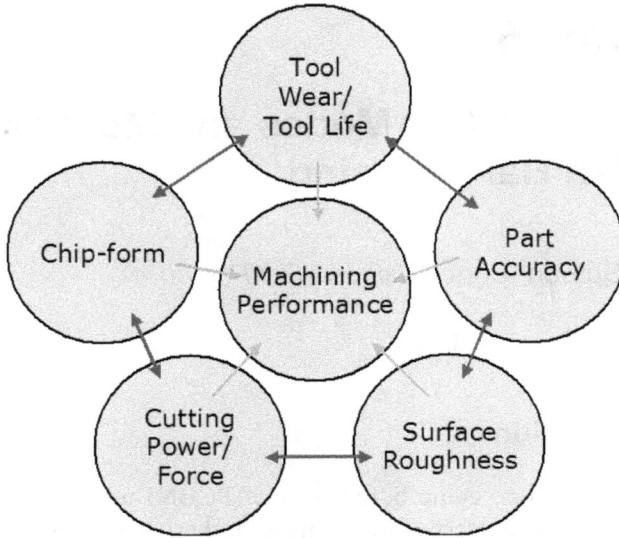

Fig. 5.1. Representation of the factors defining machining performance for Turning. Adapted from [1].

5.2. PCBN Tool Materials

PCBN is a composite material comprising cubic boron nitride (CBN) grains in a binder matrix. These materials are broadly categorised as high CBN content material or low CBN content. High CBN content grades are approximately 80–95 % CBN with a metallic-type binder. Low CBN content grades can contain from 40–70 % CBN and the majority have ceramic based binder systems such as TiC and TiN. Research to date has found that low CBN content materials provide the best performance in hard turning in terms of tool life and surface finish.

5.3. Cutting Tool Wear

PCBN tool wear is most often discussed in terms of flank and crater wear. Traditionally, tool life is defined by flank wear due to the significant influence this parameter has on the surface finish and dimensional accuracy of the machined part. Crater wear has a strong influence on process reliability as it can lead to instantaneous failure due to chipping or fracture of the tool edge (Fig. 5.2).

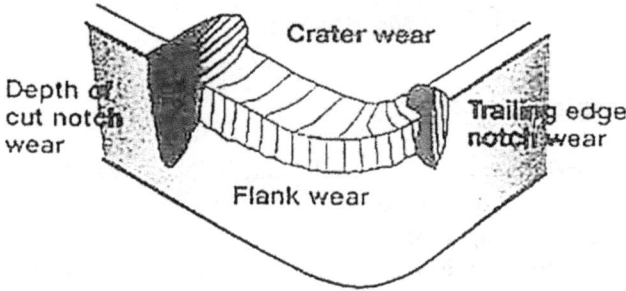

Fig. 5.2. Typical wear types observed on cutting tools. From [3].

A review of the available literature shows that while there are several different theories regarding the wear mechanisms present in hard turning with PCBN, there is general consensus that the wear is caused by a combination of several wear mechanisms. The mechanisms most commonly used to explain the wear of PCBN tools include abrasion [4–8], adhesion [4, 10-11], diffusion [4, 9–11] and chemical wear [5, 10–14, 16]. Abrasion is caused by hard particles in the workpiece and also by CBN grains from the cutting tool [4–8]. Where the binder material has been abraded by the workpiece the CBN grains are more easily removed and then contribute to further abrasion [5]. Adhesion occurs when material from the workpiece or chip melts due to high temperature and stress conditions at the cutting edge and adheres to the non-contact surfaces of the tool [5, 10–14, 16]. The area and thickness of the deposited layer depend on the cutting conditions and tool wear rate, as these factors determine the temperature in the cutting zone. The structure, composition and degree of adhesion of the layer are determined by the PCBN material [11-12]. Many researchers report that the compounds formed are not as hard as the PCBN tool material, resulting in an increase in abrasive wear [12, 16]. Diffusion is made possible by the high temperatures reached during the metal cutting process [4, 9–11]. The binder in PCBN cutting tools is reported to be most susceptible to this form of wear and some of the phases react quite readily with the workpiece material, resulting in structural changes [11]. This can make the binder less wear resistant and lead to an increase in abrasive wear. Similar to diffusion, the high temperatures at the cutting edge promote chemical reactions in that area. The built up layer which is frequently observed on PCBN tools after metal cutting is due to a chemical reaction occurring in the contact zone between the workpiece and the tool or the atmosphere [5, 10–14, 16].

157

5.4. PCBN Tool Wear Mechanisms

5.4.1. Abrasion

Some of the earliest research on tool wear, by Narutaki and Yamane [4], identified the workpiece material as having a strong influence on the wear of PCBN cutting tools. They found that the composition of the workpiece and to a lesser extent, the hardness, greatly influenced tool life. The performance of high and low CBN content materials on different steel types was evaluated. The high CBN content material with a metal binder gave the longest tool life in machining high speed steel (HSS) which is extremely abrasive due to the presence of very large carbides. This ranking was reversed for tool steels, which are less abrasive and on case-hardened steels, with the low CBN content material giving the best performance. They also reported decreasing tool wear rate and cutting temperature with increasing hardness for workpiece hardness values above 40 HRC. Later experiments by Luo, *et al.* [5] found that this trend continued up to 50HRC after which the wear rate increased. König, *et al.* [6] confirmed the effect of workpiece composition on tool life and related PCBN tool performance to the size, type and composition of the carbide phases and the percentage of martensite in the steel. Machining tests on different case-hardened, through hardened and tool steels, all at the same hardness (55 HRC) found that the lowest tool life was obtained on the tool steels due to their high carbide content. These tests also showed that for the steels with a primarily martensitic matrix, tool wear rate increased with increasing martensite content.

The influence of carbides in the workpiece was further investigated by Davies, *et al.* [7] who showed a correlation between flank wear rate and carbide size, with the smallest carbides resulting in the lowest flank wear. This experiment used M50 steel workpieces of the same chemistry but different microstructures. Luo, *et al.* [5] examined the behaviour of a PCBN material with a ceramic based binder ($TiC+Al_2O_3$) in turning hardened AISI4340 alloy steel. They proposed that the main wear mechanism for PCBN is abrasion of the binder by hard particles in the workpiece. SEM examination of the tools showed the presence of grooves on the tool flank, which is typical of abrasive wear. It is suggested that the binder of the tool is abraded by hard particles of the workpiece material, which leads to CBN grains being detached from the bond. Similar to the work by König, *et al.* [6], Poulachon, *et al.* [8]

performed machining tests with PCBN on four different steels of the same macrohardness hardness (54HRC), but with different microstuctures. Two of the steels contained large amounts of carbide, one with a relatively large carbide size (10–15 mm), the other with extremely fine carbides (1 mm). The remaining steels used in the experiment consisted primarily of martensite with one of them also containing a small amount of extremely fine, carbides. A grooved surface was observed on the flank of the tools after machining. The authors reported a correlation between the size of the primary carbides (when they were present) and the size of the grooves, and in the case of the martensitic steel, the grooves may be linked to the martensite grains. Carbides were determined to be more abrasive than martensite as the resultant grooves are deeper.

5.4.2. Diffusion and Adhesion

While it is intuitive that the workpiece will cause the cutting tool to wear due to abrasive mechanisms, there is also strong evidence that the workpiece contributes to tool wear by diffusion and adhesion. The work by Narutaki and Yamane [4] included an investigation of the chemical interactions between the tool and the workpiece. Diffusion was not observed between CBN grains and pure iron. In tests with ceramic binder PCBN material and a carbon steel, boron diffused into the steel and the concentration in the PCBN was depleted to a depth of 50 mm. This depletion was not obvious from examination of the microstructure and was not detected by Electron probe microanalysis (EPMA) and it is thought that the grain boundaries are the boron source rather than the CBN grains. In the case of metal binder PCBN material, cobalt was observed to diffuse from the tool but only from a depth of 10 mm or less. The rate of diffusion was found to increase with increasing temperature but the authors [4] concluded that since that the cutting temperature with PCBN is relatively low, measured to be less than 900 1C, this wear mechanism is only considered significant when extremely severe cutting conditions are used. Depletion of boron from the PCBN material was also found on a low CBN grade with a ceramic TiC binder after machining a case-hardened steel at 240 m/min by Zimmermann, *et al.* [9]. EPMA showed that in the crater region, the concentration of boron is lower than that in the bulk PCBN material. The authors suggest that this may be caused by tribochemical reaction in which BN dissolves into the flowing chip. Such a reaction is possible due to the extremely high pressures and temperatures in this region. It is assumed that the temperature at the rake face is higher than the flank (4800 1C). Following dissolution or diffusion of the boron nitride (BN), the remaining titanium-rich region on the rake has significantly lower hardness than the bulk tool and is rapidly abraded. Farhat [10] investigated PCBN tool wear when machining P20 mould steel at cutting speeds of 240, 600 and 1000 m/min. At all three speeds, iron deposits were presenton the flank and rake faces of the tools and also present on the surface of the crater at 600 and 1000 m/min. Here, it is proposed that the steel workpiece melts in the area of contact with the tool due to the high temperatures there and is then expelled and deposited on the non-contact surfaces of the tool. In addition, after machining at 1000 m/min there is a Mn and Cr rich layer evident on the tool flank, close to the cutting edge, which is attributed to the extremely high cutting temperatures generated. At these temperatures, B and N in the tool material dissociate and diffuse into the molten Fe present in the contact area and are subsequently deposited on the tool surfaces. The Mn and Cr rich layer is thought to be formed due to reaction between the workpiece and the PCBN tool material, which reduces the strength of the tool. This layer is not formed at the lower cutting speeds because the required high temperature is not generated. Investigations by König and Neises [11] using two PCBN materials; one high CBN material with an Al-based binder and a low CBN material with a ceramic TiC binder, found that the composition and percentage binder used in the PCBN material determines the thermal stability. Diffusion tests were carried out with these PCBN materials in contact with the steel workpiece at a temperature of 950 °C. Subsequent evaluation using a model abrasion test found a reduction in wear resistance of the PCBN materials which was attributed to thermally induced recrystallisation of the binder phases as there was no evidence of any reaction between the PCBN and the steel and no change observed in the integrity of the CBN grains. In the model abrasion test (using a Rockwell shaped indenter) the wear scar on both PCBN materials increased with increasing temperature but was larger on the material with the TiC binder. However, this ranking was reversed in the continuous turning tests with the TiC material having a lower rate of flank and crater wear. From this, the authors conclude that in addition to the structural changes in the binder there is some other mechanism contributing to the wear of the PCBN tool during hard turning.

5.4.3. Built Up Layer (Chemical Reaction)

Many researchers have reported the presence of a layer on the surfaces of the PCBN tool, outside the area of contact between it and the workpiece [5, 10–14, 16]. There is general consensus that a chemical reaction occurs between the tool and the workpiece in the area of contact, which is facilitated by the high temperature conditions. Due to the relatively high forces, the products of this reaction are expelled into the surrounding area and deposited on the tool surfaces. The area and thickness of the layer that is deposited depends on the cutting conditions and the tool wear rate as these factors determine the temperature in the contact zone. While opinions differ regarding the composition of the layer, iron and iron oxide have been identified most frequently. On a high CBN tool, Chou, *et al.* [12] found that the layer also contained silicon oxide, possibly due to the high affinity of the cobalt binder to silicon compounds in the steel. Analyisis of the layer deposited in the experiments performed by Klimenko, *et al.* [13] found evidence of a range of elements found in the tool and the workpiece (B,C, N, Si, Al, Cr, Fe) and products of their reaction with atmospheric oxygen. The layer generated in the machining test by Luo, *et al.* [5] also contains elements from both the workpiece (Fe, Ni, Mn) and the binder of the CBN tool (Al and Ti). In contrast, Barry and Byrne [14] associate the composition of the layer with elements present in small quantities in the steel workpiece. They also found a correlation between the tool wear rate and the Al and S content of the workpiece, and after examination of the layer at the trailing edge concluded that the Al content influences the tool wear rate.

König and Neises [11] reported the presence of layers consisting primarily of aluminium at the borders of the contact zone. As Al is present only in parts per million (ppm) in the 100Cr6 workpiece, this is not thought to be the source of the deposited layer. It is suggested that the chip may remove Al from the binder of the PCBN and deposited on the tool surfaces. A conclusion reached by many of the researchers is that the compounds formed by the chemical reaction in the contact zone are not as hard as the PCBN tool material, and are therefore more easily abraded. The structure and composition of the adhered layer are determined by the PCBN material. Chou, *et al.* [12] reported that the layer on low CBN material is uniform and smooth with a flake-like structure and some grooves present at the highest cutting speeds. On high CBN material the layer is rough and grooved. This feature becomes more pronounced as the cutting speed increases and the thickness also increases. The layer on the high CBN material was much more difficult to remove. SEM examination of the tools following removal of the transferred layer found that on the low CBN material there is a smooth surface on the CBN grains and shallow pockets where grains have been pulled out. This is attributed to adhesive wear, after which fine scale attrition dominates the wear because of the strong bond between the CBN and the ceramic binder. There is a grooved surface on the high CBN tool. It is suggested that bond failure between the CBN and the binder matrix results in the CBN grains being pulled out. The grooves may be caused by abrasion by the pulled- out CBN grains and hard particles in the workpiece. In contradiction to this theory, there is also evidence that the BUL can protect the tool from wear. Luo, *et al.* [5] found that PCBN tool life increased with increasing cutting speed until a critical value was reached after which tool life decreased. It is suggested that the adhered layer protects the tool until a temperature is reached at which the layer becomes soft and is removed and tool wear rate then increases. During examination of worn tools, Barry and Byrne [14] found that the CBN grains had suffered more wear than the binder phase leading the leading the authors to conclude that the BN and certain work material inclusions react, forming products which offer protection to the TiC ceramic phase against diffusion/dissolution wear. Klimenko, *et al.* [13] suggested that the formation of a liquid phase in the contact zone could be used to explain the low coefficient of friction when machining with PCBN. It was also noted that the presence of the layer on the tool surfaces can affect the dissipation of heat and hence the cutting temperature.

5.5. Factors that Influence PCBN Tool Wear

5.5.1. PCBN Tool Material Composition

Eda, *et al.* [15] proposed that the bond strength between CBN grains and ceramic binders is greater than that between CBN grains and metallic binders, which results in increased toughness. This study also attributed the increased toughness to the sub-micron size of the ceramic binder grains. Based on the superior performance of a low CBN content, ceramic binder material in machining soft steel where adhesion is the primary wear mechanism, Klimenko, *et al.* [13] also observed that the metal–CBN bond is weaker than the ceramic–CBN bond. It was also noted that in this test that the percentage of binder in the ceramic based PCBN (40 %) is much greater than that in the metal based material (10 %), suggesting that the amount of binder is significant, rather than any chemical bonding issues. Hooper, *et al.* [16] compared the performance of high and low CBN content material in finish turning of hardened steels and attributed the superior tool life of the low CBN material to the formation of a protective layer. The high CBN content material has higher thermal conductivity and therefore a lower cutting temperature. The higher temperature generated by the low CBN content material allows the formation of a protective layer on the tool. The temperatures generated during cutting with the high CBN content material are not sufficiently high to form a uniform layer. Any layer that is deposited is removed with the chip which contributes to attrition wear. It was also established that TiC is more resistant than CBN to the chemical wear ('atmospheric oxidation') that these authors [16] use in their analysis of PCBN tool wear. The CBN grains in the high content material have been plastically deformed during the synthesis process but this is not a feature of the CBN grains in the low content material. It was also found that individual CBN particles contain defects that contribute to tool wear. Low CBN content materials contain less of these structures/sites. Bossom [2] also cited the difference in thermal conductivity between high and low CBN content materials as the reason for the relatively poor performance of high CBN content materials in finish hard turning. Because of higher thermal conductivity, high CBN content materials dissipate heat more quickly. Heat generated at the cutting edge facilitates plastic deformation of the workpiece in the shear zone. The low CBN content material has lower thermal conductivity and hence more heat is retained in the cutting tip and shear zone, softening the workpiece and promoting deformation and shearing. As discussed in the previous section, several researchers have found evidence of chemical reactions involving the PCBN tool material that contribute to tool wear. Both the PCBN material composition and microstructure determine the reactions that occur and therefore influence the degree of chemical wear.

5.5.2. Tool Edge Geometry

The cutting tool edge geometry, by which is meant the chamfer angle, chamfer width and edge hone has a significant influence on tool life and to a large extent determines the surface finish and integrity of the machined part. Tool wear, on both flank and rake face constitutes a change in edge geometry. Tool life is usually defined by flank wear due to the significant influence this parameter has on the surface finish and dimensional accuracy of the machined part. Surface roughness increases as flank wear increases. Crater wear has a strong influence on process reliability as it can lead to instantaneous failure due to chipping or fracture of the tool edge. The feeds and depths of cut used in finish hard turning are relatively small (0.2 mm), and generally of the same magnitude as the tool edge geometry. As a result, cutting is confined to a small area on the nose radius and edge. Due to the extreme hardness of the workpiece in hard turning, a negative rake angle with strong edge geometry with a chamfer and hone is employed, in order to withstand the high cutting forces, stresses and temperatures that are generated.

5.5.2.1. The Effect of Tool Edge Geometry on Surface Roughness

Shintani, *et al*. [17] conducted continuous cutting tests on SCM420, carburised Cr–Mo steel (600–720 Hv) with a PCBN tool material containing approximately 60 % CBN, and with a TiN-AlN binder. They determined that the optimum tool geometry for this operation was a 0.8mm nose radius, a 30–351 negative land of width greater than the tool-chip contact length and an edge hone of 50 mm. Tool life as defined by flank wear (VB), increased with increasing chamfer width until a critical point was reached after which tool life remained constant. This critical chamfer width was found to be equal to the chip–tool contact length. It was also shown that tool life increased with increasing nose radius up to a value of 0.8 mm and then remained constant even as the nose radius was further increased. This, they attributed to the high temperatures generated when machining with a small nose radius due to the narrow tool–chip contact area and increased chip thickness. When surface finish (Ra) was the parameter used to define tool life, the maximum was also reached at a nose radius of 0.8 mm. When a smaller nose radius is used there are grooves present on the tool flank which have a negative effect on the workpiece surface finish. Similarly, maximum tool life as defined by both VB and Ra was achieved at the same edge hone size of 50 μm. There was considerable scatter observed in results when testing with smaller edge hones. This was attributed to the presence of small chips at the tool edge due to the grinding operation which are removed by the application of a 50 μm edge hone but are still present with a smaller size hone. Such chips lead to premature tool failure. Tool fabrication constraints of this nature are not likely to be an issue with modern grinding technology. Thiele and Melkote [18] found that surface roughness increases with increasing edge hone size but this influence becomes less significant as workpiece hardness increases. Their test work on AISI 52100 indicated that feed rate is the primary influence on surface roughness with edge geometry playing a secondary role. However, edge geometry becomes more important as the feed decreases.

5.5.2.2. The Effect of Tool Edge Geometry on Cutting Forces

The experiments by Shintani, *et al.* [17] found that in general, the cutting force increases with increasing chamfer width. Bossom [2] investigated the effect of tool geometry on the radial force and tool wear in machining D3 cold work tool steel with low CBN content material. It was found that the application of a chamfer to a negative tool significantly increases the radial forces but considers the values to be relatively low. The application of a chamfer results in a stable radial force that produces minimal flank wear progression. A positive geometry caused accelerated tool wear and unstable cutting conditions but the radial forces are similar to those generated with a negative tool geometry. Similarly, the work by Thiele and Melkote [18] shows that axial and radial forces are largely determined by the edge geometry and increase with increasing edge hone and/ or the addition of a chamfer. The influence of the feed and the hardness are less significant on these force components but however, the feed rate is the strongest influence on the tangential force. The increase in cutting forces due to a chamfered edge was also reported by Özel [19] in the orthogonal cutting of H13 tool steel who found that PCBN tools with an edge hone only (no chamfer) result in lower cutting forces than chamfered edges. His work also shows that a chamfered edge results in reduced rake face temperatures. Similarly, experiments by Kurt and Şeker [20] in turning AISI 52100 steel found that all components of the cutting force, but in particular the passive force increases as the size of the chamfer angle increases. Kountanya, *et al.* [21] performed orthogonal cutting tests on 100Cr6 at 60HRC over a range of cutting conditions. Again it was found that an increase in the size of the edge hone will increase both the cutting and the thrust forces but these forces decrease with increasing cutting speed. The forces increase as the chamfer becomes more negative and the proportion of the thrust force to the cutting force also increases. Klocke and Kratz [22] presented a novel chamfer design where the chamfer angle and width vary around the nose radius to accommodate the changing conditions at the cutting edge. The benefits of this tool edge design are improved workpiece surface finish and residual stress conditions due to optimised chip formation conditions and the generation of crater wear in a location such that the cutting edge is more stable and tool life is increased.

5.5.3. Machine Tool Requirements

All the components in a machining system contribute to the quality and repeatability of the process. This includes the cutting tool, tool holder, tool and workpiece clamping, machine spindle and machine tool bed [23]. PCBN is a relatively brittle tool material and requires a rigid machine setup to achieve optimum performance. Any instability in the machine tool will have a negative effect on tool wear and cutting forces, which in turn determine the surface quality and dimensional accuracy. Experiments by Chryssolouris and Toenshoff [24] investigated the relative effects of compliance in the toolholder and the workpiece. He demonstrated that in an unstable process, cracking of the tool material occurs almost immediately, leading to reduced tool life and reliability. When the setup is sufficiently rigid, PCBN tools experience flank and crater wear, similar to conventional tool materials. These experiments also found that the quality of the machined part is more sensitive to vibrations in the tool holder than the workpiece. It is evident that the simultaneous development of machine tools is critical in order to support the increasing use and applications of PCBN cutting tools. High-precision lathes are now available and in widespread use, providing hard-turned parts of extremely high dimensional accuracy and surface finish quality [1, 23, 25]. These machines are designed to address the requirements for geometrical and kinematic accuracy and system stiffness. They provide a high degree of process monitoring due to the strong influence of tool wear on part accuracy and can also compensate for the effects of thermal expansion of the workpiece, tool and toolholder. This technology allows the production of high grade parts such as gear components and roller bearings. In addition to the ever-present demand for higher productivity, flexibility and quality, there is the necessity to consider environmental issues such as the minimization of power consumption and pollution. These requirements ensure that cutting tool materials, machine tools and production methods will continue to develop and evolve.

5.6. Summary

The use of PCBN cutting tools in finish hard turning of alloy steels is now a well established production operation which continues to grow and develop. A significant amount of research has been dedicated to understanding the mechanics of this metal cutting process and this work has contributed to the successful use of PCBN in industry. It has been established that low CBN content materials provide the best performance in hard turning in terms of tool life and surface finish [2, 4, 11–13, 16]. Similar to more conventional cutting tool materials, the wear of PCBN is discussed in terms of flank and crater wear. Flank wear has a strong influence on the surface finish, integrity and dimensional accuracy of the machined part while crater wear affects process reliability. While there are different theories regarding the tool wear mechanisms involved, there is general agreement among researchers that PCBN tool wear is complex and no single mechanism alone provides a satisfactory explanation. Abrasion makes a significant contribution to flank wear and is caused by hard carbide particles and martensite in the workpiece and also by CBN grains from the cutting tool [4–8]. Where the binder material has been abraded by the workpiece, the CBN grains are more easily removed and then contribute to further abrasion [5].

Diffusion between the tool and the workpiece is made possible by the high temperatures reached during the metal cutting process and these high temperatures at the cutting edge also promote chemical reactions in that area [4, 9–11]. The binder in PCBN cutting tools is reported to be most susceptible to this form of wear and some of the phases react quite readily with the workpiece material, resulting in structural changes [11]. This can make the binder less wear resistant and lead to an increase in abrasive wear. Evidence of diffusion and chemical reactions is very often found in the crater that develops on the rake face. The built up layer which is frequently observed on PCBN tools after metal cutting is due to a chemical reaction occurring in the contact zone between the workpiece and the tool or the atmosphere [5, 10–14, 16]. Adhesion occurs when material from the workpiece or chip melts due to high temperature and stress conditions at the cutting edge and adheres to the non-contact surfaces of the tool [5, 10–14, 16]. The area and thickness of the deposited layer depend on the cutting parameters and tool wear rate, as these factors determine the temperature in the cutting zone. The structure, composition and degree of adhesion of the layer are determined by the PCBN material [11-12]. Many researchers report that the compounds formed are not as hard as the PCBN tool material, resulting in an increase in abrasive wear [12, 16]. The conditions at the contact area between the tool and the workpiece determine the wear mechanisms and there are a number of variables contributing to this. While the composition of the PCBN tool material and the steel workpiece are obviously very important, other factors such as the tool edge geometry and machine tool stability are critical to tool performance [23–25]. A negative geometry with a chamfer and edge hone has been found to increase tool life as defined by both flank wear and workpiece surface finish [2, 17–21]. The research reviewed here has shown that an in-depth understanding of the tool wear mechanisms and the parameters that control them can lead to significant gains in PCBN tool performance and process capability.

References

[1]. Byrne G., Dornfeld D., Denkena B., Advancing cutting technology, *Annals of the CIRP*, Vol. 52, 2003, pp. 483–507.

[2]. Bossom P. K., Finish machining of hard ferrous workpieces, *Industrial Diamond Review*, Vol. 90, Issue 5, 1990, pp. 228–232.

[3]. Brandt G., Ceramic cutting tools, state of the art and development trends, *Mater Technol*, Vol. 14, 1999, pp. 17–22.

[4]. Narutaki N., Yamane Y., Tool wear and cutting temperature of CBN tools in machining of hardened steels, *Annals of the CIRP*, Vol. 28, Issue 1, 1979, pp. 23–28.

[5]. Luo S. Y., Liao Y. S., Tsai Y. Y., Wear characteristics in turning high hardness alloy steel by ceramic and CBN tools, *Journal of Material Processing Technology*, Vol. 88, 1999, pp. 114–121.

[6]. König W., Komanduri R., Tönshoff H. K., Ackershott G., Machining of hard materials, *Annals of the CIRP*, Vol. 33, Issue 2, 1984, pp. 417–427.

[7]. Davies M. A., Chou Y., Evans C. J., On chip morphology, tool wear and cutting mechanics in finish hard turning, *Annals of the CIRP*, Vol. 45, Issue 1, 1996, pp. 77–82.

[8]. Poulachon G., Bandyopadhyay B. P., Jawahir I. S., Pheulpin S., Seguin E., Wear behaviour of CBN tools while turning various hardened steels, *Wear*, Vol. 256, Issue 3-4, 2004, pp. 302–310.

[9]. Zimmermann M., Lahres M., Viens D. V., Laube B. L., Investigations of the wear of cubic boron nitride cutting tools using Auger electronspectroscopy and X-ray analysis by EPMA, *Wear*, Vol. 209, 1997, pp. 241–246.

[10]. Farhat Z. N., Wear mechaniam of CBN cutting tool during high-speedmachining of mold steel, *Materials Science and Engineering*, Vol. A361, 2003, pp. 100–110.

[11]. König W., Neises A., Wear mechanisms of ultrahard, non-metallic cutting materials, *Wear*, Vol. 162–164, 1993, pp. 12–21.

[12]. Chou Y. K., Evans C. J., Barash M. M., Experimental investigation on CBN turning of hardened AISI 52100 steel, *Journal of Material Processing Technology*, Vol. 124, 2002, pp. 274–283.

[13]. Klimenko S. A., Mukovoz Yu. A., Lyashko V. A., Vashchenko A. N., Ogorodnik V. V., On the wear mechanism of cubic boron nitride base cutting tools, *Wear*, Vol. 157, 1992, pp. 1–7.

[14]. Barry J., Byrne G., Cutting tool wear in the machining of hardened steels Part II: cubic boron nitride cutting tool wear, *Wear*, Vol. 247, 2001, pp. 152–160.

[15]. Eda H., Kishi K., Hashimoto H., Wear resistance and cutting ability of a newly developed cutting tool, in Hashimoto H., Editor, in *Proceedings of the International Conference on Cutting Tool Materials*, Kentucky, USA: American Society for Metals, Ft. Mitchell, 1980, pp. 265–280.

[16]. Hooper R. M., Shakib J. I., Parry A., Brookes C. A., Mechanical properties, microstructure and wear of DBC50, *Industrial Diamond Review,* Vol. 4, Issue 89, 1989, pp. 170–173.

[17]. Shintani K., Ueki M., Fujimura Y., Optimum tool geometry of CBN tool for continuous turning of carburized steel, *International Journal of Machine Tools and Manufacture*, Vol. 29, Issue 3, 1989, pp. 403–413.

[18]. Thiele J. D., Melkote S. N., Effect of cutting edge geometry and workpiece hardness on surface generation in the finish hard turning of AISI 52100 steel, *Journal of Material Processing Technology*, Vol. 94, 1999, pp. 216–226.

[19]. Özel T., Modeling of hard part machining: effect of insert edge preparation in CBN cutting tools, *Journal of Materials Processing Technology*, Vol. 141, 2003, pp. 284–293.

[20]. Kurt A., Şeker U., The effect of chamfer angle of polycrystalline cubic boron nitride cutting tool on the cutting forces and the tool stresses in finish hard turning of AISI 52100 steel, *Materials & Design*, Vol. 26, Issue 4, 2005, pp. 351–356.

[21]. Kountanya R., Varghese B., Al-Zkeri I., D'Anna L., Altan T., Study of PCBN tool edge preparation in orthogonal hard turning, in *Proceedings of the First International Diamond at Work Conference*, 2005.

[22]. Klocke F., Kratz H., Advanced tool edge geometry for high precision hard turning, *Annals of the CIRP*, Vol. 54, Issue 1, 2005, pp. 47–50.

[23]. Klocke F., Brinksmeier E., Weinert K., Capability profile of hard cutting and grinding processes, *Annals of the CIRP*, Vol. 54, Issue 2, 2005, pp. 552–580.

[24]. Chryssolouris G., Toenshoff H. K., Effects of machine-tool-workpiece stiffness on the wear behaviour of superhard cutting materials, *Annals of the CIRP*, Vol. 31, Issue 1, 1982, pp. 65–69.

[25]. Tönshoff H. K., Arendt C., Ben Amor R., Cutting of hardened steel, *Annals of the CIRP*, Vol. 49, Issue 2, 2000, pp. 547–566.

Chapter 6

Simulation Study of a Constant Time Hybrid Approach for Large Scale Terrain Mapping Using Satellite Stereo Imagery

A. Sarkar, R. Reiger, D. Chatterjee, S. Patranabis, H. Singh and P. Mukherjee

6.1. Introduction

Recent work in the field of obtaining real time solutions to estimating robot's poses rely on approaches that scale with the environment without using any probabilistic framework, such as RSLAM. On the other hand, approaches like FastSLAM that are based purely on particle filters and other similar approaches using EKF filters involve parameters that grow with time. However, the filter based approach gives us a distribution of each pose rather than a single point in the global environment. In this work, we attempt to merge the advantages of both approaches by combining the particle based framework with a continuous relative pose representation. Our approach involves factorization of the SLAM posterior over the robot's path, in which each individual particle follows a constant time stereo SLAM approach and the particle distribution is harnessed by the algorithm to estimate the optimal trajectory. The result is a bundle distribution instead of a single bundle, which is then adjusted to estimate the current pose. For accurate bundle adjustment, especially when landmarks are sparsely encountered, fuzzy pose similarity technique is leveraged. We further ensure that the robot works in non-controlled environments too using an appropriate loop closure mechanism.To avoid problems with scale propagation, we use a stereo pair of images at each time step which are used to extract stereo features via SURF. This also avoids extensive computations for feature depth estimation that a monocular approach would otherwise demand. Our work also lays stress on data association across frames via a rigorous

A. Sarkar
Department of Mathematics, IIT Kharagpur, India

three stage matching process - this step plays a critical role in accurate bundle adjustment and final pose estimation.The contribution of the paper lies in the innovative use of particles for constant time pose estimation. Fig. 6.1 describes the work flow of the methodology.

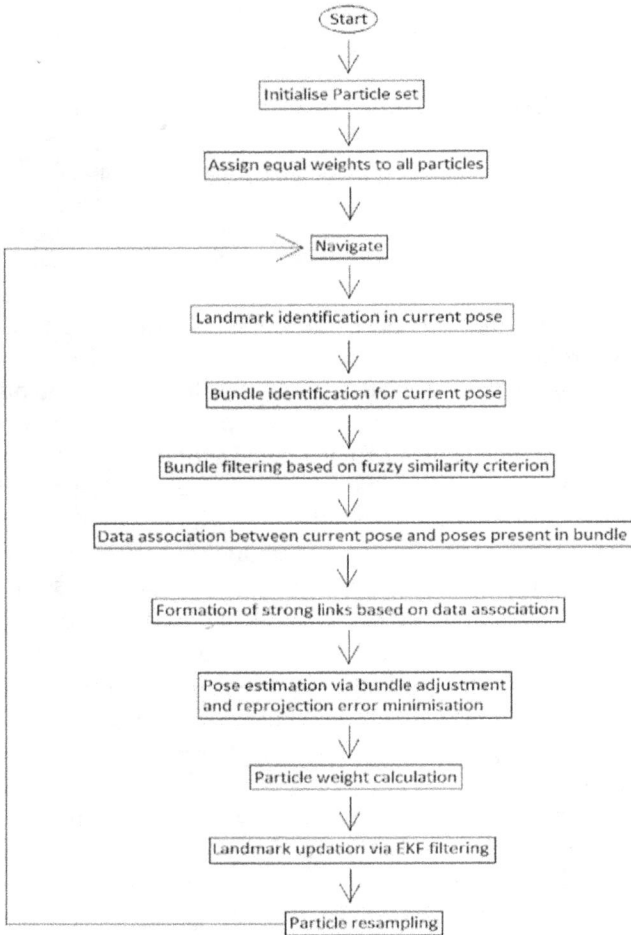

Fig. 6.1. Flow Diagram.

As a future work, we also propose to include auxiliary particle filters (APF) so that in particle filter step, the particles potentially likely to be compatible with the observation can have a better chance to survive as well as ensuring the algorithm to be of constant time as currently it is.

6.2. Related Work

In this section, we review some of the major advances in visual SLAM that use stereo systems. Surveys on SLAM approaches are available in Thrun, *et al.* (2005) [35], and Durrant-Whyte and Bailey, Part1 (2006) [11] and Bailey and Durrant Whyte, Part2 (2006)) [2]. Fast SLAM 2.0 comprises Rao-Blackwellized particle filters for pose estimation and Extended Kalman Filter (EKF) for landmark estimation. The method has a major limitation - the number of particles necessary for a specific environment is difficult to estimate and affects the computational complexity. Zhu, *et al.* (2011) [39] among others has attempted to solve this problem by proposing a method to estimate the required number of particles according to the uncertainty of sensors with some success.

However, these methods are still prone to inconsistent estimates, leading to the recent trend of local (Mouragnon, *et al.* (2009) [25]) or global (Lu and Milios (1997) [22], Klein and Murray (2007) [19], Eade and Drummond (2008) [12], Borrmann, *et al.* (2008) [12]) bundle adjustment as the underlying estimator. More recent approaches intend to provide real time solutions using stereo pair of images. This includes the work of Nister, *et al.* (2006) [26] that harnesses local bundle adjustment using global pose estimates. However the work does not address the problem of loop closure which is essential for drift reduction. The work of Jeong, *et al.* (2012) [16] may be noted for many facet of bundle adjustment. Special mention must be made of the FrameSLAM system (Konolige and Agarwal (2008) [20]) and constant time RSLAM framework (Mei, *et al.* (2010) [24]) that use bundle adjustment for constant time pose estimation. While the former aims for reduction of large-scale solution complexity, the latter focuses on locally accurate map and trajectory using relative bundle adjustment.

The innovation in our work is that it combines the relative bundle adjustment technique of RSLAM with the concept of Rao-Blackwellized particle filters. We also leverage on the concept of fuzzy pose similarity for bundle adjustment which is particularly useful when landmarks are sparsely encountered. Also, unlike other EKF filter based approaches where the time complexity worsens due to growing parameters at each time step, our approach achieves constant time complexity for each time step. Also, the complexity of the current approach is independent of environment size making it suitable for online SLAM applications over extremely large environments.

6.3. Problem Formulation

6.3.1. Simulation Environment

We simulate the robot's path and test our methodology on stereo-pair satellite image environment to demonstrate that such an UAV with stereo-pair sensor aboard can be planned in practice. Two large stereo-pair satellite images of size 12000×12000 (covering area of 30 km by 30 km), are considered from Panchromatic cameras AFT (after ward) and FORE (forward), which are at angles -5° and +26°, respectively, along the track with respect to nadir and are very helpful in deriving the depth information of the terrain imaged (Srivastava and Krishna, 2005 [34]). A series of sub-scenes corresponding to the robot's sensors' intake at different time steps is cropped from these images as if the robot is in motion and is taking such images at different time steps. The robots' path is also simulated by incorporating errors in the input control information.

A continuous relative representation (CRR) is used to represent the environment (Mei, *et al.* (2010) [24]). The advantage of this approach is that it incorporates standard bundle adjustment and does not require complex map-merging for loop closure. However, it aims mostly to provide stable motion estimates without laying too much stress on the mapping aspect. In our work, we maintain multiple copies of the same map and at each step choose the best map to represent the environment.

6.3.2. Representation of the Environment

In our representation, the set of poses are treated as nodes of a graph. Each new pose is treated as a newly added vertex in the already existing graph. The new pose is connected to other existing poses by two kinds of links:

1. Each adjacent pair of poses is connected by an odometric link which represents the required transformation to travel from one pose to another.

2. Each pair of poses that have matching features will have another additional link called a matching link. This link essentially represents a set of constraints imposed on the relative location of the poses, manifested in the form of common stereo features between them.

For each new pose, a new vertex is introduced in the graph along with a set of links as described above. Next, for this newly introduced pose, we define a bundle, which is the set of poses with the strongest links to this pose. Usually these include the poses that are either have maximum number of matching landmarks or spatial overlap or spatial proximity with the current pose. In most cases, the most recent poses will constitute the bundle. However, in situations like loop closure, some old poses may also be included in the bundle. In our work, we use a maximum overlap criterion to construct the bundle for each pose.

Furthermore, we maintain landmark coordinates with respect to their base frames, which is the frame where they were observed for the first time. For all later poses where this landmark is observed, we may easily re-project it to the corresponding pose by a series of transformations along the path in the graph connecting the home pose of the landmark with the target pose. This representation of landmarks avoids blindfold searching of match correspondences between newly observed features and the existing set of landmarks. This is because for each current pose, it is sufficient to search for correspondences among the landmarks whose home poses belong to the bundle for that pose.

At each time step, our approach tries to achieve a pose estimate that minimizes the energy of the graph via bundle adjustment as described in the following section. We use local adjustment because a global estimate would cause the number of parameters (number of landmarks and poses) to increase with each passing time step, thus degrading performance considerably.

6.3.3. Bundle Adjustment Technique

Bundle adjustment is the problem of refining a visual reconstruction to produce jointly optimal structure and viewing parameter estimates from a set of corresponding image projections (Triggs, *et al.* (1999) [36]).

6.3.3.1. Feature Matching

Given the set of poses B in the bundle for the current time-step and a set of landmarks M, our cost function aims at minimizing the re-projection error between each landmark m_i and its measurement in pose j given by z_i^j, measured over all landmarks in M and all poses in B.

As mentioned before, each landmark m_i is maintained with respect to its base pose. Thus, if a landmark m_i with base frame j_1 is observed from another frame j_2 then the re-projection of m_i in frame j_2 can be easily computed by a series of transformations along the path from frame j_1 to j_2. Let the transformation be $Z_i^{j_2}$.

We formulate the re-projection error for common landmarks between the poses j_1 and j_2, with $M_{(j_1, j_2)}$ being the set of common landmarks, as:

$$F_{j_1, j_2} = \sum_{i=1}^{M_{(j_1, j_2)}} (z_i^{j_2} - Z_i^{j_2})^T \Sigma_{i, j_2}^{-1} (z_i^{j_2} - Z_i^{j_2}) \qquad (6.1)$$

Let at the current time step the set of old poses in the bundle B be $\{j_1, j_2, \ldots, j_n\}$ and the new pose at the current time step be j_t. Then, the energy of the bundle B denoted by E_B is given by:

$$E_B = \sum_{i=1}^{n} F_{j_i, j_t} \qquad (6.2)$$

6.3.4. Graph Weight Computation

As mentioned earlier, each particle i maintains its own CRR graph comprising the set of poses and landmarks, along with the strong and weak links connecting the poses. At each time step, the new pose is introduced into each of these graphs, followed by bundle adjustment to estimate the new updated graph for the current time step.

Let the CRR graph maintained by particle P_i after time-step t be G_t^i. When the new pose at each time-step $t+1$ is introduced, we perform bundle adjustment as described above.

Let the bundle at time-step t for particle P_i be given by B_t^i. Then the energy of G_t^i is given by:

$$E_{G_t^i} = E_{G_{t-1}^i} E_{B_t^i} \qquad (6.3)$$

with

$$E_{G_1^i} = 1 \qquad (6.4)$$

We now define the particle weight at time step t W_t^i as follows:

$$W_{G_t^i} = 1 / E_{G_t^i} \qquad (6.5)$$

Finally, the particles are re-sampled according to their weights.

We now incorporate loop closure using the bag-of words model developed for image retrieval systems (Sivic and Zisserman (2003) [33]).

6.4. Loop Closure

For loop closure we use the fast appearance based mapping approach (Cummins and Newman (2008) [7]). This approach represents each place using the bag-of-words model developed for image retrieval systems in the computer vision community (Sivic and Zisserman (2003) [33]; Nistâ´er and Stewenius (2006) [26]). At time-step t the appearance map consists of a set of S_t discrete locations, each location being described by a distribution over which appearance words are likely to be observed there. Incoming sensory data is converted into a bag-of-words representation; for each location, a query is made that returns how likely it is that the observation came from that locationâ€™s distribution or from a new place. This allows us to determine if we are revisiting previously visited locations.

There remains a possibility of incorrect matches from loop closure leading to catastrophic results. However this can be avoided by applying feature matching between the currently observed region and the already observed pose with which it appears to share a strong correspondence. If the match is incorrect, most of the features of the current pose will be registered as new landmarks. In such a case, the incorrect loop closure can be easily detected.

If a correct loop closure is indeed detected, a new link is introduced to the graph corresponding to the new estimate, which is calculated by taking the weighted average of the new pose location and the old pose it corresponds to, based on the probability of match. This helps reduce drift to a large extent.

6.5. Theoretical Justification of the Hybrid Approach - Integrating RSLAM with Particle Filters

6.5.1. Desired Asymptotic Properties

Solution to SLAM (or FastSLAM) problem has two basic approaches. One of these is estimation-theoretic while the other is numerical or computational (Dissanayake, *et al.* 2001 [8]). A major advantage of the first approach is that it is possible to develop a complete proof of the various properties of the SLAM problem and to study systematically the evolution of the map as well as the uncertainty in the map and vehicle location.

On the other hand, given a set of measured image feature locations and correspondences, the goal of bundle adjustment is to find 3D point positions and camera parameters that minimize the re-projection error. This optimization problem is usually formulated as a non-linear least squares problem, where the error is a squared L_2 norm of the difference between the observed feature location and the projection of the corresponding 3D point on the image plane of the camera (Agarwal, *et al.* 2010 [1]). We note here that if the error is Gaussian, then bundle adjustment is, in fact, the maximum likelihood (ml) estimator.

Considering estimation-theoretic SLAM, Dissanayake, *et al.* (2001) [8], have given a proof for solution of SLAM problem that the estimated map converges monotonically to an actual map with zero uncertainty. We show here that similar results will hold for our model, the hybrid approach, and the particle filter will only enhance the convergence rate. Further, at least theoretically, our approach follows similar convergence with consistency. In fact it follows from the convergence properties of ml estimator. We consider different kind of weight function as opposed to the usual particle weight that exists in literature (Thrun, *et al.* 2005 [35]). And we justify the validity of such weight function by considering that it satisfies both:

1) The desired properties similar to the weight of particle filter;

2) The desired convergence properties similar to the usual weight function.

Slutsky's theorem (Cramer, 1946 [6]) then ensures the validity of algebraic operations between convergence of two random phenomena viz., bundle adjustment and particle filters and justifies the convergence properties of our hybrid approach.

The weight function used to estimate resample the particle filter in FastSLAM was essentially the ratio of two densities. Here it is worth mentioning that though we are considering a different weight function which is in terms of bundle energy, it is not conceptually different from the former and can be written as ratio of two densities precisely because our weight is actually a maximized ratio of likelihoods.

6.5.2. Properties of Maximum Likelihood (ml) Estimators in Bundle Adjustment

It is well known that the ml estimate has zero bias asymptotically and the lowest variance that any unbiased estimator can have. So in this regard, ml estimation is at least as good as any other method.

Apart from their obvious simplicity and intuitive justification, ml estimators satisfy certain statistical properties viz., *consistency* and *asymptotic normality*.

Case of Incorrect Modelling and Robustness of ml

Incorrect modelling of the observation distributions is likely to degrade quality of the ml estimate. Such incorrect modelling is, to some extent, inevitable because error distributions stand for influence that we can not fully predict or control (Triggs, *et al.* 1999 [36]). The important problem encountered here is the failure to account for the possible outliers. But, as described in B. Triggs, *et al.* (1999 [36]), the above properties of ml estimation including asymptotic minimum variance remain valid even in the presence of outliers, due to robustness of ml estimator.

Case of Distribution Mismatch

A crucial assumption of our approach is that the observations follow a Gaussian distribution. When an observation differs from Gaussian, the ml estimate converges asymptotically to the model parameter whose predicted observation distribution has minimum relative entropy (Triggs, *et al.* 1999 [36]). Hence for large samples, the problem of mismatch can be at least theoretically eliminated, and it is only the rate of convergence that is affected in this situation.

So, in conclusion, asymptotic properties of ml estimator, which was used for bundle adjustment, are comparable to the asymptotic properties that were mentioned in Dissanayake, *et al.* (2001) [8].

6.5.3. Advantages and Asymptotic Properties of Particle Filter

In particle filters, the samples of a posterior distribution are called particles and are denoted by

$$\chi_t = \left\{ x_t^{[1]}, x_t^{[2]}, \ldots, x_t^{[M]} \right\}.$$

The weight set is

$$\left\{ w_t^{[1]}, w_t^{[2]}, \ldots, w_t^{[M]} \right\}.$$

Here M denotes the number of particles in the particle set χ_t. This approach is actually similar to Bayes Filter. So, like all other Bayes filter algorithms, the particle filter constructs the belief $bel(x_t)$ recursively from the belief $bel(x_{t-1})$, from a step earlier. One of the most crucial steps in using particle filters is resampling or importance sampling. The probability of drawing each particle in the resampling step is given by its importance weight (Thrun, *et al.* 2005 [35]). By incorporating the importance weights into the resampling process, the distribution of the particles change as follows:

$$bel(x_t) = \eta P(z_t \mid x_T^{[M]}) \overline{bel}(x_t),$$

where $\overline{bel}(x_t)$ is the prior, $bel(x_t)$ is the posterior belief.

The samples of posterior distribution (particles) is accumulated by particle set χ_t. If D_i denote the accumulated data at i-th time point, then particle filter adopts the following update scheme.

$$\cdots \rightarrow bel(x_i \mid D_i) \rightarrow bel(x_{i+1} \mid D_i) \rightarrow bel(x_{i+1} \mid D_{i+1}) \rightarrow \cdots$$

As stated in Thrun, *et al.* (2005) [35] if f denote the density function of a target distribution from which we like to obtain a sample, and if g be the density function of the corresponding proposal distribution, then

$$\frac{1}{M}\sum_{m=1}^{M} I\left(x^{[m]} \in A\right) \rightarrow \int_A g(x)dx.,$$

where I is the indicator function. Further, if $w^{[m]} = \dfrac{f(x^{[m]})}{g(x^{[m]})}$, then

$$\left[\sum_{m=1}^{M} w^{[m]}\right]^{-1} \sum_{m=1}^{M} I\left(x^{[m]} \in Aw^{[m]}\right) \rightarrow \int_A f(x)dx.$$

We may note that this convergence is actually due to *law of large number*, and hence *in probability*.

In our approach we consider a seemingly different kind of weight function, which we repeat here for the sake of clarity:

Let the bundle at time-step t for particle P_i be given by B_t^i. Then the energy of G_t^i is given by:

$$E_{G_t^i} = E_{G_{t-1}^i} E_{B_t^i} \tag{6.6}$$

with

$$E_{G_1^i} = 1 \tag{6.7}$$

We now define the particle weight at time step t W_t^i as follows:

$$W_{G_t^i} = 1/E_{G_t^i} \tag{6.8}$$

Finally, the particles are re-sampled according to their weights.

It is quite evident that this new weight function follows the above convergence properties. In bundle adjustment, one basically minimizes the energy function defined as the sum of reprojection errors.

We formulate the re-projection error for common landmarks between the poses j_1 and j_2, with $M_{(}j_1, j_2)$ being the set of common landmarks as given by:

$$F_{j_1,j_2} = \sum_{i=1}^{M_{(j_1,j_2)}} (z_i^{j_2} - Z_i^{j_2})^T \Sigma_{i,j_2}^{-1} (z_i^{j_2} - Z_i^{j_2}) \qquad (6.9)$$

Let at the current time step the set of old poses in the bundle B be $\{j_1, j_2, \ldots, j_n\}$ and the new pose at the current time step be j_t. Then the energy of the bundle B denoted by E_B is given by:

$$E_B = \sum_{i=1}^{n} F_{j_i, j_t} \qquad (6.10)$$

As said earlier, in our case the error is Gaussian, hence bundle adjustment is the ml estimator. So the minimization in this regard implies that the ml estimator satisfies the convergence properties as stated above. Further, we may write the energy function in terms of ratio of likelihoods. The maximum a posteriori estimate is the parameter $\hat{I}$, that maximizes the probability of a given the data:

$$P(\theta \mid x_1, x_2, \ldots, x_n) = \frac{f(x_1, x_2, \ldots, x_n \mid \theta) P(\theta)}{P(x_1, x_2, \ldots, x_n)},$$

where $P(\theta)$ is the prior distribution for the parameter θ and where $P(x_1, x_2, \ldots, x_n)$ is the probability of the data averaged over all parameters.

We note that our weight function as given by Eq. (6.8), though stated in terms of energy of the bundle, can also be stated as maximized ratio of two Gaussian likelihood, as can be readily seen from the Eq. (6.9) and Eq. (6.10) above. Thus conceptually our weight function is not different from the usual one in particle filter, rather it is similar and hence follows similar properties.

We recall that maximum-likelihood estimator is an estimator obtained by maximizing, as a function of θ, the objective function, which is given by:

$$\hat{\ell}(\theta \mid x) = \frac{1}{n} \sum_{i=1}^{n} \ln f(x_i \mid \theta),$$

This is the sample analogue of the expected log-likelihood $\ell(\theta) = E[\ln f(x_i \mid \theta)]$, where the expectation is taken with respect to the true density $f(\cdot \mid \theta_0)$. The ml estimator satisfies a property called *efficiency*, which means it achieves the Cramer-Rao lower bound when the sample size goes to infinity. That is to say, no consistent estimator has a lower asymptotic mean squared error than the ml estimator.

By the law of large numbers, one can establish the convergence in *probability* of the log-likelihood:

$$\sup_{\theta \in \Theta} |\hat{\ell}(\theta \mid x) - \ell(\theta)| \xrightarrow{\ p\ } 0.$$

This is true for both cases of independent and identically distributed (i.i.d) and non-i.i.d observations. The dominance condition can be used to prove convergence for the case of i.i.d. observations. For convergence in probability of the non-i.i.d. case, one can show that the sequence $\hat{\ell}(\theta \mid x)$ is stochastically equi-continuous.

The above result is comparable to the convergence properties of the usual weight function that has been used so far. So, in our approach, we are taking the advantage of both convergence of ml estimator and convergence of particle filter. Moreover, the convergence, being the convergence in probability, will enhance the convergence of ml estimator, which can be shown analytically using Slutsky's theorem (Cramer, 1946 [6]). This means a mathematically valid expression of mutual convergence as in our case can easily be shown.

6.6. Landmark-Formation in Stereo Environment

6.6.1. SURF Based Feature Detection

To build our simulation environment, we have used a stereo pair of satellite images. In order for the UAV to correct its path it is of paramount importance that landmarks be identified and matched correctly, not only between time steps but also in the same time step between the stereo pair of images. Landmarks can either be closed bounded regions or individual features that are common to the stereo pair of images. While closed regions are difficult to match due to change in orientation and viewpoint in the stereo pair, merely matching individual features may not be very accurate because of the sheer number of stereo features tracked by common stereo matching algorithms, leading to the possibility of a number of spurious matches. Hence there is a need to refine the process of feature matching based on local characteristics that are scale, rotation and affine transform invariant.

As seen in the literature the most reliable stereo matching algorithm is Scale Invariant Feature Transform (SIFT) (Lowe, 1999 [21]) which performs continuous convolution of the images with Gaussian Kernels to build scale space pyramids and then build a difference of Gaussian (*DoG*) hierarchy. Features are considered as manifestations of local extrema of 1st and 2nd order derivatives in the *DoG*. Invariance with respect to scale is achieved by identifying such extrema across multiple *DoG*s.

SIFT, though highly accurate in feature identification, is a computationally cumbersome procedure, principally because of the expensive operations involved in building the *DoG*. This makes it impractical for online feature identification in SLAM. A faster alternative to SIFT is SURF (Bay, *et al.* 2008 [3]). Its major advantage over SIFT lies in the fast computation of box space operators making it highly suitable for real time use in our SLAM framework.

6.6.1.1. Stereo Feature Extraction, Outlier Detection and Identification of Landmarks

For matching between the stereo pair of snapshots taken in the same time step an exhaustive comparison of the SURF descriptors of the features identified from each image is performed. To speed up the process we match the sign of the Laplacian which automatically discards several pairs as infeasible for matching. The actual match uses the Euclidean distance of the descriptor vectors and subsequently validation is done based on a suitable threshold.

Detection of outlier matches is an important issue. We adopt a planar motion segmentation based approach (Zhang, *et al.* (2010) [38]). This approach is RANSAC-based (Fischler and Bolles (1981) [13]) and uses Kanade-Lucas-Tomasi (KLT) tracker (Lucas and Kanade (1981) [23]) along with epipolar constraints to reject outlier matches. This effectively eliminates incorrect matches that might result from image noise, repeated structures and image distortion.

6.6.2. Landmark Characterization

The feature selection process follows assumption that we desire features with a uniform distribution in the image. In view of that, once a feature has been identified as a valid one, landmark formation is carried out in the three phases as described below.

6.6.2.1. Height of Landmark

In phase one we estimate the height of the landmark from the concept of binocular vision by correlating its disparity in the X and Y directions. Since the snapshots (images) are taken at different angles, a feature may be shifted in both X and Y directions and so we consider the Euclidean distances between the features location in the two snapshots as a probable measure of its disparity. Since disparity is a good indication of the closeness of the feature to the camera, we may use it as a measure for the height of the feature.

6.6.2.2. Segmentation Based Characterization

In phase two we identify the class location of the feature. This will provide two-pronged advantages. First, all spurious features that have location mismatched will be eliminated. Secondly, the class-wise match features will ensure that features that are used for the analysis are evenly spread over the image. Before feature extraction is carried out from the AFT and FORE images we segment the snap shot of AFT image (which is closer to nadir) and classify the image into a number of classes so that the entire environment has K fixed distinct classes. The technique used here is an Markov Random Field (MRF) model based approach (Sarkar, *et al.* 2002 [31]) for determining the optimal number of segments followed by an unsupervised scheme of classification (Sarkar, *et al.* 2012 [32]) to classify the segments. Each class is characterized by its estimated mean ($\hat{m}_i$), estimated variance (S_i^2) and number of pixels (n_i). We use this information in the landmark association stage where two landmarks can be said to match if and only if they belong to the same class.

6.6.2.3. Topographic Labelling

In the third and final phase, we aim to topographically classify the feature. Our approach involves estimation of the underlined structural component of the region surrounding the feature pixel. This approach is particularly suitable for images with dominant tonal component such as indoor images. However since our aim is to characterize the feature pixel by a few local region statistics, it is expected to work for images with dominant textural component as well. Since pixel information gathered from a digital image often has noise associated with it, it is essential to be able to extract the exact signature of the feature irrespective of the noise level variation across different snap shots, which is what topographic classification aims to achieve. Thus topographic classification helps to further uniquely identify a feature.

Haralick, *et al.* (1983) [15] have used a cubic polynomial fit to estimate ideal image surface which may not always assure adequate representation of the underlined structure. On the other hand approach of Watson, *et al.* (1985) [37] to use generalized splines as basis functions for surface estimation does not account for the noise variance and as such may not be appropriate. We have used a statistical technique based approach to estimate the optimal underlying structure. In order to take into account the local topographic variation, we consider a 15×15 area surrounding the feature point. It is essential to maintain orientation invariance; so the rectangular region surrounding a particular feature should have the same orientation across different time steps with respect to the global co-ordinate system. We consider the pixel values in the region surrounding the feature to be governed by two components, viz. structural and noise. The noise is assumed to follow a bivariate Gaussian distribution in both X and Y directions with zero mean and finite covariance. To represent the noisy image surface adequately, we use an orthogonal polynomial basis (Kartikeyan and Sarkar 1995 [18]).

$$\Phi = \{\varphi_{ij}(x,y) = \pi_i(x)\pi_j(y), \quad i,j = 0,1,2,...,n-1\}, \qquad (6.11)$$

where $\pi_k(u)$ is the orthonormal polynomial of degree k in u and $n \times n$ is the size of the window around the feature that we have considered. However, the orthonormal polynomials should be so chosen such that they satisfy the following conditions,

1. $\displaystyle\sum_{(x,y)\in W} (\varphi_{ij}(x,y)\varphi_{kl}(x,y)) = 0$ $i,j,k,l < n$ and $(i,j) \neq (k,l)$

2. $\displaystyle\sum_{(x,y)\in W} \varphi_{ij}^2(x,y) = 1$

We present the elements of the basis for $n = 3$

$$\varphi_{0,0} = \frac{1}{3} \quad \varphi_{0,1} = \frac{y}{\sqrt{6}} \quad \varphi_{1,0} = \frac{x}{\sqrt{6}} \quad \varphi_{1,1} = \frac{xy}{2}$$

$$\varphi_{0,2} = \frac{3y^2 - 2}{\sqrt{18}} \quad \varphi_{2,0} = \frac{3x^2 - 2}{\sqrt{18}}$$

$$\varphi_{1,2} = \frac{x(3y^2 - 2)}{\sqrt{12}} \quad \varphi_{2,1} = \frac{y(3x^2 - 2)}{\sqrt{12}} \quad \varphi_{2,2} = \frac{(3x^2 - 2)(3y^2 - 2)}{6}.$$

Similar calculations may be performed for higher values of n.

So the underlying function of the $n \times n$ window can be uniquely expressed as

$$g(x, y) = \sum_{i=0}^{n-1} \sum_{j=0}^{n-1} A_{ij} \varphi_{ij}(x, y), \tag{6.12}$$

where φ_{ij}'s are the sources of variation and the coefficients A_{ij} are the effects of such variations. Of the n^2 components in the above summation it is essential to identify the ones that are purely due to noise. Using Nair's test of homogeneity with 1 degree of freedom, we separate the structural components from the noisy ones. The estimate of noise variance is then given by

$$\hat{\sigma}_n^2 = \sum_{\hat{A}_{i,j}^2 \in N} \frac{\hat{A}_{ij}^2}{C(N)} \tag{6.13}$$

and the structural component, on the other hand, is estimated by

$$\hat{f}(x, y) = \sum_{\hat{A}_{ij}^2 \notin N} \hat{A}_{ij} \varphi_{ij}(x, y), \tag{6.14}$$

where $\hat{A}_{ij}$'s are the estimated effects of the variation, N is the set of noisy components and $C(N)$ is the cardinality of N.

The primary quantities which have to be estimated are f_x, f_y, f_{xx}, f_{yy} and f_{xy}. These can be easily computed by performing partial differentiations of the estimated function $\hat{f}(x,y)$. Using these derivatives we compute the Hessian Matrix at the feature point and perform standard statistical tests to determine certain topographical characteristics. In particular we concentrate on testing the following:

(a) The sign of the gradient value at the pixel of interest;

(b) The signs of the eigenvalues of the Hessian matrix at that pixel;

(c) The presence or absence of zero crossings along the eigen-direction of the Hessian within the area of the pixel.

For the first characteristic, we test the multiple null hypothesis $f_x = 0$ and $f_y = 0$. Using the estimates for these derivatives $\hat{f}_x$ and $\hat{f}_y$ and their variances $V(\hat{f}_x)$ and $V(\hat{f}_y)$, we test the following statistic: $F_1 = (\hat{f}_x)^2 / (V(\hat{f}_x)/C(N)) + (\hat{f}_y)^2 / (V(\hat{f}_y)/C(N))$ which is F-distributed with $(2, C(N))$ degrees of freedom.

For the second characteristic, the null hypothesis tested is $\lambda = 0$ where λ is an eigenvalue of the Hessian matrix. We estimate $T_1 = \dfrac{\partial^2 f}{\partial \omega^2}$ (ω being the unit vector in the direction of the eigenvector of the Hessian) as $\hat{T}_1 = (\omega_x)^2 \hat{f}_{xx} + (\omega_y)^2 \hat{f}_{yy} + 2\omega_x \omega_y \hat{f}_{xy}$. The test statistic used is:

$F_2 = (\hat{T}_1)^2 / (V(\hat{T}_1)/C(N))$ which is F-distributed with $(1, C(N))$ degrees of freedom.

For the third characteristic, the null hypothesis tested is $T_2 = \nabla f . \omega = 0$. The estimate of T_2 is given by $\hat{T}_2 = (\omega_x)\hat{f}_x + (\omega_y)\hat{f}_y$. We use here the following test statistic: $F_3 = (\hat{T}_2)^2 / (V(\hat{T}_2)/C(N))$ which is F-distributed with $(1, C(N))$ degrees of freedom.

Thus, in our SLAM implementation, a landmark consists of the 2D location of a feature and its height viz., (x, y, z) along with a list of descriptors, class information, the sign of Laplacian at its corresponding pixel and its responses to the aforementioned topographic tests.

6.7. Data Association Based on Landmark Matching

We perform the data association process between the set of previously observed and newly observed landmarks in three steps. In each step, we progressively refine the association obtained in the previous step.

Our initial association is purely based on the information obtained from SURF in the form of the sign of Laplacian and the SURF descriptor list associated with each feature. Since the list is invariant to scale, rotation and illumination, two or more descriptor lists corresponding to the same feature, even if it occurs across different time steps, are expected to be within a certain Euclidean distance limit. Based on this assumption we create an initial probable match list for each newly observed landmark in the current time step. For each new landmark, we compute the Euclidean distance of its descriptor list from that of each old landmark. If the distance is less than a certain threshold we add the corresponding old landmark to the list of probable matches for the new landmark. Following this procedure, we end up creating an adjacency list for each new landmark. However, there exists a chance that two or more features from different regions may have similar local characteristics and hence similar descriptor lists. Using a stricter threshold may work in some cases, but to the contrary in most cases it creates a danger of leaving out some correct matches as well, while retaining some spurious matches. Evidently some other independent tests are necessary.

In the next step, we attempt to refine the adjacency list created thus far by pruning a few outlier matches that may have crept in due to spurious matching of descriptors. We use the class information that we have already obtained via unsupervised classification of the image closer to nadir. If two landmarks are identical, irrespective of the time step at which they are identified, they are expected to belong to the same class. We now simply test the hypothesis that a new landmark (L_i) and an old map feature (m_j) appearing in its adjacency list belong to the same class based on the mean, variance and number of pixels of the respective classes to which these landmarks belong. The test statistic used is Q_{ij} and has the following form

$$Q_{ij} = \frac{(\hat{m}_i - \hat{m}_j)^2 \times v_{ij} n_i n_j}{(n_i S_i^2 + n_j S_j^2)(n_i + n_j)}$$

It follows Fisher's F distribution with $(1, v_{ij} = n_i + n_j - 2)$ degrees of freedom. Any outlier matches that do not belong to the same class are thus removed.

In a third and final attempt at removing extraneous matches that might have survived the previous two steps, we turn to compare the local topographic characteristics of a new landmark with those of its potential matches. As stated in the previous section, three statistical tests are carried out to identify the necessary topographic characteristics of each landmark. While validating potential candidates for matching, we simply check for similar responses to these tests. For a valid matching pair, the responses should be identical. If they are not, the corresponding old landmark is removed from the adjacency list.

These tests have been so designed that they ruthlessly do away with all invalid/erroneous associations.

There exist still quite a few probable matches for each newly observed landmark due to the sheer density of stereo features that SURF may provide us with. The challenge is to avoid overlapping matches, that no two new landmarks must be matched to the same old landmark. Fortunately, the aforementioned association process has in all probability made the adjacency list manageable in terms of size. So we have designed a procedure that gives us the best non-conflicting mapping between the set of new and old landmarks. Each association between a pair of landmarks is given an initial probability based on the inverse Euclidean distance between the pair in terms of global coordinates. Then we apply a matching algorithm to obtain the final association list.

When landmarks are sparsely encountered it is easier to use fuzzy similarity based procedure as described in the following section.

6.8. Matching Landmarks Using Fuzzy Similarity

6.8.1. Fuzzy Landmarks

The objective consists of comparing an egocentric representation of a *small* area around the robot pose with some prototypes representing the features considered as the basis for the environment description.

Note that this step is only used when landmark distribution is sparse, and landmark matching is prone to errors. With a dense landmark distribution this step is too computationally taxing and is hence avoided.

Let T_i and U_i be two landmarks with current mean $\overline{T_i}$ and $\overline{U_i}$ respectively, denoted by $\times$ and Δ respectively as in Fig. 6.2. The current covariance matrices of the landmarks give rise to the probability contour ellipsoids (under the assumption of Gaussian distribution) which has been drawn by red and blue respectively.

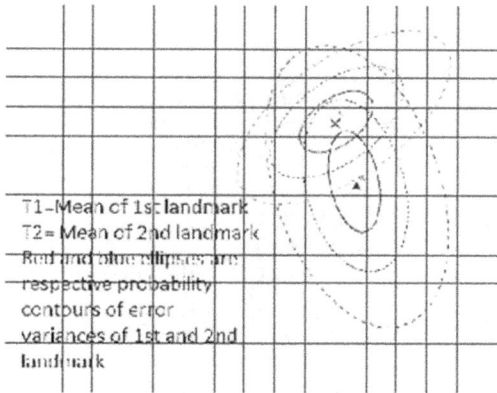

Fig. 6.2. Gaussian probability contours.

6.8.2. Fuzzy Set Terminologies

We recall that a fuzzy set is a pair (U,μ) where U is a set and $\mu:U \rightarrow [0,1]$. For each $x \in U$, the value $\mu(x)$ is called the grade of membership of x in (U,μ). For a finite set $U=\{x_1,...,x_n\}$, the fuzzy set (U,μ) is often denoted by $\{\mu(x_1)/x_1,...,\mu(x_n)/x_n\}$.

Let $x \in U$. Then x is called not included in the fuzzy set (U,μ) if $\mu(x)=0$, x is called fully included if $\mu(x)=1$, and x is called a fuzzy member if $0<\mu(x)<1$. The set $\{x \in U \mid \mu(x)>0\}$ is called the support of (U,μ) and the set $\{x \in U \mid \mu(x)=1\}$ is called its kernel. The function μ is called the membership function of the fuzzy set (U,μ). Cardinality of a non-fuzzy set, Z, is the number of elements in Z. On the other hand, the cardinality of a fuzzy set A, the so-called *sigma count*, can be expressed as a sum of the values of the membership function of A, i.e, $card(A) = \mu_A(x_1) + \cdots + \mu_A(x_n) = \sum_{i=1}^{n} \mu_A(x_i)$.

6.8.3. Similarity Metric: Fuzzy Similarity

Several authors have proposed similarity indices for fuzzy sets that can be viewed as generalizations of the classical set-theoretic similarity functions (Dubois (1980) [10], Zwick, *et al.*(1987) [40]).

There are many indices proposed in the literature as measure of similarity between fuzzy subsets. We define an index suitable for our purpose which is another version of the similarity measure due to Sokal and Michener:

$$S(M,L) = \frac{card(M \cap L) + card(\overline{M} \cap \overline{L})}{card(\overline{M} \cap L) + card(M \cap \overline{L}) + card(U)},$$

where U is the universe of discourse.

Let T_i and U_i be two landmarks with current mean $\overline{T_i}$ and $\overline{U_i}$ respectively, denoted by $\times$ and $\triangle$ respectively, which is same as $\overset{\circ}{a}$ and $\overset{\circ}{a}$ respectively as in Fig. 6.3.

We can directly use a suitably chosen ellipsoid probability contour of the current probability of each landmark, as determined by covariance matrix. In Fig. 6.3, the blue region will be assigned to zero probability, and inside the ellipsoids, the probability is as usual. Then by using the straight forward fuzzy set algebra, we may obtain the similarity index.

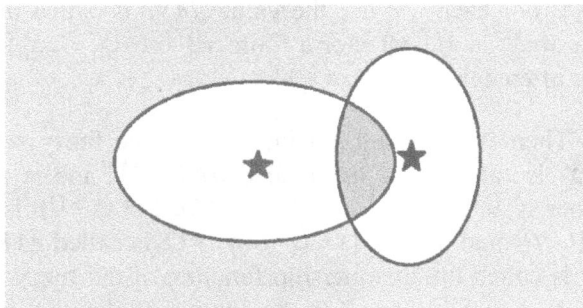

Fig. 6.3. Landmarks in fuzzy context.

We consider multiple ellipsoid probability contours and take a weighted version of the above mentioned indices. The weights are in accordance with the value of probability inside the contours, which may vary according to the scenario.

We propose using the cell (discrete)version of the ellipsoids as in Fig. 6.4 in the sense that the probability is constant on each pre-defined rectangular cells. This can be useful in the situation where it is necessary to use computationally non-intensive procedure. In particular we can divide the whole map into some cells, use $\mu(x)$ in each of them according to the coordinate and covariance matrix associated with each landmarks. $\mu(.)$ gives the underlying fuzzy map. Then use the above to reduce the whole set of landmarks to smaller one. This increases the efficiency and reduces the complexity of the algorithm. We then follow the union, intersection rules as usual.

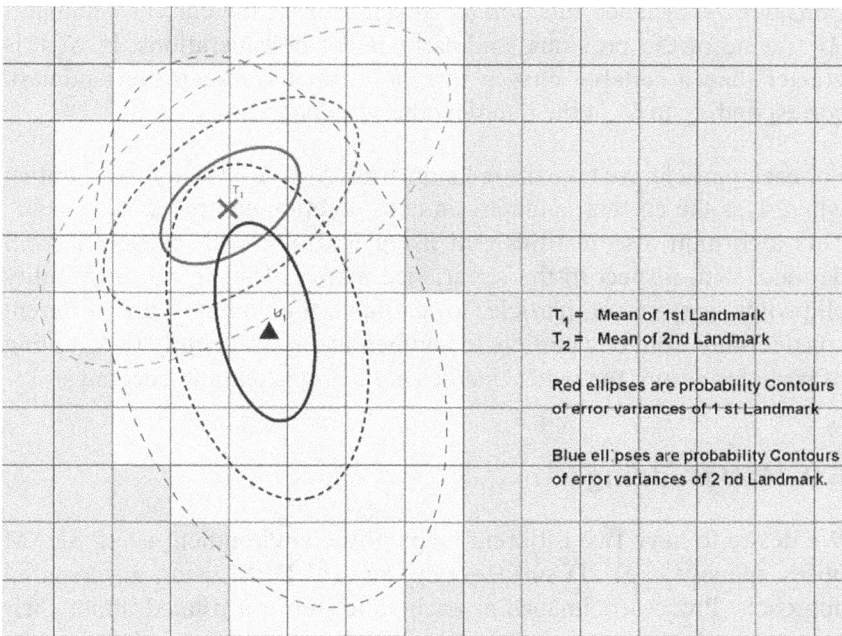

$T_1 = $ Mean of 1st Landmark
$T_2 = $ Mean of 2nd Landmark

Red ellipses are probability Contours of error variances of 1 st Landmark

Blue ellipses are probability Contours of error variances of 2 nd Landmark.

Fig. 6.4. Discrete representation of fuzzy landmark.

6.8.4. The Algorithm for Fuzzy Landmark Matching

In RSLAM, the cost function has been used to minimize the reprojection between the landmark m_i and its measurements z_i^j in image j for all M_i images and all N landmarks. So in the cost function, basically $N \times M_i$ variables are to be used for minimization. Instead of that, we reduce the N landmarks by N^δ which is specific to each particles, and $N^\delta < N$, thereby reducing the computational cost also. It consists of computing the degree of matching between the measured landmarks and the current one, interpreted as fuzzy sets. The comparison may be performed by using one out of the many fuzzy similarity functions defined in literature. In our case the similarity function due to Sokal and Michener is suggestive, and proper motivation behind this lies in the fact that we are taking Gaussian error into account, and secondly, we are using the precise information available from the mean and the covariance matrix into account.

Consider the distance function as $S(M, L)$. L is the current landmark, M is one of the previous landmarks under considerations. If $S(.)$ is greater than a suitably chosen threshold, then we keep the landmark corresponding to M in the bundle, otherwise not.

For each particle, we take those landmarks, $S(., L)$ of which is tolerable, where L is the current landmark in consideration and proceed as usual. This algorithm also justifies that using RSLAM with particle is well defined, with respect to the covariance matrix. That is, the probability ellipsoids vary from particle to particle, the bundles are different considerably from one particle to another, and subsequently re-sampling of particles ensure the better bundles are being taken into account.

6.9. Map Building

We desire to have two different maps of the environment using SLAM poses, namely - (a) 2D surface map, and (b) DEM using a sequential approach. Processed images at each time-step are rotated about their center so that their orientation is in line with the robot's pose. To do so, the ith and the jth pixel of the current image are moved to $(i - w/2) * sin(\psi_t) - (j - l/2) * cos(\psi_t)$ and $(i - w/2) * cos(\psi_t) + (j - l/2) * sin(\psi_t)$ respectively, where w and l are the width and length of the image. The images are then placed on the map by using the pose of the robot as center of each image. To do this x_t and y_t are added to i^{th} and j^{th} pixel of image respectively. So finally,

$$i \Rightarrow (i - w/2)*cos(\psi_t) - (j - l/2)*sin(\psi_t) + x_t \qquad (6.15)$$

$$j \Rightarrow (i - w/2)*sin(\psi_t) + (j - l/2)*cos(\psi_t) + y_t \qquad (6.16)$$

At every time step this is done to add an image to the global map. In this way the final global map is constructed with origin at the center of the first image.

6.9.1. Disparity Computation

We implement a computationally efficient and robust method (Radhika, *et al.* 2007 [29]) for stereo image matching of remote sensing images. The key features of our method are essentially hierarchical registration of image and a robust technique for detecting outliers based on local statistics. The proposed technique takes two images: the AFT and the FORE images with along-track disparity as input. The detailed DEM generation procedure follows the block diagram shown in Fig. 6.5.

6.9.1.1. Hierarchical Decomposition

A pyramidal image registration scheme is followed. The large image is progressively resized into smaller ones upto four stages. The DEM generated for a lower resolution image aids the development of the DEM of higher resolution images.

6.9.1.2. Feature Selection Matching

Relevant feature points have been identified using SURF in one of the images, say AFT, followed by an area based matching procedure. We implement a block-matching algorithm with normalized correlation coefficient as matching criterion to identify a match point in the FORE image for each feature selected in the AFT image. An intensity image (**R**) is created by matching the patch/template image (**T**) in shifting windows on the source image (**I**). In our case, the source is taken from the FORE image, and template is from the AFT image. The resulting intensity image is created using the following operation:

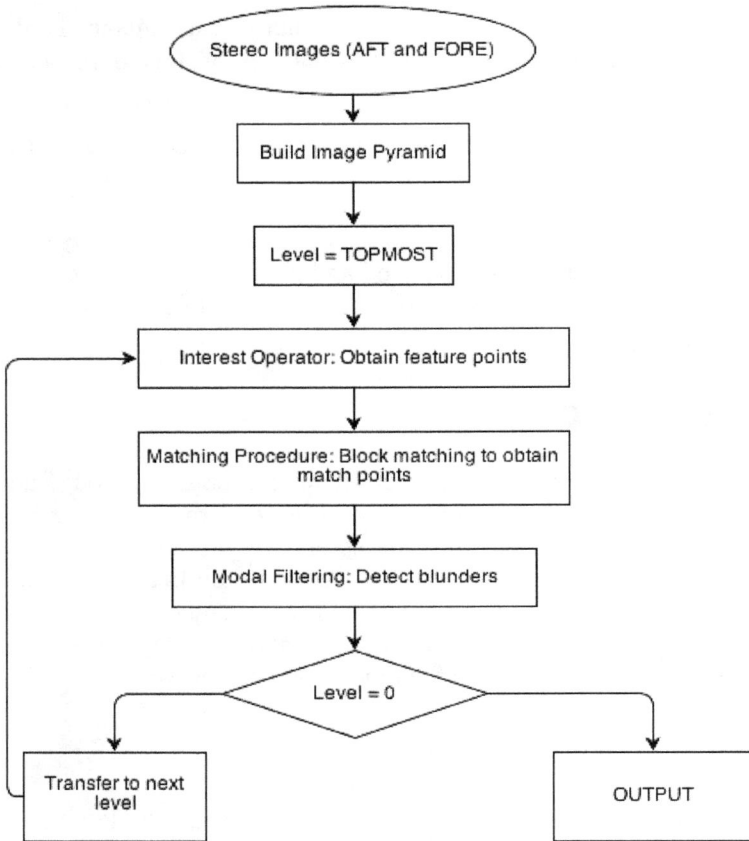

Fig. 6.5. Work Flow for Disparity Generation.

$$R(x, y) = \frac{\sum_{x',y'} (T(x', y') \cdot I'(x + x', y + y'))}{\sqrt{\sum_{x',y'} T(x', y')^2 \cdot \sum_{x',y'} I'(x + x', y + y')^2}}, \qquad (6.17)$$

where

$$T'(x', y') = T(x', y') - 1/(w \cdot h) \cdot \sum_{x'',y''} T(x'', y'') \qquad (6.18)$$

$$I'(x + x', y + y') = I(x + x', y + y') - 1/(wh) \sum_{x'',y''} I(x + x'', y + y'') \quad (6.19)$$

196

The maxima in the resultant intensity image is chosen as the appropriate match. In the next step we propose an algorithm to detect incorrect matches.

6.9.1.3.Blunder Detection

Despite there being a high number of reliable matches, we do get some spurious matches which may potentially skew the distribution. We detect blunders through local statistics, such as modal filtering.

We consider windows of size 25×25 pixels in each image, and calculate the modal disparity value within that window. All disparity values that are greater than a fixed distance away from the modal value are considered as outliers. The centroid of selected features in each window is then used. This effectively removes the outliers, and the centroid aids in providing sub-pixel resolution of the disparity image.

Subsequently a cubic spline is fitted to interpolate unknown values. The algorithm (Haber, *et al.* 2001 [14]) for this is linear time and is able to handle irregular distributions.

It should be noted that the disparity generation algorithm may be run completely offline. This is just to generate a DEM along the robot's track.

6.10. Experimental Results

The robot's environment used for our simulation has been exhibited in Fig. 6.6. The successive images captured by robot has an overlap of about 50 % or more. In the proposed methodology each pair of captured image has been processed individually and landmarks are identified as discussed in Section 6.6.

Sensor noise was simulated by adding random intensity values drawn from a Gaussian distribution.Changes in illumination were also simulated by adding a random illumination offset to each image, drawn once again from a Gaussian distribution. Here we present some illustrations.

Fig. 6.6. Robot's environment. The test environment
for the proposed methodology.

6.10.1. Effect of Topographic Labelling on System Performance

Fig. 6.7 and Fig. 6.8 show a SLAM simulation using the proposed methodology over a sequence of 53 frames with a spacing of 20 cm along the line joining the center of consecutive frames, with and without topographic labelling of landmarks. Fig. 6.7 shows the trajectories with and without topographic labelling along with real path. Fig. 6.8 exhibits pose error for the two cases. These figures appropriately captures the precision obtained using topographic labelling of landmarks. The average error without topographic labelling is 1.53 while the average error with topographic labelling is 0.72. Without the topographic labelling, the error has increased by a factor of 2.13. This is quite apparent from the error comparison in either case. This simulation confirms that topographic labelling adds in obtaining precise landmark estimates.

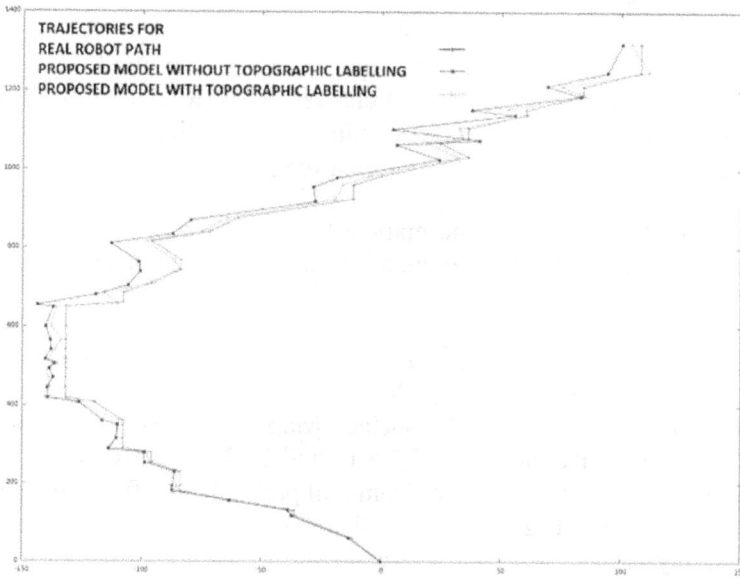

Fig. 6.7. 2D trajectory comparison with real path (red), proposed method with Topographic Labelling (green) and without (blue).

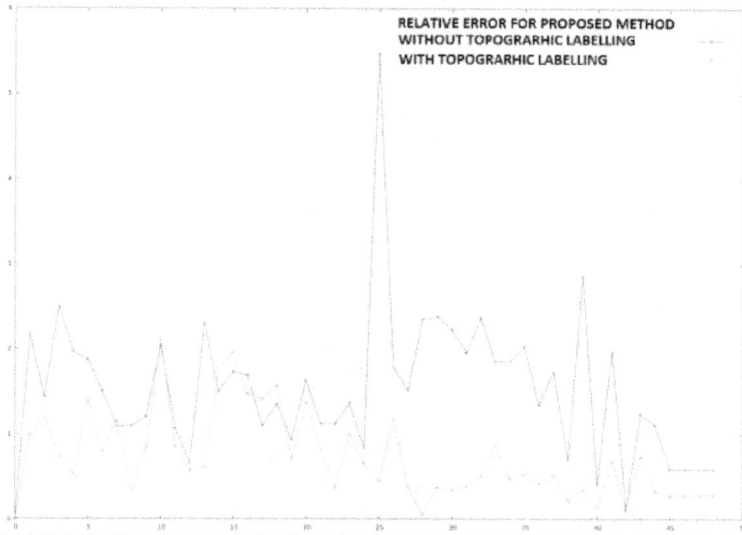

Fig. 6.8. Pose error comparison for proposed method with (green) and without (red) Topographic Labelling.

199

6.10.2. Example-1

In Example-1 we demonstrate our proposed methodology with 117 frames using 1, 20 and 50 particles respectively, over a distance of approximately 182 km. We consider in our simulation a percentage of overlap as more than 50 % for consecutive images.

The information regarding the map-features and robot poses contained in the particle with highest weight, after the final time step, has been used to build the *map* of the environment.

The 2-D (XY) and the 3-D (XYZ) trajectories of the robot's path along with control path and SLAM path have been exhibited in Fig. 6.9 and Fig. 6.10 respectively. The 2D surface Map and DEM built by Robot with 50 particles are shown in Fig. 6.11 and 6.12 respectively. Fig. 6.13 presents the pose error estimate (from real path) of each frame in each of the 3 scenarios for 1, 20 and 50 particles in red, green and blue colour respectively.

Table 6.1 summarizes the experimental set up.

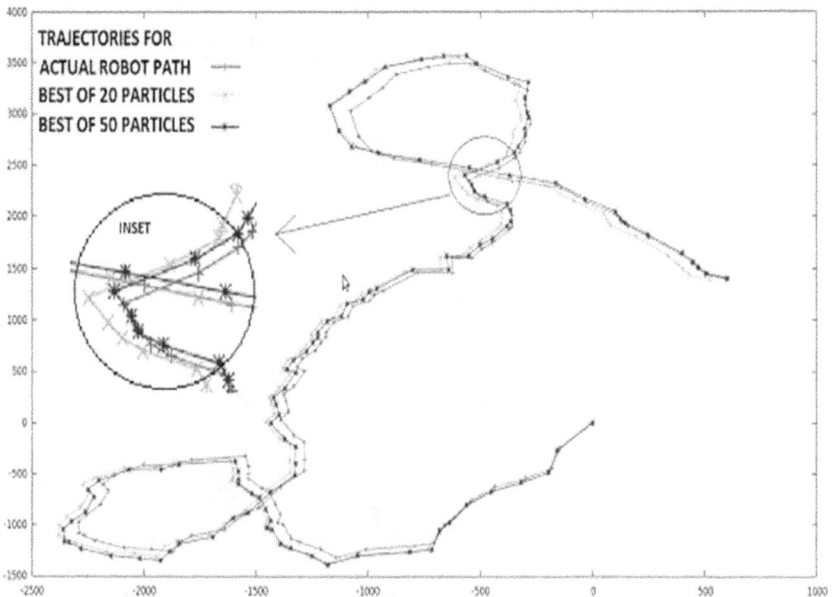

Fig. 6.9. The robot's trajectory in 2D (XY) plane.

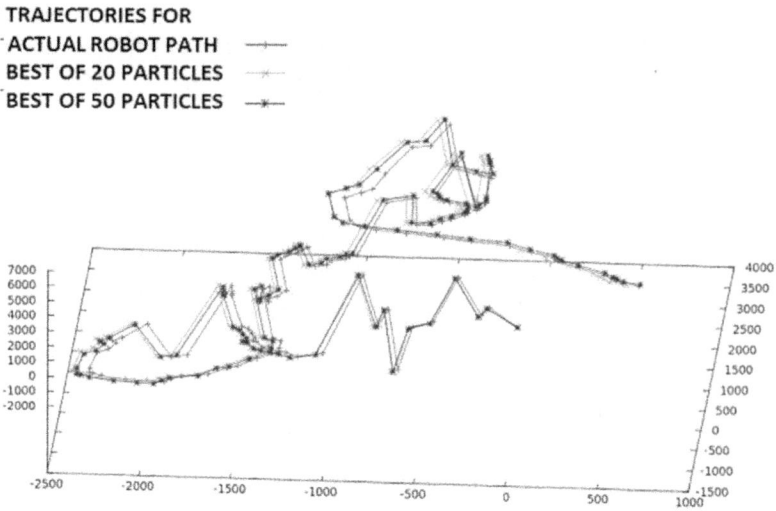

Fig. 6.10. The robot's trajectory in 3D space.

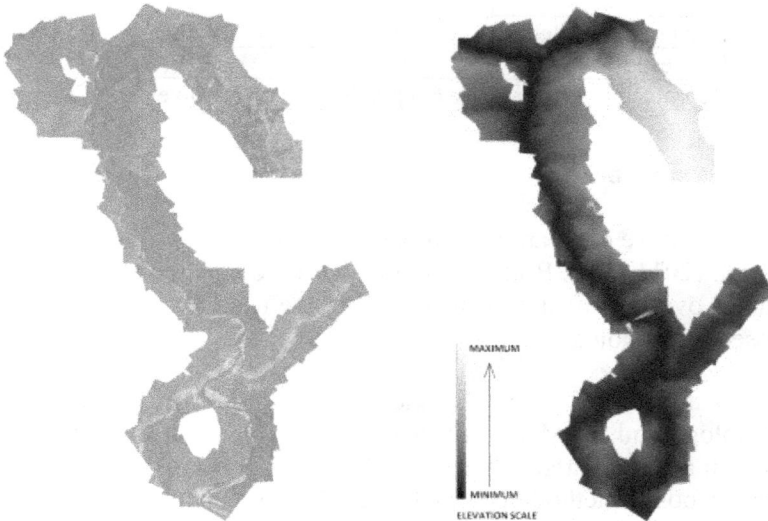

Fig. 6.11. 2D Surface Map build by Robot with 50 particles.

Fig. 6.12. 3D Map (DEM) build by Robot with 50 particles.

201

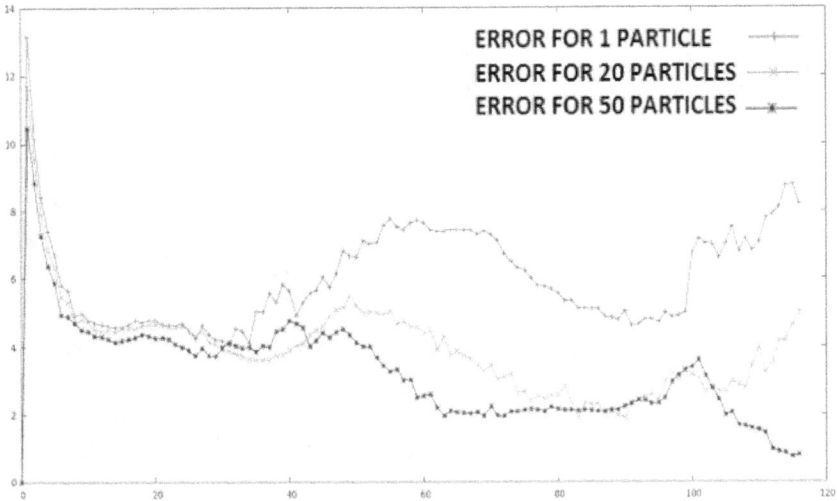

Fig. 6.13. Pose Error Analysis for proposed method.

Table 6.1. Simulation features for the proposed model.

...	20 particles	50 particles
Distance Travelled	182 km	182 km
Frames Processed	117	117
Reprojection Error	6.0 pixels	3.9 pixels

6.10.3. Example-2

In Example-2 we compare our proposed methodology with the work of Mei, *etal.* (2011 [24]). Both methodologies are compared with same set of frames over identical trajectories. We have used 50 particles for the proposed methodology.

Fig. 6.14 and Fig. 6.15 exhibit the 2-D maps built by the proposed methodology and by Mei, *et al.* respectively. 2-D Trajectories for the robot path are compared in Fig. 6.16. Fig. 6.17 presents the pose errors for the proposed methodology and RSLAM approach in green and red curves respectively.

Table 6.2 summarizes the experimental set up.

202

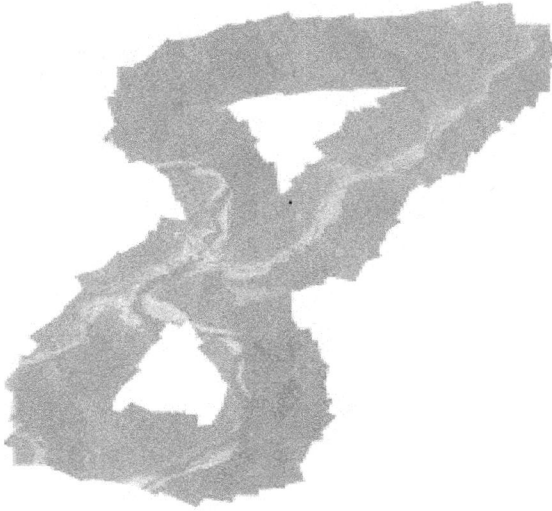

Fig. 6.14. 2D Surface Map of proposed methodology (50 paricles).

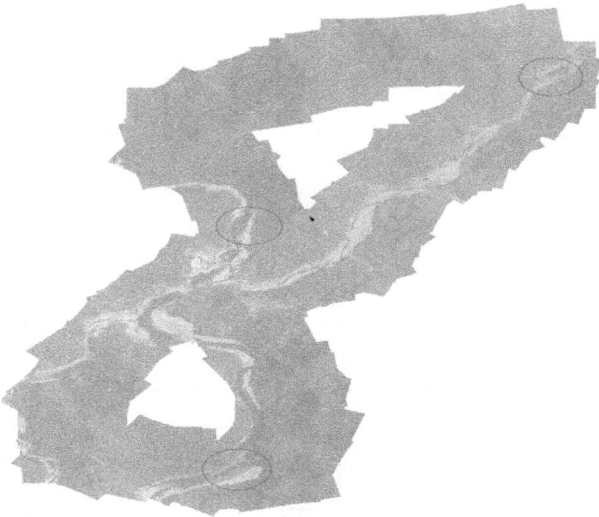

Fig. 6.15. 2D Surface Map of RSLAM.

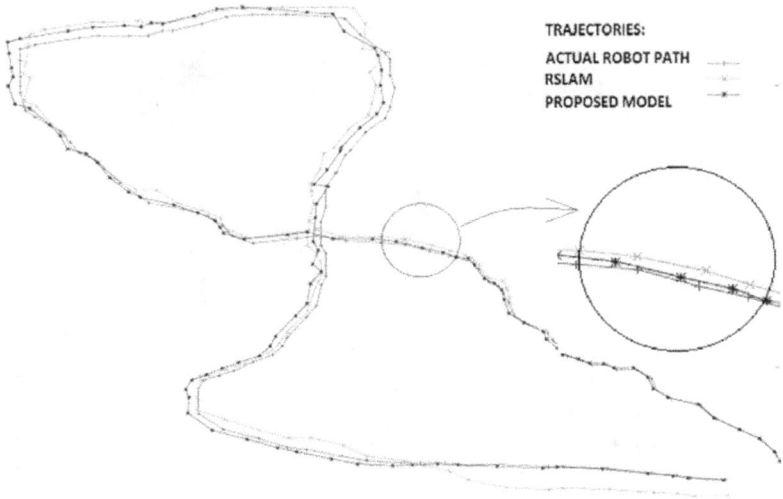

Fig. 6.16. Robot Trajectories in 2D Plane.

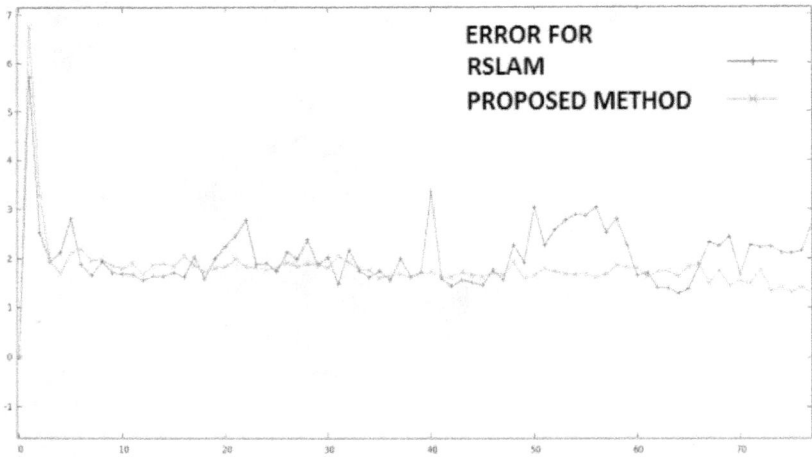

Fig. 6.17. Error Analysis for proposed method.

Table 6.2. Comparison with RSLAM.

...	**RSLAM**	**Proposed Model**
Distance Travelled	20 km	20 km
Frames Processed	118	118
Reprojection Error	4.2 pixels	1.9 pixels

6.10.4. Example-3

Example-3 shows a real sequence over a very long distance of approximately 4 km covered in 53 frames, using both our proposed methodology and the standard particle filter based FastSLAM 2.0. Fig. 6.18 and Fig. 6.19 exhibit 2D surface maps for proposed methodology and FastSLAM 2.0 respectively. Fig. 6.20 gives a comparison of the 2-D trajectories for the two methods. Fig. 6.21 gives the error analysis for FastSLAM (in red) and the proposed methodology (in green).

Table 6.3 summarizes the experimental set up.

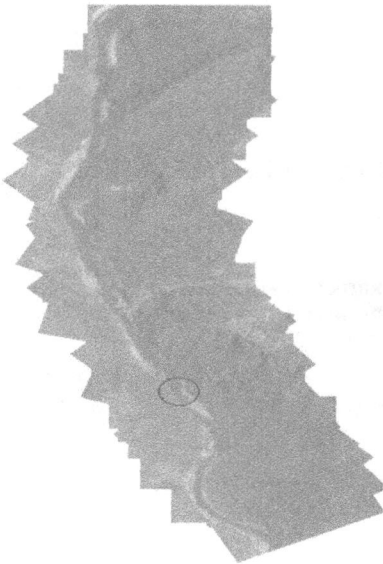

Fig. 6.18. 2D Surface Map build by proposed method.   **Fig. 6.19.** 2D Surface Map build by FastSLAM2.0.

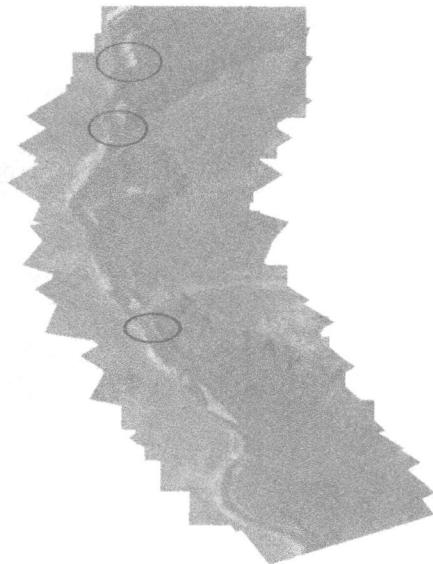

Table 6.3. Comparison with FastSLAM 2.0.

...	FastSLAM 2.0	Proposed Model
Distance Travelled	4 km	4 km
Frames Processed	53	53
Reprojection Error	4 pixels	0.5 pixels

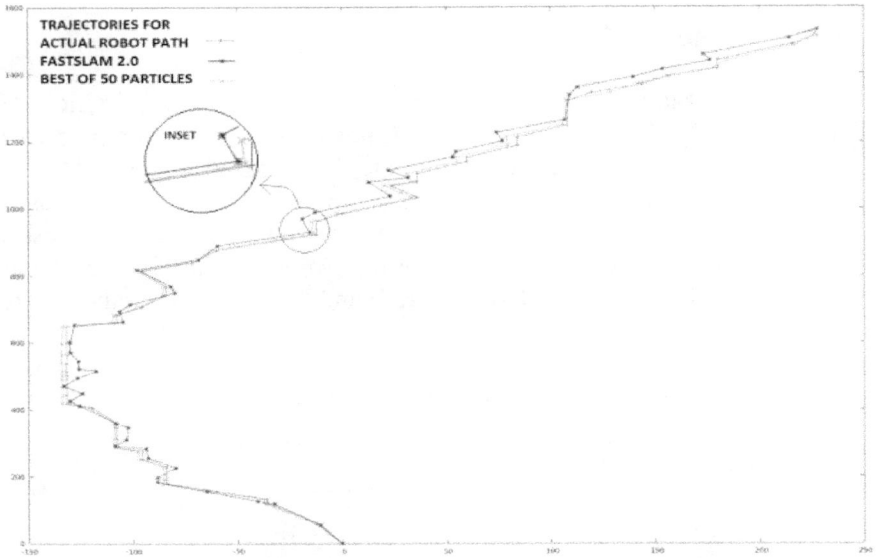

Fig. 6.20. 2D Trajectory for proposed model vs FastSLAM2.0.

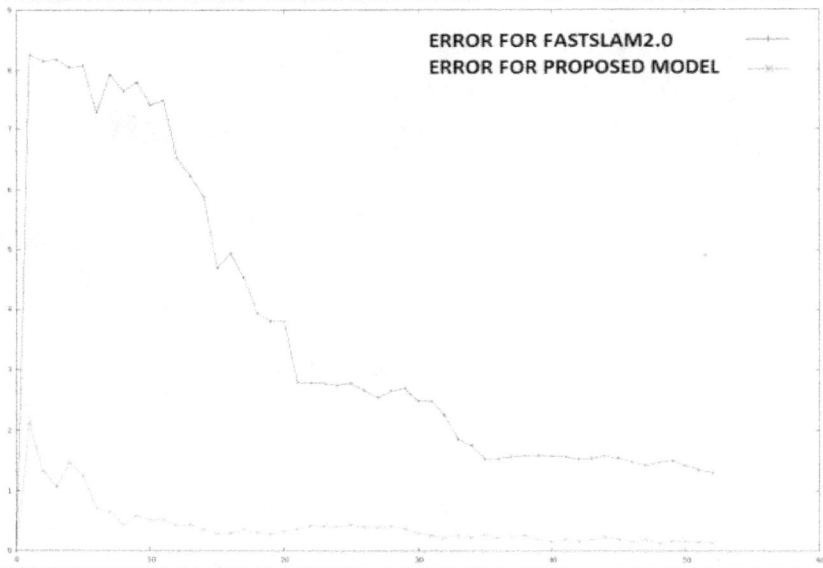

Fig. 6.21. Error Analysis for proposed method vs FastSLAM2.0.

6.10.5. Simulation with Proposed Model

See Table 6.1.

6.10.6. Comparison with Existing Models

See Table 6.2 and 6.3.

6.11. Future Work: Improvement of the Hybrid Approach Using Auxiliary Particle Filter

The usual particle filter re-samples the particles at the end of iteration $(t-1)$ before the tth observation is available, which is somewhat imprudent and wasteful. For more details about general APF algorithm, we refer to Douc, *et al.* (2009) [9] and Johansen, *et al.* (2008) [17]. The proposed hybrid approach can further be improved with the help the auxiliary particle filter algorithm (APF), originally proposed by Pitt, *et al.* (1999) [28]. It improves some deficiencies when dealing with tailed observation densities. As mentioned in Pitt, *et al.* (1999) [28], The particle filter works fine for standard problems where the model is a good approximation to the data along with conditional densities being reasonably flat. On the contrary the particle filter behaves badly when there is possibility of having critical outliers. Unlike the traditional particle filter the APF allows to affect the particle sample allocation by designing freely a set of first-stage importance weights for the selection procedure. Using that, it then assigns large weight to particles whose successors are likely to land up in zones of the state space having high posterior probability. There are many essential asymptotic properties of APF which has been stated in Douc, *et al.* (2009) [9]. the convergence properties of APF ensures fast convergence to the true distribution under standard regularity condition. The auxiliary variable step of APF is equivalent to multinomial re-sampling. So, no additional computational cost is being introduced as there is no need to re-sample before the pre-weighting. In other words, this modification ensures the constant time algorithm as before. Moreover, using APF, we can ensure the same convergence rate even for model relaxation upto non linear Gaussian. It is worth mentioning that the APF method is particularly suitable when the sensor observation is of high dimensional, which is very much relevant in our current situation.

6.12. Summary and Conclusion

We first note that Fig. 6.7 and 6.8 demonstrate the advantage of topographic labelling along with SURF features for landmark identification.

In Example-1 it is observed that the path traced by the proposed method with 50 particles is much closer to the actual robot path than the path traced by the robot taking 20 particles and 1 particle as indicated in the Fig. 6.9 (see inset). The reason is obvious; that the use of more number of samples expectedly gives a better estimate of the distribution.

However it must be pointed out that despite the use of particles, the pose estimation algorithm still depends heavily on bundle adjustment and consequently, on feature matching across frames. If sufficient number of features are not detected in some time step, the entire particle distribution will be less accurate. This explains why, even with 50 particles, the pose estimates in some cases are inaccurate.

Example-2 is a direct demonstration of the advantage of using bundle adjustment in the particle framework. Using particles enables our methodology to choose the best particle that gives a low average deviation error over the entire robot path, especially when very few features are encountered. RSLAM on the other hand shows a greater fluctuation in error along the robot path. It is interesting to note that initially RSLAM gives lesser error than the proposed methodology because at this stage the distribution of poses has not yet stabilized fully. But later on, the average error encountered in our method is much lesser and there are very few fluctuations, indicating that the distribution of particles now accurately traces the actual path. Of course, the proposed method takes longer time than RSLAM but the average error is found to be lower. In building 2D maps we observe a couple of improper alignment (marked in red circle in Fig. 6.15) for the approach due to RSLAM while the proposed methodology generated the map with proper alignment.

In Example-3, a comparison reveals that our proposed method gives much more precise estimates than FastSLAM 2.0, suggesting that the proposed method has an edge over FastSLAM 2.0. Also it is evident from the 2-D map built by FastSLAM 2.0 that there are three places (marked in red circle in Fig. 6.19) where alignment is not in proper position whereas in Fig. 6.18, built by proposed methodology a minor error has been found.

From these illustrations it appears that the work successfully merges the concepts of relative bundle adjustment with particle filters to achieve a constant-time, precise pose estimation algorithm. Also, the incorporation of topographic labelling and loop closure makes the system highly robust and efficient as demonstrated in the examples. As future work, inclusion of the APF algorithm will ensure much more efficient handling in the adverse scenario, particularly when there are chance of having severe outliers or the sensor observation becomes high dimensional.

Acknowledgements

We gratefully acknowledge European Aeronautics Defence and Space Company (EADS N.V) for the financial support of this work. (IIT Kharagpur approval number with EADS IIT/SRIC/MA/LNU/ 2009-10/138).

References

[1]. Agarwal S., Snavely N., Seitz S. M., Szeliski R., Bundle adjustment in the large, in *Proceedings of the 11th European Conference on Computer Vision: Part II* (ECCV'10), *Springer*, 2010, pp. 29-42.

[2]. Bailey T., Durrant-Whyte H., Simultaneous localisation and mapping (SLAM): Part II - state of the art, *Robotics and Automation Magazine*, Vol. 13, No. 3, 2006, pp. 108-117.

[3]. Bay H., Ess A., Tuytelaars T., Van Gool L., Speeded-up robust features (surf). Computer vision and image understanding, Vol. 110, No. 3, 2008, pp. 346-359.

[4]. Beg I., Ashraf S., Similarity measures for fuzzy sets, *Applications & Computational Mathematics,* Vol. 8, No. 2, 2009, pp. 192-202.

[5]. Bormann D., Elseberg J., Lingemann K., Nuchter A., Hertzberg J., Globally consistent 3D mapping with scan matching, *Robotics and Autonomous Systems*, Vol. 56, No. 2, 2008, pp. 130-142.

[6]. Cramer H., Mathematical Methods of Statistics, *Princeton University Press*, 1946.

[7]. Cummins M., Newman P., FAB-MAP: Probabilistic localization and mapping in the space of appearance, *International Journal of Robotics Research*, Vol. 27, No. 6, 2008, pp. 647-665.

[8]. Dissanayake M. W. M. G., Newman P., Steven C., Durrant-Whyte H. F., Csorba M., A solution to the simultaneous localization and map building (SLAM) problem, *IEEE Transactions on Robotics and Automation*, Vol. 17, No. 3, 2001, pp. 229-241.

[9]. Doucet A., Johansen A., Crisan D., Rozovesky B., A tutorial on particle filtering and smoothing: Fifteen years later, *Oxford University Press*, 2009.

[10]. Dubois D. J., Fuzzy sets and systems: theory and applications, *Academic Press*, Vol. 144, 1980.

[11]. Durrant-Whyte H., Bailey T., Simultaneous localisation and mapping (SLAM): Part I - the essential algorithms, *Robotics and Automation Magazine*, 2006, pp. 1-9.

[12]. Eade E., Drummond T., Unified loop closing and recovery for real time monocular SLAM, *BMVC*, 2008, pp. 1-10.

[13]. Fischler M., Bolles R., Random sample consensus: A paradigm for model fitting with applications to image analysis and automated cartography, *ACM*, Vol. 24, 1981, pp. 381-395.

[14]. Haber J., Zeilfelder F., Davydov O., Seidel H. P., Smooth approximation and rendering of large scattered data sets, in *Proceedings of the IEEE Conference on Computer Society (VISUALIZATION' 01)*, 2001, pp. 341-348.

[15]. Haralick R. M., Watson L. T., Laffey T. J., The topographic primal sketch, *The International Journal of Robotics Research*, Vol. 2, No. 1, 1983, pp. 50-72.

[16]. Jeong Y., Nister D., Steedly D., Szeliski R., Kweon I. S., Pushing the envelope of modern methods for bundle adjustment, *IEEE Trans. Pattern Anal. Mach. Intell.*, Vol. 34, No. 8, 2012, pp. 1605-1617.

[17]. Johansen A M., and Doucet A., A note on auxiliary particle filters, *Statistics and Probability Letters.*, Vol. 78, No. 8, 2008, pp. 1498-1504.

[18]. Kartikeyan B., Sarkar A., The assignment of topographic labels by a statistical model-based approach, Signal Processing, Vol. 42, No. 1, 1995, pp. 71-86.

[19]. Klein G., Murray D., Parallel tracking and mapping for small AR workspaces, *Mixed and Augmented Reality*, 2007, pp. 225-234.

[20]. Konolige K., Agrawal M., Real-time localization in outdoor environments using stereo vision and inexpensive GPS, in *Proceedings of the IEEE18th International Conference on Pattern Recognition*, Vol. 3, 2006, pp. 1063-1068.

[21]. Lowe D. G., Object recognition from local scale-invariant features, in *Proceedings of the Seventh IEEE International Conference on Computer Vision*, Vol. 2, 1999, pp. 1150-1157.

[22]. Lu F., Milios E., Globally consistent range scan alignment for environment mapping, *Autonomous Robots*, Vol. 4, No. 4, 1997, pp. 333-349.

[23]. Lucas B., Kanade T., An iterative image registration technique with an application to stereo vision, in *Proceedings of the 7th International Joint Conference on Artificial Intelligence (IJCAI '81)*, 1981, pp. 674-679.

[24]. Mei C., Sibley G., Cummins M., Newman P., Reid I., RSLAM: A system for large-scale mapping in constant-time using stereo, *International Journal of Computer Vision*, Special Issue of BMVC, 2010, pp. 1-17.

[25]. Mouragnon E., Lhuillier M., Dhome M., Dekeyser F., Sayd P., Generic and real-time structure from motion using local bundle adjustment, *Image Vision Computing*, Vol. 27, No. 8, 2009, pp. 1178-1193.

[26]. Nister D., Stewenius H., Scalable recognition with a vocabulary tree, in *Proceedings of the Conference on Computer Vision and Pattern Recognition*, Vol. 2, 2006, pp. 2161-2168.

[27]. Paz L. M., Pinies P., Tardos J. D., Neira J., Large-scale 6-dof SLAM with stereo-in-hand, *IEEE Transactions on Robotics*, Vol. 24, No. 5, 2008, pp. 946-957.

[28]. Pitt M. K. and Shepard N., Filtering via simulation: Auxillary particle filters, Particle filters for tracking applications. Artech House, 1999.

[29]. Radhika V. N., Kartikeyan B., Krishna B. G., Chowdhury S., Srivastava P. K., Robust stereo image matching for spaceborne imagery, *IEEE Transactions on Geoscience and Remote Sensing*, Vol. 45, No. 9, 2007, pp. 2993-3000.

[30]. Russell B., Freeman W. T., Efros A., Sivic J., Zisserman, A., Using multiple segmentations to discover objects and their extent in image collections, *CVPR'06*, Vol. 2, 2006, pp. 1605-1614.

[31]. Sarkar A., Biswas M. K., Kartikeyan B., Kumar V., Majumder K., Pal D., A MRF model-based segmentation approach to classification for multispectral imagery, *IEEE Transactions on Geoscience and Remote Sensing*, Vol. 40, No. 5, 2002, pp. 1102-1113.

[32]. Sarkar A., Vulimiri A., Paul S., Iqbal J., Banerjee A., Chatterjee R., Ray S. S., Unsupervised and supervised classification of hyperspectral imaging data using projection pursuit and Markov randomed segmentation, *International Journal of Remote Sensing*, Vol. 33, No. 18, 2012, pp. 5799-5818.

[33]. Sivic J., Zisserman A., Video google: A text retrieval approach to object matching in videos, *Computer Vision*, Vol. 2, 2003, pp. 1470-1477.

[34]. Srivastava P., Krishna B., Geometrical processing of cartosat-1 satellite data, *Bull. Nat. Natural Resour. Manag. Syst.*, No. (B)-30, 2005, pp. 7-16.

[35]. Thrun S., Burgard W., Fox W., Probabilistic Robotics, *MIT Press*, 2005.

[36]. Triggs B., McLauchlan P. B., Hartley R. I., Fitzgibbon A., Bundle adjustment - a modern synthesis, in *Proc. ICCV Workshop on Vision Algorithms, Springer-VerlagLNCS*, 1999, pp. 298-372.

[37]. Watson L. T., Laffey T. J., Haralick R. M., Topographic classification of digital image intensity surfaces using generalized splines and the discrete cosine transformation, *Computer Vision, Graphics, and Image Processing*, Vol. 29, No. 2, 1985, pp. 143-167.

[38]. Zhang G., Dong Z., Jia J., Wong T.-T., Bao H., Efficient non-consecutive feature tracking for structure-from-motion, Computer Vision – ECCV 2010, *Springer,* 2010, pp. 422-435.

[39]. Zhu J. N. Z., Yuan Z., Du S., Adaotuve skan akgirutgn wutg sanokubg vased ib state ybcertaubttm, *Electronics Letters*, Vol. 47, No. 4, 2011, pp. 284-286.

[40]. Zwick R., Carlstein E., Budescu D. V., Measures of similarity among fuzzy concepts: A comparative analysis, *International Journal of Approximate Reasoning*, Vol. 1, No. 2, 1987, pp. 221-242.

Chapter 7

An Overview of Systems, Control and Optimisation (SCO) in Recent European R&D Programmes and Projects (2013-2017) under the Emergence of New Concepts and Broad Industrial Initiatives

Alkis Konstantellos

7.1. Introduction

7.1.1. The Broad R&D landscape for Systems, Control and Optimisation (SCO)

The envelope and intensity of R&D efforts has significantly increased, worldwide[1] the last decade, through substantial renewal of the topics proposed for funding and by several new governmental programmes and parallel independent work at universities, research centres, enthusiasts' groups[2], in industrial and service sectors and through their associations

Alkis Konstantellos
European Commission (retired), Brussels, Belgium: Complex Systems and Advanced Computing Unit. <u>Disclaimer</u>: The views expressed in this paper and the selection of projects and their grouping are those of the author and do not necessarily reflect the position of the European Commission on the topics discussed.

[1] A Supplement to the R&D Magazine, Winter 2017: R&D Forecast 2017, page 3 and 14
http://digital.rdmag.com/researchanddevelopment/2017_global_r_d_funding_forecast?pg=1#pg1
[2] E.g. Open source Robotics Foundation (OSROF):
https://www.osrfoundation.org/wordpress2/wp-content/uploads/2015/11/rft-boyer.pdf

[2-9]. In Europe, there are several examples of new large scale programmes and public-private initiatives funding classic scientific fields[1], or addressing contemporary Science and Technology (S&T) challenges in proactive[2] and open[3] ways. Concerning Systems, Control and Optimisation (SCO) topics, although not a priority per se, they are encountered in several places in the published European Union (EU) Work Programmes and Roadmaps, especially in Cyber-Physical Systems, Industry 4.0, Aeronautics, Robotics and Energy [10-16] and eventually in the funded projects. The EU-wide efforts are occasionally combined with, or complemented by National [4]research in individual countries.

There have also been European projects supporting international cooperation in challenging systems topics of common interest[5]. Certain R&D topics in systems and control seem to re-appear after couple years in more advanced form today (e.g. AI, DSPs, Data compression/Fusion, Decision Making). Concerning the new Artificial Intelligence (AI) see footnote 2 of the last paragraph of the concluding remarks of this chapter. Fig. 7.1 presents a general overview of the key SCO domains, underpinning methods and contributing areas (e.g. computing, communications). Methods are framed in dotted lines and systems in solid lines. Applications cut across, hinting at possible future extensions. Mixing systems and methods is certainly not proper, but is done here only for schematically positioning the topics discussed. <u>Comment</u>: Obviously, a typical R&D project does not address all aspects mentioned above, although some projects deal with combinations of topics in varying intensities. Large-scale projects, tend to tackle more than one challenge and in such cases, they usually provide an integration environment/platform for the whole system.

---

[1] i.e. mathematics, engineering, social and life sciences
[2] (i.e. indicating a-priori attractive themes), for example in FET Proactive in H2020 (http://ec.europa.eu/programmes/horizon2020/en/h2020-section/fet-proactive)
[3] (i.e. left up to the proposers to suggest), for example in FET Open in H2020 http://ec.europa.eu/programmes/horizon2020/en/h2020-section/fet-open), MSCA, ERC
[4] E.g. The role of EU funding in UK research and innovation, by Technopolis Group, 2017 https://royalsociety.org/~/media/policy/Publications/2017/2017-05-technopolis-role-of-EU-funding-report.PDF
[5] (EU with e.g. Brazil, China, Egypt, Japan, Korea, Russia, USA)

Chapter 7. An Overview of Systems, Control & Optimisation (SCO) in Recent
European R&D Programmes and Projects (2013-2017) under the Emergence of New
Concepts and Broad Industrial Initiatives

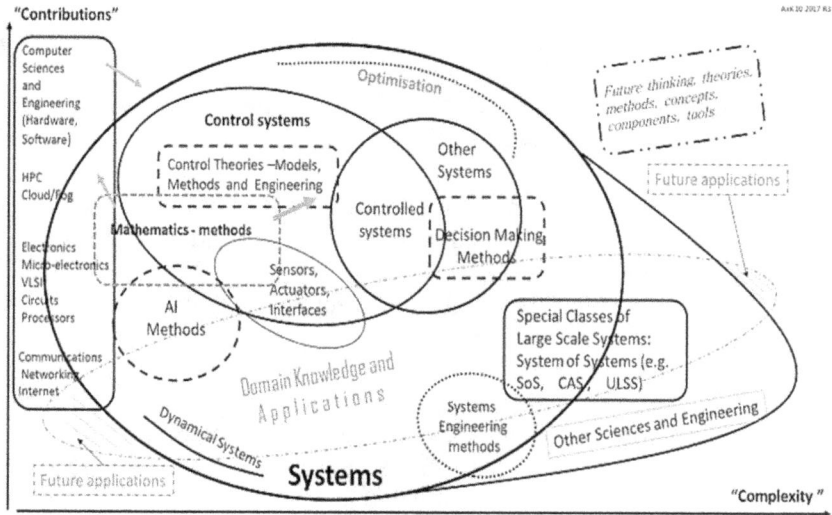

Fig. 7.1. General overview of the key SCO domains, underpinning methods and contributing areas. R&D&I = Research, Development and Innovation (H 2020), AI = Artificial Intelligence, SoS =System of Systems, CAS = Complex Adaptive Systems, ULSS = Ultra Large-Scale Systems. **Note**: Applications include the broader Robotics fields.

7.1.2. Structure of the Paper and Profiling of Projects

The chapter is structured as follows: (1) first we present the systems, control and optimisation landscape (SCO), (2) give some explanations of the concepts and terminology, (3) introduce the emerging topics (e.g. CPS, IoT, Ind.4.0), (4) discuss the R&D&I[1] programmes mainly from the SCO point of view (both theoretical and applied), (5) emphasise the transformations of the programmes designed to meet current policies and socio-techno-economic objectives aiming at complete innovation ecosystems and (6) we conclude with additional comments about S&T approaches which have significant exploitation potential. Since many projects deal with more than one topic, we portray only their dominant systems-oriented contributions. The Annex to this chapter (7A) provides additional information on selective project groups beyond section 7.4.2 .

---

[1] R&D&I: Research, Development and Innovation

Note: SCO methods applied to fundamental Electrical, Electronic and Microelectronic *Circuits[1] and Processor Architectures* are not addressed specifically in this paper.

7.1.3. Topics in Projects vs. Topics Supported by Key Scientific Societies

The topics in the European R&D Framework Programmes address a fraction of the themes included in the portfolio and conferences of prestigious organisations such as IETF[2], VDI/VDE, ITU, IFIP[3], IFR[4] ASME, SIAM, ACM, ESA[5] and their Technical Committees, mainly because funding agencies, except probably for basic research, reflect specifically agreed overarching R&D&I policies, identified gaps in previous programmes, stakeholders and socio-economic needs and the available R&D funds. Nevertheless, the two trajectories (topics in Programmes and topics promoted by S&T societies), despite their independence, are not totally unrelated if one considers the detailed outcomes of the funded projects. They meet around points of admittedly attractive challenges, but which may be formulated in different wordings.

7.1.4. Mapping Projects Content According to Inherent Single, or Multiple Innovations

The project examples are grouped in Section 7.4.2 along three main axes (Fig. 7.2) regarding three levels of sophistication: a) System and control modelling & design, b) Application domain-knowledge and c) The Implementation of the systems (computations, hardware, software, interfaces, I/Os and networking). There is in practice a 4th dimension addressing additional aspects, such as organization and life cycle. This means that we may have a challenging new application requiring just a simple control (at least initially) while the implementation platform may need to be the state-of-the-art. Sensors and

---

[1] Franco Maloberti, Anthony C. Davis: A Short History of Circuits and Systems, IEEE 2016, *River Publishers*, ISBN 978-87-93379-69-5.

[2] IETF: Internet Engineering Task Force, https://www.ietf.org/

[3] IFIP: International Federation for Information Processing, http://www.ifip.org/

[4] IFR: International Federation of Robotics, https://ifr.org/

[5] European Space Agency (ESA)

Chapter 7. An Overview of Systems, Control & Optimisation (SCO) in Recent
European R&D Programmes and Projects (2013-2017) under the Emergence of New
Concepts and Broad Industrial Initiatives

Actuators are part of the systems, or treated separately in R&D. Based
on these considerations, we may encounter any combination between
"simple" and "advanced" in the four axes. Examples: a) State-of-the-art
robust MPC for a new crystallization process, such as advanced ice-
cream production process, [see projects FRISBEE, MINICRYSTAL and
a more general one, CAFÉ for the food industry], b) A simple condition-
monitoring system of aircraft engines as in projects T-CREST, ARGO,
Par-MERASA, AEGART and HYPSTAIR (hybrid propulsion).

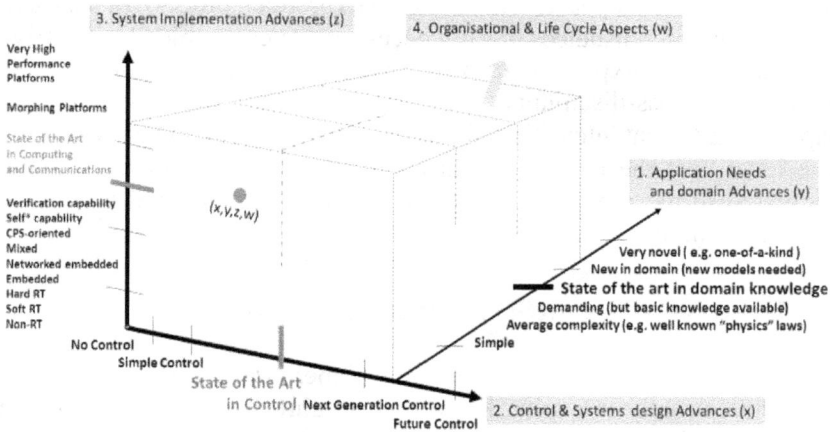

Activities and progress in x,y,z,w usually advance independently from each other

Fig. 7.2. A 4-dimensional R&D space: y-Applications, x-Systems & Control,
z-Implementation environment and w-Organisation. Each "axis", indicates
qualitatively the scale of advances and the State-of-the-Art is indicated as
reference. Therefore projects map to different levels (x, y, z, w).

Comments: In recent years funded and other R&D projects benefit from
the contributions of different disciplines and diverse expertise from
theories, methods, tool-chains, platforms and reuse, as well as from
cross-domain fertilisation. There are two points we wish to emphasise:
The first point is the dilemma between two options: (a) to do a generic
development first and then domain customisation, or (b) to start from
scratch within each domain with minimal cross-domain reuse. In
European projects and the recent work in the European Technology
Platforms and Joint Undertakings e.g. SHIFT2RAIL, CLEAN SKY,

EXCEL, SPIRE, both approaches (a, b) have been tried. They also benefit from common partners in cross-sectoral programmes, such as EXCEL/ARTEMIS and in domain-specific ones such as CLEAN SKY for aerospace, SHFT2RAIL for rail and SPIRE[1] for processing industries. Nevertheless, there are several differences in systems requirements, among apparently similar applications in different sectors. An example is the high number of different standards in systems and control used by the key sectors for similar purposes (e.g. safety). The second point, based on Fig. 7.2 and the earlier discussion, is the observation that the three levels (i.e. SCO methods, implementation and application knowledge) are most frequently progressing independently, at least in the initial phases, to reach a goal (product, process, or service). Under other circumstances it is not necessary, or economically attractive at all to apply, or expect to reach all three close to current/optimum state of the art. Cross-disciplinary cooperation is very useful, but it may happen at different intensities and times in the life cycle of a new system. The ideal "get the best of all, in all dimensions" is a matter usually addressed in the next generations of products, systems and processes, under a "continuous improvement" philosophy.

7.1.5. The Sample of R&D Projects Considered

Preliminary Remark: Projects in this paper are presented by their acronyms, usually in capital letters. A straightforward way to obtain a project description is by their acronym, or ID number from the Cordis portal[2] and its search engine. Note, that the same acronym may correspond to more than one projects, in totally different topics, but the project ID is unique.

Out of more than 1000 systems-related projects identified in the FP7, H2020 and national programmes, roughly about 10 % represent highly relevant fundamental research by the corresponding R&D project teams, or single investigators. For projects which have started at the time of writing this paper, obviously only intended activities are mentioned. Remark 1: In this chapter we concentrate on the attractiveness of the topics in the projects and the S&T domains involved, and not on the evaluation of the results of the projects, which is done through dedicated follow-up mechanisms, periodic programme impact assessments and

---

[1] SPIRE: Sustainable Process Industry through Resource and Energy Efficiency (European Public-Private Partnership) https://www.spire2030.eu/
[2] http://cordis.europa.eu/projects/home_en.html

estimations of the Return on R&D investment (ROI) by the European Commission, European Parliament and individual countries, [17-19]. Remark 2: The set of projects discussed is indicative and by no means a complete picture of all funded projects. Comment: This paper complements timely and content-wise a previous short overview paper [1].

7.2. Systems, Control, Control Systems, Control in Systems and "No-control"

7.2.1. Terminology, Evolution, Explanations, Different Views

Concerning SCO, the concept of Cyber-Physical Systems serves as a useful basis (although it does not clearly cover all possible cases and boundaries). We call "Physical" anything other than processor-based parts, that is, a chemical, or biological process, mechanical, electrical, hydraulic, pneumatic, thermal systems to mention a few. "Cyber" refers to the intelligent part (interacting with the physical) and usually having a control, monitoring, signal processing, or decision-making function. "Controlled systems are CPS". Fig. 7.3 outlines typical combinations of systems and their control. A comprehensive and rigorous presentation of CPS is given for example in the following textbooks: E. A. Lee & S. A. Seshia [20], R. Alur [21] and several edited books e.g. [22]. Some conventions and ambiguities about the boundaries between Cyber and Physical parts of a CPS model are discussed in [23], but in practice there is no unexpected design gap, assuming there is enough domain knowledge in both parts (Cy - Phy) and the whole CPS. In some models and applications, CPS is in fact a "Cyber- for -Physical" System. See also Fig.7.6 in Section 7.3.5 of this chapter.

We now consider the following nested hierarchy, from the degree of constraints point of view:

System $\supseteq$ Dynamical System $\supseteq$ Open loop Control $\supset$ Closed loop control $\supseteq$ embedded control (unsupervised) $\supset$ embedded control (supervised, HitL[1])

---

[1] HitL: Human in the Loop (frequently with no-automatic supervision)

To emphasise the interest in general systems research, and in understanding rigorous definitions, categorisations and analysis of dynamical[1] systems & control the reader may refer for example to [24-28]. All above mentioned system classes are well represented in funded projects. Examples of funded projects are given in Section 7.4.2, some of their papers and general reference publications are included in the Annex. Particularly, the following fields have been very popular in European programmes: 1) the area of Non-Linear systems [29-30] and phenomena like Bifurcations [31-32] are investigated by several groups from {physics, complexity} and {electrical, electronics, control and systems engineering}; 2) Partial differential equations (PDEs) [33] systems modelled by PDEs and their control, researched by {mathematics, computing, systems and control} communities and 3) advanced logic, automata, discrete systems and their control-optimisation.

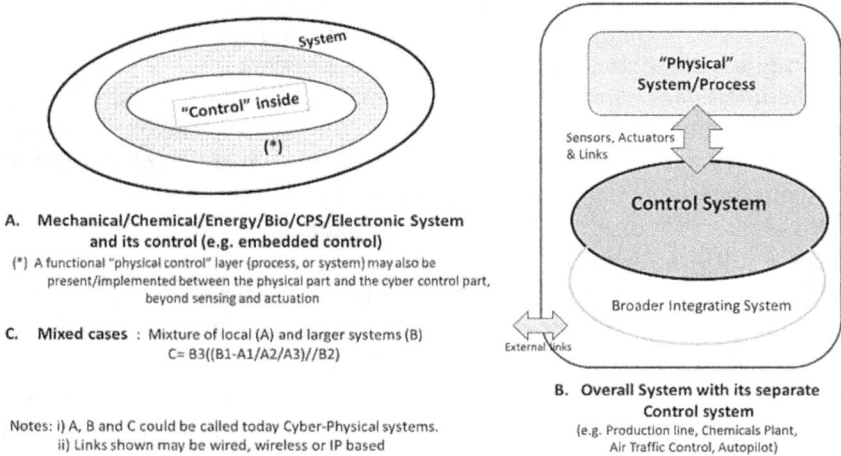

A. Mechanical/Chemical/Energy/Bio/CPS/Electronic System and its control (e.g. embedded control)

(*) A functional "physical control" layer (process, or system) may also be present/implemented between the physical part and the cyber control part, beyond sensing and actuation

C. Mixed cases : Mixture of local (A) and larger systems (B)
C= B3((B1-A1/A2/A3)//B2)

Notes: i) A, B and C could be called today Cyber-Physical systems.
ii) Links shown may be wired, wireless or IP based

B. Overall System with its separate Control system
(e.g. Production line, Chemicals Plant, Air Traffic Control, Autopilot)

Fig. 7.3. From Embedded Systems & Control to Cyber-Physical Systems (CPS) and mixed cases.

On the applied side, {robotics, cooperative & adaptive control and model-based control} are the most appealing topics. In several control systems there is a "Human in the loop" as e.g. operator, driver, pilot, captain, (or just a user) with certain overriding authority/responsibility.

---

[1] http://www.scholarpedia.org/article/Encyclopedia_of_dynamical_systems

Embedded control systems may offer, by construction, a high level of autonomy, but they also exhibit risks, for example during unattended operations. Note that, SAE J3016[1] defines six levels of automated driving[2].

System: in general, is a set of parts (also components/objects/ modules/sub-systems), or even other systems, (physical and/or cyber), the rules describing their internal states and the interactions among them, as well as means of implementing the required communications. If the states change with time, we are talking about dynamical systems. Many interacting systems, in an environment, may be: a) A set of distributed (interconnected) systems hierarchically coordinated for a concrete goal, or b) System(s) of Systems (SoS), if they comply with special conditions (e.g. M.W. Maier criteria)[3]. In the latter case, the involved systems are called constituent systems. There are several general and project-specific reports [34-36] regarding SoS. An example of optimisation approach for SoS is in [37].

Although there is no commonly accepted clear definition of Complex systems, they are studied by different communities, such as {dynamical systems, control, general systems, complexity, physics, "old and new cybernetics", non-linear systems and philosophy}. Other proposed classes of large scale systems are {Ultra Large-Scale Systems (ULSS)[4] and Complex Adaptive Systems (CAS)[5] [38-40]}. Related terms are Systems theory, Systems Engineering[6] and Systems Thinking[7]. Examples: EC Projects CRITICS (Critical Transitions in Complex Systems) and ILDS (Integrability and Linearization of Dynamical Systems). Although these topics touch upon systematic organisational requirements dealt within Systems Engineering, especially CAS approaches are used in non-engineering applications, such as in General

---

[1] SAE: Society of Automotive Engineers, www.sae.org/autodrive

[2] https://www.sae.org/misc/pdfs/automated_driving.pdf

[3] Maier M. W., Architecting Principles for Systems-of-Systems, *Systems Engineering*, Vol. 1, Issue 4, 1998.

[4] https://www.sei.cmu.edu/uls/

[5] https://en.wikipedia.org/wiki/Complex_adaptive_system

[6] http://sebokwiki.org/wiki/Guide_to_the_Systems_Engineering_Body_of_Knowledge_(SEBoK)

[7] Ross D. Arnold, Jon P. Wade: A Definition of Systems Thinking: A Systems Approach, 2016 https://www.researchgate.net/publication/273894661

systems, Health Care [40a], [40b] and, like SoS, in the Defence sectors and services [40c], [40d]. Moreover, combinations of the topics CPS, SoS and CAS have recently emerged in R&D activities and are discussed in Section 7.3.5.

Control: in the sense of control theory, method, algorithm for modelling, analysis and design. *Note*: At a mathematical level, projects address e.g. differential equations (ODE, or PDE), non- linear phenomena, automata, co-algebras and advanced logic. The jargon term "control, or controls" means, e.g. block controllers, converters, interfaces and the adjustment of control parameters. Moreover, several control topics have been transformed over the last decades and the respective scientific communities re-shaped, or mixed[1]. On the product side advanced control methods have penetrated dramatically the consumer electronics sectors: 1) "A camera today would run Kalman estimators [41], for auto focusing, or use big data for both imaging and focusing", or 2) Adaptive control achieves proper quantization for audio and mpeg video compression (Mpeg4-Mpeg21) and quality of service algorithms [42].

Control System: In short, Control system = (control laws, sensors, devices, networks, supplies). In the sense of an integrated device (small, embedded), or bigger (cabinet/rack-mounted system, including the intelligent platform e.g. hardware, RTOS, drivers, s/w layers, algorithms, communication networks if distributed, and associated/connected with inputs (from sensors) and outputs (to actuators), called I/Os for the control system. In embedded control, (some) sensors and/or actuators may be attached directly to the control system "block", or the other way around, the control is attached to the sensing/actuation "block". In large process control systems (using a control room), the sensors and actuators are remotely installed. MIMO[2] models with multiple inputs & multiple outputs (low order multivariable control was traditionally implemented by several local controllers). In practice, a control system is clearly identifiable even within a larger system environment. It is probably designed from scratch (e.g. on FPGAs/DSPs/ASICs), or purchased and

---

[1] e.g. Control, Robotics with Autonomic Computing, Autonomous systems, Data Compression, *Thirteenth International Conference on Autonomic and Autonomous Systems, ICAS 2017,* Spain, https://www.iaria.org/conferences2017/ICAS17.html

[2] MIMO in communications systems, although multidimensional [nxm] (antennas, receivers), is an open system, compared to closed loop MIMO feedback control.

Chapter 7. An Overview of Systems, Control & Optimisation (SCO) in Recent
European R&D Programmes and Projects (2013-2017) under the Emergence of New
Concepts and Broad Industrial Initiatives

programmed/configured (e.g. in commercial PLC, DCS, SCADA, PCS and embedded controllers[1]. Sub-category of control systems are the embedded control systems, but not all embedded systems execute control functions. Control systems and their instrumentation constitute an industrial sub-sector (see NACE, NAICS), which is well defined for example in process control and avionics, manufactured around the world.

Control in systems: are usually smaller parts of very broad/large systems, which are not always clearly distinguishable from other electronic, or intelligent systems, while offering control functionalities. Many small size embedded controllers[2] belong to this category. Examples: 1) Part of control inside the physical ABS block in a car (identifiable); 2) "Intelligent" control loop in Phase-Difference Detection system for auto-focusing in a modern digital camera (invisible) [43] and 3) Turbine pressure controllers (identifiable).

No-control in Systems: while imperfections in nature and human-made systems require a control system with certain characteristics, other systems include operational stages which need no control. This is the case for example in cognitive systems aiming at understanding and interpreting phenomena and situations during part of their operations, or high-level open-loop coordination applications. Examples: Condition monitoring in electric machines, alarm system, FDI/FDD and Decision support systems (DSS).

Optimisation[3]: Based on mathematical methods and numerical codes to obtain min max solutions for parameters, or dynamical (state) variables using rigorous, or heuristic approaches. Deterministic and stochastic[4] methods are popular in and beyond control engineering. Alternatively, control objectives may be included in an optimisation problem formulation, either as condition, or constraint. In this context,

---

[1] https://m.eet.com/media/1246048/2017-embedded-market-study.pdf
[2] IDC study for the European Commission, 2012,
http://cordis.europa.eu/fp7/ict/embedded-systems-engineering/documents/idc-study-presentation.pdf
[3] For example Floudas, Christodoulos A., Pardalos, Panos M. (Eds.):
Encyclopedia of Optimization, Springer, 2009.
[4] European Conference on Stochastic Optimization, ECSO 2017,
http://ecso2017.inf.uniroma3.it/

"Optimisation" issues are strongly linked to control, at low, MES and ERP levels. Projects funded in this area cover applications[1] ranging from simple parameter optimisation for processes[2], up to Mixed Integer Non-Linear Programming applications[3]. Concerning tools see e.g. recent overview paper by Andrea Callia D'Iddio and Michael Huth[4].

7.2.2. Another Consideration: Focus on Control vs. Focus on Applications (as addressed in Projects)

Beyond the above general cases (1-6), there is a different consideration, depending on the level of control compared to the scope of the application itself. We distinguish projects addressing/developing: a) theoretical methods, with/without reference to applications, b) advancing control concepts, methods, algorithms and tools and applying these, to existing, or totally new applications, c) control in very attractive, difficult to model, or poorly understood "physical" phenomena, processes, or systems, where the emphasis is not on control, but on the prime goal of "how to control the target element/ system" and d) simplistic use of the term, for example "we will use an FPGA based controllers for laser beam positioning". Here the R&D element is not evident.

A typical sketch of a controlled system is shown in Fig. 7.4, emphasising that there are several essential parts required to execute closed loop control. Example: The project STEELANOL[5] "demonstrates the production of bioethanol from emissions of the steelmaking process which has the potential to significantly reduce greenhouse gas emissions

---

[1] Sumpf P., Klemm M., Throndsen W., Büscher C., Robison R., Schippl J., Foulds C., Buchmann K., Nikolaev A. and Kern-Gillard, T., 2017. Energy system optimisation and smart technologies – a social sciences and humanities annotated bibliography. Cambridge: SHAPE ENERGY.

[2] Jan Grözinger, Dr. Andreas Hermelink, Bernhard von Manteuffel, Markus Offermann, Sven Schimschar, Optimising the energy use of technical building systems, ECOFYS,
https://www.ecofys.com/files/files/ecofys-2017-optimising-the-energy-use-of-tbs-final-report.pdf

[3] http://energy.tamu.edu/events/goc-2017/

[4] Andrea Callia D'Iddio and Michael Huth: ManyOpt: An Extensible Tool for Mixed, Non-Linear Optimization Through SMT Solving,
https://arxiv.org/pdf/1702.01332.pdf

[5] EC, H2020 project for converting Carbon (at Steelmaking) to Ethanol,
http://www.steelanol.eu/en

Chapter 7. An Overview of Systems, Control & Optimisation (SCO) in Recent
European R&D Programmes and Projects (2013-2017) under the Emergence of New
Concepts and Broad Industrial Initiatives

compared to oil-derived fuels". Such a project addresses design challenges beyond control.

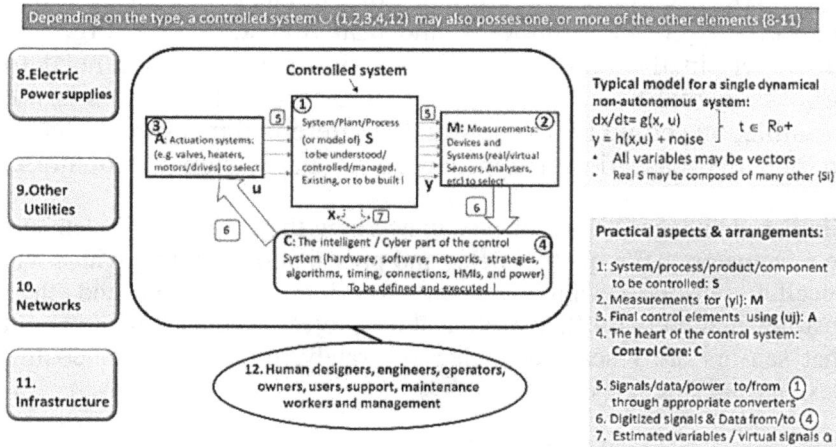

Fig. 7.4. Typical controlled System with Measurements, Actuators and the Control core (S-M-A-C).

7.2.3. The Important Role of Sensors and Actuators

In several control projects, sensors and actuators are considered as obvious elements. However, to design complete control systems (i.e. closing the loops), sensors and actuators may become components and sub-systems of paramount importance. Certainly, in many cases we have no problem to defining them, since there are suitable sensors and actuators available on the market (e.g. temperature, humidity, thickness, deflection, acceleration, MEMs-based and motors, pumps, heaters respectively). But in other cases, special, or complex measuring systems (optical, laser devices, composition analysers, mass spectrometers, mechanical/thermal detectors, chromatographs etc.), are required. Furthermore, when there is no known technology to measure a variable, then dedicated R&D is required[1]. A different challenge with sensors

---

[1] US Department of Energy: Advanced sensors and Instrumentation, September, 2017

emerged the last 10-15 years: RFIDs, Distributed and then Wireless Sensor Networks (WSNs), and WS&ANs[1] to reach IoT[2]. As far as theoretical analysis and modelling is concerned, it may be sufficient to consider sensors and actuators represented by signals y(t), y(k) (continuous and discrete/binary) and u(t), u(k) respectively, i.e. the values y and u at time t, tk. See Fig. 7.4. However, in the realisation of the control system, we need to process in real-time, if possible, variables y and u. If this is not possible, one could try to use suitable inferential measurements, or in the worst-case, design appropriate stochastic signal estimators.

Similar challenges are posed also concerning the final control elements (e.g. actuators). We quote articles and presentations dealing with some specific sensors designs and an overview of sensors and their instrumentation [44-47], as well as three project examples, pointing out that sensors are researched either separately, or jointly with control designs and strategies:

Example-1: NANOBIOTOUCH project, (Nano-resolved multi-scale investigations of human tactile sensations and tissue engineered Nano-Bio sensors)-application driven R&D.

Example-2: STARGATE[3] project (Sensors towards advanced monitoring and control of gas turbine engines, 2012-2016) it is stated that "obtaining reliable and accurate measurements from all parts of gas turbine engines during development phases is critical to optimising the design to maximise performance. The role of sensors is also fundamental to the operational performance of gas turbines. The limitations of current sensors in terms of temperature (project target is up to 1600 °C), accuracy, drift, stability, degradation with time, limit the maximum operating ceiling the gas turbine can be run". Improvement driven R&D.

---

https://www.energy.gov/sites/prod/files/2017/09/f37/NEET-%20Advanced%20Sensors%20and%20Instrumentation%20Newsletter%20-%20Issue%207%2C%20September%202017.pdf
[1] WS&ANs: Wireless Sensors and Actuators Networks
[2] Internet of Things
[3] Sensors Towards Advanced Monitoring and Control of Gas Turbine Engines
http://cordis.europa.eu/result/rcn/187512_en.html

*Chapter 7. An Overview of Systems, Control & Optimisation (SCO) in Recent
European R&D Programmes and Projects (2013-2017) under the Emergence of New
Concepts and Broad Industrial Initiatives*

Example-3: DIAGNO-RAIL[1] project combining innovative portable visual, acoustic, (NMR) portable devices. While this project task is not related to a control problem, the measurement instrumentation with new sensors is a system' challenge which will help to reduce risks associated with derailments and collisions. Novel measurement requirements.

Projects on sensors: Further to above examples: EU-ROPAS (Roll-to-roll Paper Sensors, 2011-2015), DELILAH (Diesel engine matching the ideal light platform of helicopters, 2013), HITEAS (High Temperature Energy Autonomous System, Aerospace, 2014-2016), MORGAN[2] (Materials for Robust Gallium Nitride-for High temperature& Pressure sensors; ORAMA (Oxide materials Towards a matured post-silicon electronics era), WINDSCANNER (Lidar), THERMOBOT (Autonomous robotic system for thermo-graphic detection of cracks), SCINTILLA (Scintillation Detectors And New Technologies For Nuclear Security).

Projects on actuators: examples include ACTUATION2015, RESEARCH, E-SEMA, DYNXPERTS, ADLAND

7.2.4. Challenges for Systems and Control Other Than Explicit Feedback Arrangements of Fig. 7.4

This is frequently the case at higher level, where the "System" may represent other than core control environments e.g. general ICT for Decision Making, Management Information support, or Dynamic knowledge systems for real time strategy switching concerning raw material utilisation. Examples: SoS Modelling and architectural aspects in DANSE and COMPASS projects. Regarding Road-mapping see completed projects T-AREASoS and CPSoS and the International organisation INCOSE[3]. Concerning applications in Energy, Internet, Transportation and Mobile Telephony, the reader may consult the global technology and business reports e.g. by MIT [48]. Similarly, several detailed studies on Energy-related SCO have been completed in Europe [49-51].

---

[1] In-situ diagnostics for rail and subways:
http://cordis.europa.eu/result/rcn/148677_en.html
[2] http://www.morganproject.eu/
[3] http://www.incose.org and http://www.incose.org/AboutSE/WhatIsSE

7.3. The European R&D scene: European Commission, National & Other Programmes

7.3.1. Outline

The European Research, Development and Innovation landscape, consists of a very broad[1], multidimensional[2] and dynamic[3] set of strategic and supporting programmes and initiatives. Table 7.1 presents an overview of the period 2007-2017 and announced plans. *Note*: Every year, several business and R&D oriented "Predictions", "Hot-topics" and "Foresight" reports are published, some of which are regularly updated e.g. [52]. The EC has also presented visions for medium and long-term socio-techno-economic scenarios [3], [53] and the European Parliament among others, a specific report on Ethics related specifically to CPS, 2017 [54].

Table 7.1. EU R&D programmes 2007-2017 and outlook beyond 2020.

EU Framework Programmes / Issues	FP7 (2007-2013)	H2020 (FP8) (2014-2020)	Provisional name : FP9 (2021-2027) Under construction. (*) Very preliminary expectations
High level EU Objective(s) relevant to Systems	Industry- Academia Cooperation, Societal challenges, Small Businesses support	Cover the complete Innovation Ecosystem, Establish Institutional and Contractual Platforms (PPPs/ETPs/JUs/cPPPs)	(*) to be defined. *Expected to include a limited number of very ambitious "Moon-shot"-type projects.*
Control R&D focus	Networked Embedded Systems & Control, Autonomous Vehicles, Complex Systems - Emergence	Cyber-Physical Systems, Systems of Systems, Safety and security	(*) to be defined
Systems R&D focus	Wireless Technologies, Real-time Systems Design, SoAs, Multi-Many Core Processors, Mixed-Criticality systems	Internet of Things (IoT), Industrial IoT, HPC, Cloud/Fog, Novel Transportation Systems	(*) to be defined . *Expected to include Integrated platforms to support global R&D collaborations.*
World-wide S&T trend(s) in the period. Expectations for 2020-2030	Nanotechnology, Smart Grids, Green energy, New Materials, Ultra Large Scale Systems, Affordable Mobile communications, Energy efficiency	Industry 4.0 (4IR), Data Efficiency, Digital Societies, 5th Generation Mobile Networks Decision Making Support	*Prepare for the next generation R&D workforce. S&T to help tackle economic and employment crises, Natural Catastrophes, Epidemics, other challenges and conflicts . Achieve Responsible Connectivity*

S&T = Science and Technology, SoAs = Service Oriented Architectures, cPPPs= contractual public-private partnerships (PPPs), ETPs = European Technology Platforms, JUs = Joint Undertakings, HPC = High Performance Computing, Industry 4.0 = Germany and EU (4IR in the UK) = 4th generation industrial Revolution –Initiative

---

[1] Central EU, national/countries, public, private, individual and bi/multi-lateral partnerships
[2] (i.e. several types and modalities of projects/initiatives)
[3] (e.g. periodically revised/updated)

Chapter 7. An Overview of Systems, Control & Optimisation (SCO) in Recent
European R&D Programmes and Projects (2013-2017) under the Emergence of New
Concepts and Broad Industrial Initiatives

7.3.2. The Main R&D&I Programmes in the European Union (EU)

For the EU of 28 Member states and certain non-EU participating countries, such as Switzerland, Norway, Israel and Turkey, the main R&D&I environment is the Framework programme (FP), managed by the European Commission (EC) and external entities/agencies (institutional, or contractual) currently: 7 Joint undertakings (JUs), more than 40 Technology Platforms (ETPs) and Public-Private Partnerships (PPPs). Cross-cutting ETPs are: Nano-futures, Industrial Safety, ConXEPT (Consumer issues). "European Technology Platforms (ETPs) are industry-led stakeholder fora recognised by the European Commission as key actors in driving innovation, knowledge transfer and European competitiveness".

Note also the Future and Emerging Technologies (FET) is part of H2020 offering opportunities for proactive and open research in systems and control, for example Quantum systems, Complexity and High-Performance Computing. Especially the quantum systems domain is inviting control theories and estimation methods to help with phenomena towards Quantum Computers [80]. Other funding mechanisms of interest to SCO, include:

- The European Research Council (ERC) managed by the ERC Executive Agency ERCEA which funds projects under the leadership of individual researchers and is also open to international researchers;

- The Marie Skłodowska-Curie Actions (MSCA) are open to all domains of research and innovation, from fundamental research to market take-up and innovation services. Research and innovation fields are chosen freely by the applicants (individuals and/or organisations) in a fully 'bottom-up' manner. MSC supports mobility and training of researchers. See examples of funded actions [1];

-The European Institute of Innovation and Technology (EIIT) and its Knowledge and Innovation Communities (KICs) and Industrial associations ARTEMIS[2], EPoSS[3];

---

[1] https://en.wikipedia.org/wiki/Marie_Sk%C5%82odowska-Curie_Actions

[2] http://www.artemis-ia.eu

[3] http://www.smart-systems-integration.org

- ITEA3 cluster of EUREKA[1] (different members from EU FP);

- The European Science Foundation ESF[2], recently with a new orientation and mandate for supporting R&D excellence.

See Table 7.2 for a summary of programmes addressing topics in SCO.

Table 7.2. Summary of main European R&D Programmes related to control and systems (2017).

Programme (*) EC = European Commission	Sub-programmes, platforms, "initiatives"	Notes
1. Framework Programme (FP), EC Currently running: Horizon 2020 (H2020)	e.g. LEIT-ICT, FET, Robotics, Cognitive Systems, Cyber-Physical Systems, "Smart Anytime Everywhere (SAE)", Energy, Aero, Security, High Performance Computing (HPC)	Duration of 7-years. Broad participation including international possibilities
2. European Research Council (ERC), EC	Fundamental Sciences and Engineering fields, (Various: senior researchers, early careers), similar to NSF (USA)	e.g. Mathematics, CPS, Control, Systems sciences
3. Marie Skłodowska-Curie Actions (MS-C), EC	Fundamental Sciences – R&D (incl. systems and control)	Single investigators including Training support
4. European Technology Platforms (ETPs), EC	e.g. Smart Grids, Rail, Robotics, Bio-fuels, Photonics , Wind Energy, Forests, Industrial Safety, Software, Water, Nano	More than 40 running Platforms (Public-Private-Partnerships)
5. European Joint Undertakings (JUs), EC	e.g. ARTEMIS/ECSEL (Embedded systems, CPS), SESAR (Air Traffic Control), CLEAN SKY (Aero), IMI (Medicine)	EC Institutional entities (similar to ETPs)
6. Contractual Public-Private-Partnerships, EC [Also: KICs, EIPs and the EIIT]	Factories of the Future (FOF), Energy-efficient Buildings, European Green Vehicles Initiative, Sustainable Process Industry (SPIRE) - including Process control and Automation	Abbreviated PPPs, KIC = Knowledge Communities EIP = Eur. Innovation Partnerships EIIT = European Institute of Innovation & Technology
7. European Defence Agency (EDA), EC	e.g. Navigation ,Guidance and Control, Autonomous Vehicles (Sea Surface and Underwater) dual use technologies	EDA calls for proposals, Independent from FP
8. European Science Foundation (ESF)	Support services for Research activities (new)	Independent from the EC
9. EUREKA Clusters (Europe and beyond)	e.g. ITEA3 (incl. Embedded systems and Control)	e.g. EU, Canada, Turkey
10. R&D programmes of individual countries. National Organisations (examples)	e.g. ANR (France), BMBF (Germany), GSRT (Greece), TEST/RAZUM(Croatia), EPSRC (UK), BMWFW, ABA (Austria), MIUR (Italy), SRC, VINOVA (Sweden), SNSF (Switzerland)	Independent R&D, or precursors/complementary to certain EC funded projects

(*) Names of programmes and/or sub-programmes may change from period to period to reflect new R&D policies and objectives

Compiled AvK 10-201

TRL: Since the start of the H2020 programme, the Technology Readiness Level (TRL)[3] indicator was introduced by the EC and the external agencies (dealing i.e. with ETPs). Although there are some ambiguities concerning its applicability to multiple technologies, encountered in large systems and control projects TRL is a useful tool to help proposers to realise, or to indicate the intended level of maturity projects results will eventually (in the English meaning of the word) reach. See [23] for an introductory discussion of possible extensions to heterogeneous systems. Comments: In principle, the FP funds low TRL

---

[1] http://www.eurekanetwork.org/eureka-clusters

[2] http://www.esf.org/

[3] EARTO: The TRL Scale as a Research & Innovation Policy Tool, 30 April 2014, http://www.earto.eu/fileadmin/content/03_Publications/The_TRL_Scale_as_a_R_I_Policy_Tool_-_EARTO_Recommendations_-_Final.pdf

Chapter 7. An Overview of Systems, Control & Optimisation (SCO) in Recent
European R&D Programmes and Projects (2013-2017) under the Emergence of New
Concepts and Broad Industrial Initiatives

(3-5) activities, while the other external Organisations/ Agencies/ Undertakings fund larger, more integrated and higher TRL (≥ 5) projects, but there may be exceptions.

<u>Joint Undertakings (JUs):</u> are entities, primarily to fund ambitious large-scale projects. Many of them are related to SCO (point 5 in Table 7.2). They include (by 2017): Clean Sky 2(CS2) for aircraft themes, Fuel Cells and Hydrogen 2 (FCH2), Innovative Medicines Initiative 2 (IMI2), Electronic Components and Systems for European Leadership (ECSEL replacing ARTEMIS and ENIAC), Bio-based Industries (BBI), Air Traffic Management Research (SESAR) and Shift2Rail. Another type of external EC agencies to run thematic R&D are the so called cPPPs - Contractual PPPs: such as the Electric Vehicles Initiative (EGVI); (5G PPP) 5G Infrastructure; Sustainable Process Industry (SPIRE); Robotics; Photonics; High Performance Computing; Big Data and Cybersecurity. In the energy field, there is the SET plan with specific actions.

The Word-cloud of frequency of keywords from Horizon 2020 projects, is shown in Fig. 7.5. The terms system, control, model, dynamics are very visible, optimisation not. Note that this is not reflecting the text of the issued work programme, but the keywords of the individual funded projects, thus represents the actual R&D work completed, planed, or underway (projects up to about the end of 2016) as declared by the project teams and their publications.

Example of H2020, addressing SCO topics, is the current draft[1] (rev.5i): "Leadership in Information Technologies (LEIT). ICT-2019-2020 addresses e.g. "Computing technologies and engineering methods for cyber-physical systems of systems Specific Challenge: Cyber-physical Systems of Systems (CPSoS), like transport networks or large manufacturing facilities, interact with and are controlled by a considerable number of distributed and networked computing elements and human users. These complex and physically-entangled systems of systems are of crucial importance for the quality of life of the citizens and for the European economy. At system level the challenge is to bring a step change to the engineering techniques supporting the design-operation continuum of dynamic CPSoS and to exploit emerging technologies such as augmented reality and artificial intelligence. At

---

[1] https://ec.europa.eu/programmes/horizon2020/sites/horizon2020/files/h2020-leit-ict-2018-2020_pre_publication.pdf

computing level, the challenge is to develop radically new solutions overcoming the intrinsic limitations of today's computing system architectures and software design practices".

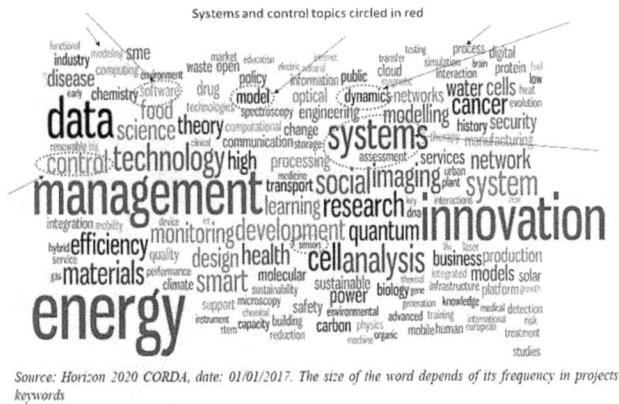

Source: Horizon 2020 CORDA, date: 01/01/2017. The size of the word depends of its frequency in projects' keywords

Fig. 7.5. A Word-cloud of frequency of keywords (H2020 funded projects) by end 2016 (EC data).

7.3.4. National and Other Programmes

Parallel to the above-mentioned FP and related EU-wide activities, all European countries run their own R&D&I programmes, frequently in conjunction with, or mutually extending and supplementing parallel FP projects. Concerning Systems and Control, see for example ANR in France, BMBF, BMWi and DFG in Germany. See point 10 in Table 7.2 and additional country R&D plans for example in: Austria [55], Croatia [56], Denmark [57], Greece [58], Spain [59] and Sweden [60]. Furthermore, certain dual-use projects (Defence-Civil) are funded by the European Defence Agency (EDA)[1] R&D, for example in SCO, projects in Guidance and Control of Underwater Vehicles and coordinating control for fleets of surface and underwater systems. An example is the French Schools of Discrete Systems Synthesis, Synchronous Languages, Model Checking, Automata and Verification, which started in the 80 s, supported by certain EC programmes and still contributing strongly

---

[1] European Defence Agency Role in Research & Technology, 2016
https://www.eda.europa.eu/docs/default-source/eda-publications/eda-r-t-2016-a4---v09

Chapter 7. An Overview of Systems, Control & Optimisation (SCO) in Recent
European R&D Programmes and Projects (2013-2017) under the Emergence of New
Concepts and Broad Industrial Initiatives

today, with significant impact on forefront science, technology and industry. Concerning cooperation beyond Europe, it includes complementary work, for example: EU-USA in Nanotechnology, CPS, Embedded Systems & Control design, Meta-modelling; EU- Russia about Complex systems, Multi-core Architectures and Aerospace; EU-Brazil about IoT and Aerospace. See a new report [61] reviewing the EU-US R&D cooperation in CPS. Other projects supporting EU-US cooperation, in pre-competitive R&D, are in block 7) of Section 7.4.2 in this chapter.

7.3.5. Renewal of R&D Programmes

Public R&D Programmes frequently change, unlike most fundamental fields of science and engineering, as far as the orientation and key objectives is concerned. Over a typical period of typically 2-5 years, several topics in the programmes, sub-programmes and the headings of the calls for proposals may be renewed, while to achieve continuity certain high interest specific longer-term topics are retained for longer. Note that the headings of most pubic programmes are usually attractive, high level, general terms, to enable innovation at project level (e.g. in Europe H2020 incudes "Smart Anytime Everywhere (SAE[1])", proactively suggesting topics of interest to work upon. Some relatively recent, internationally recognised and promoted themes for R&D have been: Internet of Things (IoT), Cyber-Physical Systems(CPS), Industry 4.0 (4IR), 5G Mobile, Systems of Systems (SoS), SoSE for SoS Engineering and Complex Adaptive Systems (CAS). Table 7.3 gives a compact identification of these topics, their nature, origins, key emphasis and contributions to. Note that SoS and SoSE are not new, but today are back on the R&D arena together with CPS and IoT as important classes of very large systems. As mentioned elsewhere in this chapter, combinations of terms are also emerging reflecting multiple challenges e.g. (CPS+SoS -> CPSoS, CAS[2]+SoS -> CASoS). Similarly, several

---

[1] Page 13 in
https://ec.europa.eu/programmes/horizon2020/sites/horizon2020/files/05i.%20
LEIT-ICT_2016-2017_pre-publication.pdf
[2] CAS: Complex Adaptive Systems, SoS: Systems of Systems, CPS: Cyber-Physical Systems

extended terms e.g. *C*P*S and*SoS[1] have been proposed in projects to reflect mixed configurations. Note that research is already undergoing in this direction, addressing e.g. large scale wireless CPS, see [62]. For a primitive macroscopic concept of a "System of the Future", see some food for thought in [23].

Table 7.3. New concepts, classes of systems, technologies and initiatives.

Issue / Topic	Broad Class of Systems	Specific Class of Systems	Generic or Specific R&D	Addresses New Methods & Design	Drives Technology -Engineering -Products -Processes -Standards	Origin (initial Idea)	Key Supporting Organisations, Associations Governments
CPS	X conceptual		S Foundational	X	Expected to address all	R&D Mixed	NSF (USA) EC H2020 (EU) Many & Expanding
SoS	-	X	g	N/A	Prc	Complexity	IEEE Systems Council,
SoSE	-	X for SoS S	X	E	S.E. Defence Industry	INCOSE, DOD/MOD/OMG Other countries
IoT	(X)	(X)	s	X	T,E,Prd,S	Wireless sensing	ITU, IEEE, IETF Many
Ind 4.0 (4IR)	Initiative (not a class)	-	G&s (security)	X	X Multi-sectoral)	ACATECH (DE) Manufacturing	German Gov. Manufacturing Industry, Many
5G-Mob	-	X (sectoral)	s	X	T,Pr,E (on-going)	4G+	ITU, Mobile Industry, IEEE
CAS	-	X	g	M	Pr	S.E. Complexity	Sandia Lab, Santa Fe, FET (EU programme)

Left margin key:
CAS= Complex Adaptive Systems
IETF=Internet Engineering Task Force
ITU = International Tele-communications Union
T= Technology
E = Engineering
Prd = product
Pr = Process
S= Standards
s = specific
g = generic
AxK Oct 2017

Right margin:
In **bold X** the key emphasis
Note: Situation of status may change in the future denoted by (_)

Despite some initial hype and ambiguities concerning the definitions and scope of the above terms (especially IoT and CPS), today there are several expected tangible and intangible benefits, such as enabling stronger multidisciplinary activities than before and a push towards swift exploitation of the R&D results. For example, CPS is nowadays endorsed by the Industry 4.0 initiative together with IoT and I2oT (Industrial IoT[2]). As a simplified historic reference, the ingredients of CPS from the past and its potential role for effective cross-disciplinary systems R&D in the future are schematically depicted in Fig. 7.6.

---

[1] * denotes a qualifier such as smart, networked, wireless, advanced, distributed.
[2] Industrial IoT Consortium: Reference Architecture, 2017, https://www.iiconsortium.org/IIC_PUB_G1_V1.80_2017-01-31.pdf

Chapter 7. An Overview of Systems, Control & Optimisation (SCO) in Recent
European R&D Programmes and Projects (2013-2017) under the Emergence of New
Concepts and Broad Industrial Initiatives

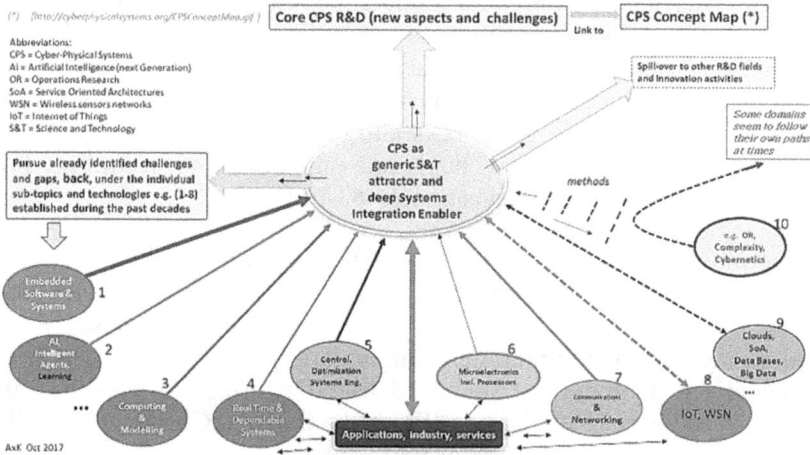

Fig. 7.6. A holistic schematic mapping of emerging CPS concept
and underpinning S&T domains.

Internet of Things (IoT)[1], is a tangible, close to market concept,
integrating complementary technologies of distributed sensors & devices
and wireless communications networks and protocols. It has evolved
dramatically upwards, in terms of capabilities, potential applications,
S&T intensions, industrial and worldwide[2] institutional and
governments[3] support. IoT will benefit a lot from 5G, Cloud/Edge/Fog
infrastructures with help from High Performance Computing (HPC). IoT
uses a different (and lighter) architecture[4] than that of a typical
networked control system. It is also at the crossroads of two research
communities, namely Control and Real-time Device Networking,

---

[1] ITU: Harnessing the Internet of Things for Global Development, Geneva
2016, https://www.itu.int/en/action/broadband/Documents/Harnessing-IoT-
Global-Development.pdf

[2] Internet Society: Global Internet Report 2016,
https://www.internetsociety.org/globalinternetreport/2016/wp-
content/uploads/2016/11/ISOC_GIR_2016-v1.pdf

[3] http://wireless.ictp.it/school_2017/Slides/IoT_ICTP_2017.pdf

[4] Thaler Dave, Hannes Tschofenig, Mary Barnes, Architectural Considerations
in Smart Object Networking, IETF 92 Technical Plenary - IAB RFC 7452. 6
Sept. 2015. Web. https://www.ietf.org/proceedings/92/slides/slides-92-iab-
techplenary-2.pdf

recently debating their S&T coverage, for example IEEE Internet Society[1] referring to CPS/Control and IEEE Control Society[2] referring to IoT. We consider that, as pointed out by Tariq Samad, from the side of Control, in practice there is a less ambiguous separation and many synergies these two domains can benefit from. IoT is also an interesting example of penetration of networks into systems and vice versa. As far as European policies, general landscape and funded projects in IoT is concerned, the reader may consult e.g. [63-65].

Concerning Industry 4.0, it was first initiated by the German Government and several German Organisations (e.g. VDMA[3], ZVEI[4], BITKOM[5]). Industry 4.0 was supported by the European Commission as key R&D goal in the H2020 programme. Similar large programmes have been launched e.g. in the US (Advanced Manufacturing Partnership 2.0), under "4th Industrial Revolution (4IR) in the Catapult Clusters in the UK, "Industrie du Futur"[6] in France, Manufacturing Innovation 3.0[7] in Korea, "China 2025"[8] in China and "New Robots Strategy"[9] plus "Society 5.0", in Japan.

There is a rich R&D activity in CPS around the world, not only on foundations and integrated design approaches, but also addressing innovative applications (as was done at earlier times in the so called networked control systems domains). Beyond the first European themes

---

[1] IEEE Internet Initiative: Towards a definition of the Internet of Things (IoT)–May 2015,
https://iot.ieee.org/images/files/pdf/IEEE_IoT_Towards_Definition_Internet_of_Things_Issue1_14MAY15.pdf
[2] Tariq Samad, Control Systems and the Internet of Things [Technical Activities], *IEEE Control Systems,* Vol. 36, Issue 1, Feb. 2016.
[3] VDMA: Leitfaden Industrie 4.0, Orientierungshilfe zur Einführung in den Mittelstand, www.industrie40.vdma.org
[4] ZVEI: Zentralverband Elektrotechnik- und Elektronikindustrie, www.zvei.org
[5] https://www.bitkom.org/industrie40/
[6] IESF: Développement de l'Industrie du Futur, 2015, http://home.iesf.fr/offres/doc_inline_src/752/C3_Industrie%2Bdu%2Bfutur.pdf
[7] http://www.oecd.org/science/sci-tech/Session%203c%20-%20LEE%20Hangkoo.pdf
[8] Made in China 2025, State Council, July 7, 2015,
http://www.cittadellascienza.it/cina/wp-content/uploads/2017/02/IoT-ONE-Made-in-China-2025.pdf
[9] Japan's Robot Strategy - Vision, Strategy, Action Plan, METI 2015,
http://www.meti.go.jp

Chapter 7. An Overview of Systems, Control & Optimisation (SCO) in Recent
European R&D Programmes and Projects (2013-2017) under the Emergence of New
Concepts and Broad Industrial Initiatives

proposed around 2013 – linking Internet and Embedded Systems [66] and in H2020 [10], it should be mentioned, for example, the latest NIST CPS architectural proposals and a guide for designers [67-68] and the NSF solicitations recommending the research communities to strengthen their efforts in CPS [69]. Interesting fundamental research and applications come also from other regions, e.g. Russia. For example, CPS for civil engineering, self-organising systems and supply chain management [70-72] see also in the Annex.

7.3.6. Views and some criticism about the Control Domain

It is worth mentioning that the Control domain is occasionally criticised, for example by neighbouring fields (e.g. Complexity, Adaptive Systems[1], OR[2]), or some of its sub-domains (e.g. robotics[3]), as being a mature domain since the 70 s. Often also, control is ignored, i.e. by the typical expression that "we have inputs and outputs" and "something in-between" (implying a processor, software, agent, decision mechanism, or emergence phenomena, but omitting the quantitative nature of core control functions). While the foundations of few rigorous control theories and applied control design methods have indeed been advanced to a maturity level the last 40 year, today, the field is still very active, successfully addressing all new systems challenges.

Moreover, teams from different disciplines contribute, cooperate and share know-how beyond core control theories and general engineering. In practice, the complexities encountered are much bigger when for example we are simultaneously confronted with a combination of several scientific and technological questions. In particular, concerning complex "physical" objects we must understand, measure and control. A concrete example:

Example: Renewable Energy (systems oriented) applications strongly emerged during the last decade. Conception, design, engineering, testing and operation of large off-shore wind farms (typical 80-200 Turbines). This is an enormous multidimensional challenge involving among

---

[1] https://en.wikipedia.org/wiki/Complex_adaptive_system
[2] OR: Operations Research domain
[3] E.g. implied in The Robot Companion for Citizens Manifesto: www.robotcompanions.eu

others, mechanical design and anchoring of the towers, composite construction of the blades, sensors for monitoring blade conditions, dynamic wind models and seasonal variations, impact of wake vortices behind each turbine on the other turbines, electric generators design, maintenance issues, control of the farm and of individual turbines, optimisation for control and other parts, implementation of the control per turbine and for the whole farm, decision support systems, organisation as well as theoretical development and simulation of the control strategy, trials, testing plus reliability and life cycle considerations.

These challenges are reflected in more than 50 projects 2007-2017. Here are some examples: CMSWIND (Condition monitoring), CLUSTERDESIGN and EERA-DTOC (Design tools[1]), HPC-BLADE (Composites turbine blades), RELIAWIND (Reliability), GEOWAVES (Geotechnical design), CORETO (Maintenance), AEOLUS (Advanced distributed control for large scale wind farms), WINDSCANNER (Lidar-based sensors system), AEOLUS4FUTURE (materials and engineering), TOTALCONTROL (Optimisation and Control), WINDRIVE (Brushless Drivetrain), SAFEWIND (Multi-Scale Modelling), ACTIVEWINDFARMS (Optimization and Control, Atmospheric Boundary Layer analysis), SMARTERSHIELD (Smart erosion shield for electro-mechanical de-icers- actuators), IAQSENSE (Nanotechnology based gas multispectral sensing system for environmental control and protection), NANODEVICE (Novel Concepts, Methods, and Technologies for the Measurement and Analysis of Airborne Engineered Nanoparticles in Workplace Air).

In R&D, the strong cooperation between control, computing and communications fields is an additional enabler to address several multi-disciplinary challenges in the future. This cooperation is well emphasised in excellent papers such as: "Impact of Control Report" of the IEEE CSS- IoC Report v2,v1, 2011 and 2014 [73], the recent IFAC Strategic Report 2017 [74] and a comprehensive and deep survey by K. J. Åström and P. R. Kumar [75]. As historical reference, let us also quote [76]. Comment: The new popular topics do not eliminate, nor displace the traditional systems disciplines. The coexistence of control and CPS, for example, is synergistic and an enabler for co-actions and co-thinking, as Janos Stzipanovits, Eduard Lee and Panos Antsaklis comment: "for a tighter integration of the cyber and physical worlds".

---

[1] http://www.cluster-design.eu, www.eera-dtoc.eu

*Chapter 7. An Overview of Systems, Control & Optimisation (SCO) in Recent
European R&D Programmes and Projects (2013-2017) under the Emergence of New
Concepts and Broad Industrial Initiatives*

Additional perspectives for the domain of CPS - Systems & Control provides the work done in robotics, life sciences and complex swarm systems (see Annex) [77-79].

7.3.7. What Communities Deal with Systems and Control R&D ?

Fig. 7.7 below lists, on the left-hand side the main topics (in SCO) which are encountered in the tasks of funded projects, the right-hand side presents the well-established S&T communities involved. The point we want to emphasise here is that today there is not a 1-1 mapping- we had years ago, but is rather n-to-m and in some instances, it may be "everybody can do almost everything". This S&T plurality demonstrates the positive proliferation of methods (originating in systems, control and optimisation) to almost all other fields and disciplines, but also poses questions with respect to the crucial necessity of involvement of the core disciplines to be involved in specific challenging problems.

Main Systems & Control Topics in funded R&D Projects (*) in Europe, advancing one, or more of the following :

- Fundamental Control *Theories* (including mathematical methods for e.g. dynamical systems)
- Control *methods and tools* (modelling, design, analysis, synthesis, verification, testing - mainly generic)
- Control *systems* (e.g. embedded, non-embedded local/global, small/large, distributed/networked)
- Control *part in other Systems* (domain-specific applications)
- *Systems* (Real time, Programable, Integrated, Networked, Languages, Operating systems, Non-functional aspects)
- *Simulation*, Simulators, Human/Entity in the Loop
- *Cyber-Physical Systems (emerging* methods and platforms)
- *Internet of Things* and Industrial IoT (in conjunction with Control)
- *High Performance* (Computing) Systems
- *Complex Systems* (e.g. Adaptation, Swarm intelligence, Self*)
- Systems of Systems (e.g. SoS Engineering and ULSS)
- Quantum systems control (e.g. communications)
- Robotics and cognitive systems, Autonomous systems
- Optical systems, Photonics and Lasers
- "Physical control" methods and processes (non-ICT)
- Within recent initiatives (e.g. Industry 4.0, New AI, Smart A-E)
- "Control" term used in its simplistic meaning, or without further elaboration ("an FPGA controller")

(*) Topics addressed in the projects are not necessarily mentioned in the Work Programmes

Scientific and Technological communities usually involved in Systems and Control Projects

- Control engineering
- Systems sciences & Engineering
- Computer sciences & Embedded Systems   *Emerging CPS*
  (Software engineering/Computational Mechanics)   *Community*
- Communications/Networks/Internet
- Micro and Nanoelectronics
- Physics, Materials, Biology and Chemistry, Mathematical Physics
- Complexity Sciences/Cybernetics/Large scale systems
- Operations Research, Decision Making & Support
- Manufacturing, Production Engineering
- Other Engineering domains
  (e.g. Electrical/Mechanical/ Aeronautics/Photonics/
  Transportation/Automotive/Shipbuilding-Ocean/
  Chemical-Pharma-Bio/Food Processing/ Forestry/
  Agricultural/ Health and medical)
- Reliability, Security, Safety Engineering
- Sensors, Instruments, Actuators, Interfaces and HMIs
- Non-Engineering domains
  (e.g. Management and Organisational fields)
- Other sciences, disciplines and domains
- Organisations, Standardisation bodies, Governments
- New, Emerging, Mixed, or Unclassified Constituencies

mapping is today not 1:1   AvA 10-2017

Fig. 7.7. Systems and Control topics in funded projects vs. the S&T Communities involved.

7.4. Project Categories and Examples Regarding Systems, Control and Optimisation

7.4.1. Preliminary Remarks

The descriptions of the projects here are not necessarily the project titles, but reflect the key challenges related to systems, control and optimisation research. A summary of representative for example foundational and forefront research topics is shown in Table 7.4.

Table 7.4. Examples of rigorous Systems-Control-Optimisation (SCO) methods in projects.

Rigorous Methods researched in funded projects	Examples of Topics in projects and Notes
Algebraic , geometric Methods	Grassmann Determinantal analysis, Geom. measure theory
Automata and advanced logic	Automata-based scheduling, Game automata, Proof adaptation/ Code-Refactoring, CPS compositional certification
Cooperative control	Swarm intelligence, Collective decisions, Consensus filtering/control
Discrete Events (DEVs), Hybrid Systems Verification	Collaborative verification, Event recognition, Event-triggered control
Dynamical Systems, Differential Equations (PDE), Stochastic Differential Equations, Non-equilibrium dyn.	Quasi linear PDEs, hyperbolic PDEs and , PDEs and Control, PDES- CFD, PDEs in Hydro-Dynamics, Random walks, Long Time Horizons control
Control & Estimation methods	(Parameter varying) LPV and qLPV synthesis, symbolic control methods
FDI-FDD-FTC methods (Identification, Detection, Control)	Fault tolerance and methods for its safeguarding, FT Control, Dependability imperatives
Large scale systems and complex phenomena,	Emergence, Predictability, Behavioural aspects, Cognitive aspects,
Model Predictive Control and advanced methods (MPC)	e.g. Multi-parametric & distributed MPC , Risk-averse MPC
MIMO control (multiple inputs-multiple outputs)	Multivariable dynamical systems control.
Non-Linear systems analysis,	Bifurcations, Stability , multiple-discontinuity, polynomial vector fields
Optimisation methods (generic), Approx dyn program.	Linear, Non-Linear, Mixed Integer–L or NL Programming, Learning-based
Quantum Systems (QS)	QS measurements, QS controllability, 2D Quantum walks
Stochastic methods and processes (e.g. Markov DM), Reinforcement Learning	Incl. Stochastic control methods, decision making (DM) processes (POMDP= Partially Observable Markov Decision Proc.), Multi-Task R, learning-based control

7.4.2. Project Examples, According to Their SCO Content, a Bottom up View

In the set of projects funded in Europe between 2013 and 2017 (some may be completed by now, others just started), we are able to conveniently use the general mapping of Section 7.3 (control, application and implementation). While the boundaries between these three aspects are not always crisp, and there are projects addressing more than one topic, the following examples indicate well the associated R&D

activities. Project acronyms are in capital letters, projects may appear also in other sections of this chapter to emphasise other tasks. A sample of selective project groups with additional information is in the Annex.

1) Theoretical and fundamental issues in control and systems. Examples:

- Automata-based systems LASSO, RSCS, PACOMANEDIA

- Co-Algebras: HOT CO-ALGEBRAS, PRO-ALG

- Control for (models with) Partial Differential Equations: Project CPDENL

- Linear Parameter Variable (LPV), quasi LPV Systems and NLPV Systems: APROCS

- Refactoring methods (for code robustification), Project SPHINX[1]

- Bifurcations in non-autonomous systems: CDSANAB, LDNAD

- Optimisation methods- mixed Integer NLP: MINO (ID- 316647)

- Verification: EQUALIS, CASSTING, VERIWARE, (BUCOPHSYS)

- Geometric methods (e.g. for networks): CONNECT (ID 734922)

- Dynamical Systems: ILDS

Other projects of this nature are for example: DEMAND (Decentralised consensus); PROCSYS (Symbolic Control); SYSDYNET, DYNSYSAPPL and CONDYS (Dynamic Stochastic systems); SYMBIOSYS, ON-TIME, ARGO-WCET and RETNET (Temporal and behavioural analysis); AD-DAP (Adaptive control); GEOMECH; CS-AWARE, EXCELL (Big data for control), MetCogCon (Metacognitive Control); Flow in specific Manifolds, Project ICCCSYSTEM (robust control); Project RWHG Study of Random Walks, Optimisation methods – for CMOS design: MANON

2) Advancing SCO design methods and algorithms. Examples:

- Complex systems: SETCOMP, FEDERATES

- Distributed reconfiguration methods: RECONFIG

---

[1] Same project under group 6 in this section

- MPC, Projects: SMPCBCSG, CONNECT, MPC-GT (in Energy systems)

- Sensitivity of control systems: SADCO

- Optimisation: OPT4SMART, MORE, MOBIL, COVMAPS, POLARBEAR, OPTICO

- Timing aspects: SYMBIOSIS, CRONOS

- Advanced Temporal Logic (LTL, LTL-MoP): MISSA, DATAVERIF, AVS-ISS

Other projects in this category, but with more applications orientation than advancing control are: System security and safety: MODESE, COCKPITCI, DEIS, CAR, SERENITI; Model-based control design, Project MOBOCON, RULE (Rule based), MULTIECS; Advanced Guidance and Control: Project ICOEUR (EU-Russia cooperation); System security and safety: MODESE, COCKPITCI, DEIS, CAR, SERENITI; Adaptive systems, Self-Adaptation: FEDERATE, AGENT; CPS design: VICYPHYSYS; CPS Optimisation: oCPS, TAPPS,

3) Control methods for challenging/difficult/novel applications/ processes. Examples:

- Cryogenic systems control: CHATT

- Nano-structure Control: CRONOS, CISSTEM

- Robust control for energy applications: DECADE

- Self-assembly: SACS (in supra-molecules), NANODIRECT, COMPLOIDS (Colloids)

- Control of Laser processing systems: MAShES

- Variable geometry thrusters for ships: AZIPILOT

- Control under CPS: SCORPIUS, CYPHERS, CPSOS, ROAD2CPS, EUROCPS

- Control systems reliability: HARMONICS (ID-269851)

- Optimisation &control, PDEs (ACTIVE WINDFARMS), AEOLUS, TOTAL CONTROL

- Control for post Wankel engines: LIBRALATO (Libralato-type engine prototype)

Chapter 7. An Overview of Systems, Control & Optimisation (SCO) in Recent
European R&D Programmes and Projects (2013-2017) under the Emergence of New
Concepts and Broad Industrial Initiatives

Other projects of control for challenging / new processes are: Water quality control (AQUALITY, SMARTAP); Turbine noise reduction (STARGATE, ORINOCO), Control of spin: HINTS; Crystallisation processes (CRYSTAL-VIS); Optimisation and control (OPTICO, IDEALVENT, HIGHTECS; Control for smart buildings (BUILDNET), see also projects under MPC in 2) above

4) Control as part of broader designs/large systems/plants. Examples:

- Automated/Autonomous ships: BRAAVOO, CASCADE, MARINECOSTABILITY

- Food processing systems: CAFÉ (Crystallisation process design and optimisation)

- Future aircraft systems: ACFA 2000, HYPSTAIR (solar planes), HIGHTECS

- Green energy production: COGENT

- Industry 4.0 (4IR) optimisation: DISRUPT, PRODUCTIVE 4.0, CONNECTED FACTORY, BEinCPPS, APS4AME

- Mechatronics Systems: i-MECH

- Monitoring & control of complete plants: CRYSTAL-VIS, MODUS

- Processing plants control & safety: OPTICO, ESCORTS (on standards)

- Rail Signalling systems: INESS, ADDSAFE, ON-TIME

- Sensors design and integration: ICOS-INWIRE, AGEN, DISIRE (in-situ), see Section 7.2.3

5) Mixed criticality Systems[1] (on multicores). Examples: COMMICS, SAFURE. A survey of "Mixed Criticality Systems in Control" projects can be found in A. Crespo, *et al.* [81]

6) Tools for systems and Control (software). Examples:

---

[1] Alan Burns and Robert I. Davis: Mixed Criticality Systems - A Review, 2017, https://www-users.cs.york.ac.uk/burns/review.pdf

- Maintenance and decision-making support: MANTIS, CARL-PdM, Real time design: CP-SETIS, ADVANCE (Verification)

- Control systems design: CONTROL-CPS, AXIOM, CPSLABS, MODESEC

- Robotics design: ROBMOSYS, Co4Robots, ROSIN (Robotics Operating System)

- Modelling and Platforms for CPS: INTO-CPS, PLATFORM4CPS, CPSLABS

- Verification of systems: UNCOVER CPS, IMMORTAL, ENABLE-3, MODUS

- Code optimisation and improvements: SPHINX with KeYmaera tools[1]

- MPC Tuning support tools: ITMPC

- Plasma Physics: NUSIKIMO

- Numerical Control Algorithms: SLICOT (older project and Library, still useful)

7) Supporting activities and projects. Examples:

- Roadmaps: Road2CPS, CPSoS, CYPHERS, PLATFORMS4CPS, Road2SoS

- EU-US: UECIMUAVS, PICASSO, TAMS4CPS, CPS SUMMIT, TAREA-SOS, COOPEUS (ID-312118)

- Research centres and Networks for control & systems: EPPIC (for production control and CPS) in Hungary

- Support for SCO in new Member States e.g. (Estonia, Romania for RT systems, Bulgaria for Control and Cybernetics)

8) Other systems and control related projects. Examples:

- Manufacturing and CPS: GOOD-MAN (Zero defects manufacturing)

- Cloud, Edge, Fog for control: FORA, RELATE, SPECS, SECCRIT

- Simulation: COSSIM, DQSIM (Quantum simulator), QUCHIP

---

[1] CMU, http://symbolaris.com/info/KeYmaera.html

- Testing, Standardisation & Certification: AMASS, CERTMILS, U-TEST (primarily CPS)

- Optimisation for systems and processes: PHANTOM, DISRUPT

- Virtualisation: EURO-MILS

- Sensors, Actuators: SENSINDOOR, ACTUATORS2020, HIGHTECS, ACCOMIN

7.4.3. SCO Topics in Large Scale and Broad Scope Projects

The set of projects in the previous Section 7.4.2 gave the extent of topics funded from theory to practical applications in systems, control and optimisation. They did not however indicate how comprehensive the activities were, nor the % of the control with respect to the scope of the overall project. Such a deeper analysis would be useful, but is however beyond this overview. Nevertheless, we are going to briefly present two sectoral activities namely (1) in Manufacturing/Production and Processing and (2) Aeronautics, because they are well coordinated, following roadmaps and strategic agendas and because they strongly inspire concrete future SCO R&D within the European programmes, or under independent work.

7.4.3.1. Manufacturing, Enterprise ICT and SCO

Manufacturing refers to product design and shop floor mass production facilities. The R&D communities and the resulted commercial grade control, coordination and integration systems used are mainly discrete automation systems with some continuous controllers, many robots and few humans in the loop. Humans are usually found in the testing and final Quality control departments. For smaller production plants the type of systems may vary. Over the past 30 years, there have been enormous transformations and improvements of the overall factory systems architectures, capabilities, performance and the machine-human environment. The trend to abandon, to some degree, the traditional hierarchical pyramid-type architecture to a better manageable, mesh-type flexible, non-hierarchical enterprise grid, is the result of several parallel developments e.g. powerful distributed real-time computing, optimisation, control and decision support tools, supported by new overall concepts such as networked embedded systems (now Cyber-

Physical systems), sensor networks leading to the Industrial Internet of Things and Service Oriented Architectures (SoAs) supporting MES (middleware) and ERP rich-semantic communications. In the EU, there are several programmes and initiatives directly supporting manufacturing through H2020, the Factory of the Future (abbreviated FoF) and the EFFRA[1] Association, developing a roadmap, as well as indirectly by several sectoral EU platforms such as Robotics, ICT LEIT, Clean Sky2, ECSEL and Rail2Shift. Links to international fora are present, such as IMS[2] (Intelligent Manufacturing Systems) and WMF (World Manufacturing Forum)[3]. The relevance of Industry 4.0 to manufacturing is analysed in [82] and the last 30 years of European R&D for manufacturing systems given in [83], the Factory of the Future in [84]. Concerning the role of sensors and control see [85]. Finally, Industry 4.0 for the process industries, in contrast to discrete manufacturing is analysed e.g. in an ARC report [86]. Fig. 7.8 shows schematically the transformation of the manufacturing enterprise architecture (1985-2015) and rough estimates to 2025. Industry has moved from CIM to CPS and Industry 4.0. The right part of the figure is based on vision documents by VDE/VDI[4] (2015), Industry 4.0 RAMI Reference Architecture (2015), PWC[5] and ARC[6] reports.

Projects in manufacturing can be found in the FoF database[7], the PPP lists[8] and in IMS[9] (International projects). Examples of those projects dealing with SCO topics include: HiPR (Monitoring and control of the condition of micro-tooling for complex high-precision 3D parts), HiMicro (quality control methods - computer-tomography (CT) metrology and digital holography), CassaMobile ('plug & produce'

---

[1] http://www.effra.eu/portal
[2] http://www.ims.org/wp-content/uploads/2017/01/2.02_Max-Blanchet_WMF2016.pdf
[3] https://www.worldmanufacturingforum.org/
[4] https://www.vdi.de/fileadmin/user_upload/VDI-GMA_Statusreport_Referenzarchitekturmodell-Industrie40.pdf
[5] https://www.pwc.com/gx/en/industries/industries-4.0/landing-page/industry-4.0-building-your-digital-enterprise-april-2016.pdf
[6] ARC report, in Siemens website: https://www.siemens.com/content/dam/internet/siemens-com/us/cyber-security/pdfs/arc-view-cyber-security-feb2016.pdf
[7] http://www.fofamproject.eu/database/projects.html
[8] http://ec.europa.eu/research/participants/data/ref/h2020/other/wp/h2020-wp1617-list-cpps-kets_en.pdf
[9] http://www.ims.org/access-projects/

Chapter 7. An Overview of Systems, Control & Optimisation (SCO) in Recent
European R&D Programmes and Projects (2013-2017) under the Emergence of New
Concepts and Broad Industrial Initiatives

architecture includes mechanical and control system), Symbionica (Reconfigurable Machine for the new Additive and Subtractive Manufacturing using advanced closed loop control), HYPROLINE: (Quality Control of High performance Production line for Small Series Metal Parts), TWIN-CONTROL(Machine tool and machining process performance simulation). INTEGRATE (ENIAC project, Integrated Solutions for Agile Manufacturing in High-mix Semiconductor Fab). A large-scale project in automotive is LIBRALATO (Libralato Engine Prototype). Additional examples:

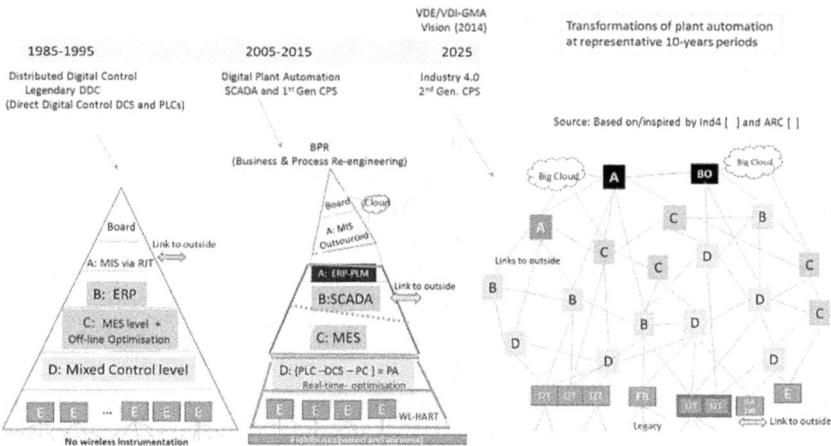

Fig. 7.8. System Hierarchies in Manufacturing, Production, Process Industries. Schemetic Evolution (1985-2025). The right hand architecture is based on RAMI (Industry 4.0)

UK-US collaboration Example: project related to Industry 4.0 (4IR): FACTORY-2050[1], French National project (ANR): MAGE (Microcontrollers for Autonomy with Highly Efficient Energy Use).

Recent projects related to process industries are presented in the SPIRE[2] portal. Examples with SCO topics: CONSENS (Integrated Control and Sensing for Sustainable Operation of Flexible Intensified Processes),

---

[1] http://www.amrc.co.uk/facilities/factory-2050
[2] https://www.spire2030.eu/projects/our-spire-projects

DISIRE (Integrated Process Control based on Distributed In-Situ Sensors into Raw Material, Energy Feedstock), CSPEC (in-line Cascade laser spectrometer for process control), MONSOON (Model-based control framework for Site-wide Optimization of data-intensive processes) and RECOBA (Cross-sectorial real-time sensing, advanced control and optimisation of batch processes saving energy/raw materials). A broad scope project is CYANOFACTORY (Design, construction, demonstration of solar biofuel production using novel (photo)synthetic cell factories). MORE (Real-time Monitoring and Optimization of Resource Efficiency in Integrated Plants).

7.4.4. Aerospace Research and SCO Topics

A substantial research effort, which is very relevant to SCO, is conducted in the Aerospace field along two main lines 1) aircraft structures, propulsion and systems and 2) Air Traffic Control and Management. The European programmes include appropriate slots in H2020 LEIT ICT (e.g. Space[1], Industrial ICT, CPS) and two Joint Undertakings: 1) CLEAN SKY[2] for aircraft designs, materials, morphing wings, flight control systems, networks, sensors and actuators and 2) SESAR[3] for Air Traffic Management and future systems, such as anti-collision ACAS-x (FAA[4], EUROCONTROL[5], EUROCAE[6], RTCA[7], FSF[8]). The challenges in 1) include noise reduction, variable thrusters, open rotor engines, advanced flight control systems, sensors & actuators, software validation, flow control and alternative propulsion such as electric and hybrid engines. The challenges in 2) address e.g. the implications of

---

[1] ICT LEIT-Space Programme under H2020, European Commission, see http://www.simplernet.it/c/document_library/get_file?p_l_id=23888371&folderId=23888312&name=DLFE-180505.pdf

[2] Example of plan 2015-17: http://ec.europa.eu/research/participants/portal/doc/call/h2020/jti-cs2-2015-cfp02-eng-01-02/1662632-cs_work_plan_2015-2017_en.pdf

[3] Fact Sheet: https://www.sesarju.eu/sites/default/files/documents/sesar-factsheet-2017.pdf

[4] https://www.faa.gov/nextgen/snapshots/stories/?slide=27

[5] http://www.eurocontrol.int/articles/history-future-airborne-collision-avoidance

[6] https://www.skybrary.aero/bookshelf/books/2390.pdf

[7] https://www.rtca.org/content/sc-147

[8] https://flightsafety.org/asw-article/acas-x/

Chapter 7. An Overview of Systems, Control & Optimisation (SCO) in Recent
European R&D Programmes and Projects (2013-2017) under the Emergence of New
Concepts and Broad Industrial Initiatives

denser skies in near future, Civil Drones integration (ICAO[1]), Sense and avoid concepts and systems, Harmonisation of international standards, regional-global transportation and Remote Airport Towers[2] conceptual design and trials (e.g. Hungary, Norway, Sweden[3]) following similar international activities[4].

Examples of aerospace SCO projects (by title only) corresponding to above topics: DELILAH (Diesel engine matching the ideal light platform of the helicopter),SPARTAN (Space exploration Research for Throttleable Advanced engine), EMA4FLIGHT (Development of Electromechanical Actuators and Electronic control Units for Flight Control Systems), CROP (Cycloidal Rotor Optimized for Propulsion), VISION (Validation of Integrated Safety-enhanced Intelligent flight control), PJ27 IOPVLD (Flight Object Interoperability VLD[5]), DROC2OM (Drone Critical Communications), PERCEVITE (Sense and avoid technology for small drones), NOVEMOR (Novel Air Vehicles Configurations- From Fluttering Wings to Morphing Flight), MAWS (Modelling of Adaptive Wing Structures), SARISTU (Smart Intelligent Aircraft Structures), SMYLE (addressing Smart alloys controlling aircraft wings- mathematical modelling the deformation of shape-memory alloys (SMAs)), RBF4AERO (Innovative benchmark technology for aircraft engineering design and efficient design phase optimisation), POLARBEAR (Production and Analysis Evolution For Lattice Related Barrel Elements Under Operations With Advanced Robustness), iFLY (on Verification).

7.4.5.Alternative and Other Interesting SCO Applications

Systems, Control and Optimisation approaches for modelling, simulation, design and analysis have been applied to interesting, but

---

[1] ICAO: UAVs integration
https://www.icao.int/Meetings/UAS2017/Documents/UAS2017_RFI.pdf
[2] http://www.sesarju.eu/sites/default/files/documents/events/remote-day2-business-persp1.pdf
[3] https://www.lfv.se/en
[4] https://www.natca.org/index.php/insider-articles/1771-april-7-2017-itf-remote-tower-services
[5] VLD: Very Large Scale Demonstrator

relatively little discussed applications. We introduce the topics as "Food for Thought" to the broader SCO communities.

7.4.5.1. Metacognitive Systems Concepts

Metacognition (in simple words thinking about one's thoughts) is studied in philosophy, psychology, complexity, learning and computer sciences. In human brain research, as discussed in psychology research and learning techniques. Metacognition is a phenomenon also known in health care, related to Compulsory Behavioural disorders (e.g. Anorexia), their diagnosis and therapy [87-91]. Psychology researchers refer to their model as a "Systems Theory" interpretation. However, due to the obvious feedback loops, it is appearing to be an appropriate example for a control modelling case study. Especially the 2-level hierarchical structure is a motivation for a potential paradigm for a coordinating control with local memories and some dynamics. Fig. 7.9 is a simplified system-inspired possible interpretation of a two-level cognitive-metacognitive thought-processing model as an alternative example of hierarchical regulatory control in engineered systems.

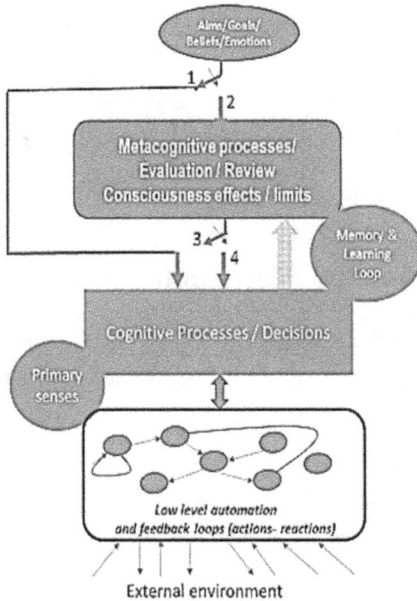

Fig. 7.9. A less common "control concept" from applied psychology and learning systems.

Chapter 7. An Overview of Systems, Control & Optimisation (SCO) in Recent
European R&D Programmes and Projects (2013-2017) under the Emergence of New
Concepts and Broad Industrial Initiatives

Few projects addressing engineering metacognition in learning, have been funded, but to our knowledge we have not found any publication dealing with modelling attempts using control theoretical "thinking" and real-time descriptions. Hybrid systems and automata-based simulation may be a first step towards quantitative modelling of mental metacognition under a cooperation across then triptych "psychology-dynamical systems and - computation". Uncertainties may also occur by noisy data transfers. Furthermore, Joëlle Proust book [92] presents a deep discussion about Metacognition from Philosophy and Complexity points of view.

7.4.5.2. Formal Methods: Improved Concepts and Approaches - New Tools

The areas of Formal Methods including Temporal logic, Model Checking[1], Controller synthesis and Verification, static analysis, especially regarding discrete and hybrid systems has been a focal point of rigorous research across the Atlantic since the mid-80s. The topics have attracted distinguished researchers from Computer sciences, Control Engineering, Optimisation, Discrete Logic and Mathematical Reasoning. One highlight of this area was the 2007 Turing[2][3] Award to Edmund M. Clarke (CMU, US), E. Allen Emerson (University of Texas, Austin, US) and Joseph Sifakis (VERIMAG/CNRS France and EPFL Switzerland).

The last decade to 2017, the interest of established and new scientific groups is still high, tackling the earlier challenges and entering recent demanding extensions such as Run time Verification, Symbolic Polyhedra and Control, Game-theoretic approaches for optimisation and control, Linear Temporal Logic MissiOn Planning (LTLMoP), Synthesis of COOL.

Controllers, Controllers for infinite state systems, Discrete Controller Synthesis, Revisiting PID controllers for complex logic and Hybrid

---

[1] Edmund M. Clarke, Thomas A. Henzinger, Helmut Veith: Handbook of Model Checking Springer International Publishing, 2016
[2] Turing Award 2007: http://www.ams.org/notices/200806/tx080600709p.pdf
[3] Allen Emerson: A Bird's Eye View http://www-verimag.imag.fr/~sifakis/TuringAwardPaper-Apr14.pdf

systems. Refactoring code for improving performance, Tools like {PRISM, ROMEO, GRAPHITE, COSYMA, FRAMA-C source code analysis, CADP}, Advanced partitioning of systems (h/w-s/w), Advanced Co-Synthesis in COOL, Hierarchical reactive controller synthesis. Funded projects include: RECONFIG. (Discrete Controller Synthesis from Individual Agent Specifications), STANCE (ID-317753) and IMMORTAL (ID-644905), MODUS (Methodology and supporting toolset advancing embedded systems quality), AVS-ISS (Analysis, Verification, and Synthesis for Infinite-State Systems). More details about this group can be found in the Annex.

7.5. Concluding Remarks and new challenges

First a bipolar situation: the process side- (the application-driven challenges and the Control and closed loop optimisation side. A concurrent engineering approach (process//control, product//control) is the ideal situation, but frequently this is not possible because of lack of sufficient knowledge. Examples: (1) In the design of future fusion reactors and their subsystems, lack of good models of the phenomena and processes, insufficient experimental data and less-repeatable measurements, make the specifications for suitable control laws and sequential operations difficult, so that instrumentation and control may not be available 100 % when then physics are ready, (2) In the more familiar case of wind farm control, while the individual turbines can be manufactured to the specs, the real-time control of each turbine and on top of that the coordinating control strategy of the whole farm may suffer e.g. from unpredictable behaviour of the actual wind pattern at a time. Therefore, better dynamic models of the wind and the aerodynamic model of the turbine blades would require additional studies. In short "Hydrodynamics and CFD meet Control Thinking and both meet Software".

A second potential challenge is the space/element(s) located between the physical and cyber parts. This may be called the "Physical" control area that is "to find how to control something". Example: A project in metal, or additive manufacturing, using photonics sub-systems is confronted with the physical-control challenge, see HYPROLINE[1] (High performance Production line for Small Series Metal Parts). Ignoring for the time being the above burden, we should stress the important role of

---

[1] www.Hyproline.eu

modelling, implementation and verification of control systems on real-time platforms[1]. Under the recent new concepts, some researchers may consider control, de facto, as element of a CPS, others may continue to avoid the term CPS, thus addressing the triple {application, control, implementation. In general, "Control" directly implies the notion of a system, but the opposite is not always true.

The set of good capabilities mentioned earlier in this section, may exhibit semi-periodic behaviour and may disappear and re-appear. Then hopefully, much better than the old ones, again to motivate new researchers and lead to stronger, more effective, highly user-confident methods, for example, we are already witnessing an interesting "come-back"[2] of some classic-modern topics in systems and control which have had a strong impact [3] [4] on S&T. Here are some examples of advanced and promising rigorous methods e.g. in hybrid control [93], distributed Temporal logic and verification e.g. [93-95] work, partly supported by the H2020 ERC BUCOPHSYS grant and by the Swedish Foundation for Strategic Research (SSF), [96], stochastic modelling [97], stochastic co-simulations and novel discussion about time in systems, and "Humans in the Internet of Things [98]". Also beyond these possible enhancements, certain early, yet unexploited, mathematical theories] may surprise everybody if they could provide straightforward solutions to some of the chronic challenging problems. For example, behaviour when it is applied to complex multi-logic governed systems, requiring non-traditional methods to design and tune [99].

Against the momentum of current developments, there are some voices raising concerns about the risks of "self-evolving autonomous intelligence" and of "super-systems". It is understood that this is not

---

[1] http://www.cleansky.eu/sites/default/files/inline-files/Skyline%2020%20-%20final_0.pdf

[2] Oded Maler, Amir Pnueli, Joseph Sifakis: On the Synthesis of Discrete Controllers for Timed Systems, http://www-verimag.imag.fr/~sifakis/RECH/Synth-MalerPnueli.pdf

[3] Eugene Asarin, Oded Maler, Amir Pnueli Joseph Sifakis: CONTROLLER SYNTHESIS FOR TIMED AUTOMATA, http://www-verimag.imag.fr/~maler/Papers/newsynth.pdf

[4] R. Alur, C. Courcoubetis, N. Halbwachs, T. A. Henzinger, P.-H. Ho, X. Nicollin, A. Olivero, J. Sifakis, and S. Yovine. The algorithmic analysis of hybrid systems, *Theor. Comput. Sci.*, Vol. 138, Issue 1, 1995, pp. 3–34.

against the useful controlled systems with verifiable logic, (even if it is based on the new AI[1]), with good sensors and actuators, which keep us happy, or make life convenient, but it is about potentially uncontrollable cyber systems, or cyber attacking actions in the future. Is it "Live 3.0" by Max Tegmark [100], [102], "Super-Intelligence" by Nick Bostrom [101], or may be "New Human Mentality and Responsibility" to be concerned, or not at all about. We trust the counter-criticism, that the bad eventuality- likelihood will never happen, because of rational humans in the front-end, middle- and high-level decision loops. Probably we may just need to enhance the concept of "correct by Design (CbD)" through "correct by evolution (CbE)" to "correct by integration (CbI)" and may ultimately be " correct by engineering responsibility". But these terms cannot easily be defined, at present and each project may use different interpretations.

Summary

Recent projects and other R&D work, have supported earlier views, that Systems, Control and Optimisation (SCO) disciplines and engineering domains, provide us with powerful methods and systems, successfully tested in many real-life applications, while meeting, or exceeding their initial goals. Despite the fact, that there are many options, a plethora of promising methods and tools (some of them being rather sophisticated), the methods for complete designs, verification, integration and testing are continuously improving, thus creating capabilities ready to fulfil future demanding engineering and socio-technical undertakings under extreme constraints. But unfortunately, this portfolio of tangible capabilities is not evolving without other obstacles, foreseeable or not, because for example, we are confronted with (1) an ever-increasing complexity of today's surrounding phenomena, new materials, systems (natural and human-made-experimental) about which we lack the deep, or perfect knowledge, (2) similarly increasing complexity and uncertainty about the definitions and requirements of the systems to be built and (3) some irrational human considerations. Certainly, programmes, funded projects, independent R&D and innovative ideas,

---

[1] IBM: Preparing for the future of artificial intelligence. IBM response to the White House Office of Science and Technology Policy's Request for information (Summer 2016), IBM Corporation 2016, http://research.ibm.com/cognitive-computing/cognitive-horizons-network/downloads/IBM_Research_CSOC_OSTP_Response%20v2.pdf

as discussed in this chapter, greatly contribute to solve problems and improve weak situations, but sound alternative thinking may also help and should be welcome in the future.

This paper includes several examples of projects, emphasising challenging points and encourages further multi-disciplinary work and cooperation, bridging the corners of the triangle: domain knowledge, systems & control and contemporary implementations. This is expected to continue stronger in the years to come, guided by leading visionaries, architects and champions in R&D& Innovation in Europe and the whole world, because the challenges may prove to be deeper than they appear to be today. References and surveys beyond the Cordis database, are included to illustrate the relevance of the topics. This short exposure, with additional personal remarks, confirms that the fascinating world of Systems, Control and Optimisation, is continuously advancing, going towards smart solutions for the complex needs of our societies.

References

[1]. A. Konstantellos, A short overview of control in European R&D programmes (1983-2013), *European Journal of Control*, Vol. 19, 2013, pp. 351-357. http://dx.doi.org/10.1016/j.ejcon.2013.06.002

[2]. OECD: Science, Technology and Innovation Outlook 2016, Megatrends affecting science, technology and innovation. https://www.oecd.org/sti/Megatrends%20affecting%20science,%20techn ology%20and%20innovation.pdf and full report http://www.oecd.org/sti/oecd-science-technology-and-innovation-outlook-25186167.htm

[3]. White Paper on the Future of Europe, Reflections and scenarios for the EU27 by 2025, *European Commission COM (2017)2025 of 1 March 2017*. https://europa.eu/european-union/sites/europaeu/files/whitepaper_en.pdf

[4]. BMBF: The new High-Tech Strategy - Innovation for Germany, August 2014. https://www.bmbf.de/pub/HTS_Broschuere_eng.pdf

[5]. Global Challenges and Data-Driven Science, in *Proceedings of the International CODATA 2017 Conference*, Oct 8-13, 2017, Saint-Petersburg, Russia. http://www.codata.org/events/conferences/codata-2017-saint-petersburg

[6]. Phil Cooper, Smart Systems Technology Support by the UK Sensors & Instrumentation Knowledge Transfer Network (KTN, UK), 2017. https://issuu.com/magazineproduction/docs/uk_tech_events_show_guide _2015_ezin and https://ktn-uk.co.uk/programmes/eu-programmes

[7]. Jan Schneider, Dirk Börner, Peter van Rosmalen, Marcus Specht, Augmenting the Senses: A Review on Sensor-Based Learning Support, *Sensors*, Basel, Vol. 15, Issue 2, 2015, pp. 4097–4133. http://europepmc.org/articles/PMC4367401

[8]. Complex Systems: Control and Modeling problems (CSCMP'2017), Sept. 2017. http://www.iccs.ru/cscmp/cscmp_topics_e.html#

[9]. Resilience and Security Recommendations for Power Systems with Distributed Energy Resources (DER) Cyber-physical Systems. IEC TR 62351-12:2016, *International Electrotechnical Commission*, April 2016.

[10]. European Commission: Horizon 2020 Work programme for 2018-2020 (Draft 5,), 2017, pp. 1-138. https://ec.europa.eu/programmes/horizon2020/sites/horizon2020/files/h2020-leit-ict-2018-2020_pre_publication.pdf

[11]. European Commission: Contractual public-private partnerships in Horizon 2020, pp. 1-16. https://ec.europa.eu/research/industrial_technologies/pdf/contractual-ppps-in-horizon2020_en.pdf

[12]. Report from the Workshop on Cyber-Physical Systems: Uplifting Europe's Innovation Capacity, Thompson Haydn (Editor), *European Commission*, Brussels, Belgium, December 2013. https://ec.europa.eu/digital-single-market/en/news/report-workshop-cyber-physical-systems-uplifting-europe%E2%80%99s-innovation-capacity

[13]. Meike Reimann, Carsten Rückriegel, *et al.*, Road2CPS Priorities and Recommendations for Research and Innovation in Cyber-Physical Systems, 2017. https://www.steinbeis-europa.de/files/road2cps_2017-01.pdf

[14]. Christian Albrecht, Meike Reimann, Roadmaps and Recommendations for Strategic Action in the field of Systems of Systems in Europe, *Steinbeis-Edition*, Stuttgart, 2015. https://www.ifm.eng.cam.ac.uk/uploads/Roadmapping/Road2SoS-Ebook.pdf

[15]. Partnership for Robotics in Europe: Robotics 2020 Multi-Annual Roadmap for Robotics in Europe, *SPARC*, 2015. https://www.eu-robotics.net/cms/upload/downloads/ppp-documents/Multi-Annual_Roadmap2020_ICT-24_Rev_B_full.pdf

[16]. SESAR J. U., The Roadmap for Delivering High Performing Aviation for Europe, European ATM Master Plan Executive View, Edition 2015. https://ec.europa.eu/transport/sites/transport/files/modes/air/sesar/doc/eu-atm-master-plan-2015.pdf

[17]. Luke Georghiou, Value of Research, Policy Paper by the Research, Innovation, and Science Policy Experts (RISE), Directorate-General for Research and Innovation 2015 Research, Innovation, and Science Policy Experts High Level Group EUR 27367 EN, *European Commission*, Report, https://ec.europa.eu/research/openvision/pdf/rise/georghiou-value_research.pdf

Chapter 7. An Overview of Systems, Control & Optimisation (SCO) in Recent
European R&D Programmes and Projects (2013-2017) under the Emergence of New
Concepts and Broad Industrial Initiatives

[18]. Commission Staff Working Document In-Depth Interim Evaluation of HORIZON 2020, European Commission, version 13.06.2017, Brussels, 29.5.2017 SWD(2017) 220 final, https://ec.europa.eu/research/ evaluations/pdf/archive/h2020_evaluations/swd(2017)220-in-depth-interim_evaluation-h2020.pdf

[19]. Republic of Estonia, Ministry of Education and Research: Increased coherence and openness of European Union research and innovation partnerships, Report by Technopolis Group. https://www.hm.ee/sites/default/files/uuringud/eu_partenrships_2017.pdf

[20]. Edward A. Lee, Sanjit A. Seshia, Introduction to Embedded Systems, A Cyber-Physical Systems Approach, Second Edition, *MIT Press*, 2017. http://leeseshia.org/

[21]. Rajeev Alur, Principles of Cyber-Physical Systems, *MIT Press*, 2015. https://mitpress.mit.edu/books/principles-cyber-physical-systems

[22]. Houbing Song, Danda B. Rawat, S. Jeschke, Christian Brecher (Editors), Cyber-Physical Systems: Foundations, Principles and Applications, *Elsevier*, 2017.

[23]. A. Konstantellos, Chapter 1 in: Kostas Siozios, Dimitrios Soudris, Elias Kosmatopoulos (Editors), Cyber-Physical Systems, Decision Making Mechanisms and Applications, *River Publishers*, 2017 Gistrup and Delft.

[24]. Jan Willem Polderman, Jan C. Willems, Introduction to the Mathematical Theory of Systems and Control (Book). http://wwwhome.math.utwente.nl/ ~poldermanjw/onderwijs/DISC/mathmod/book.pdf

[25]. Björn Birnir, Dynamical Systems Theory, Center for Complex and Nonlinear Dynamics and Department of Mathematics University of California Santa Barbara, 2012. http://birnir.math.ucsb.edu/files/ bjorn/class-documents/main.pdf

[26]. H. W. Broer, B. Hasselblatt, F. Takens (Editors), Handbook of Dynamical Systems, Vol. 3, *Elsevier*, 2010.

[27]. Stavros Anastassiou, Spyros Pnevmatikos, Tassos Bountis, Classification of dynamical systems based on a decomposition of their vector fields, *J. Differential Equations, Elsevier,* Vol. 253, 2012, pp. 2252–2262.

[28]. X. D. Koutsoukos, P. J. Antsaklis, Hybrid Dynamical Systems: Review and Recent Progress, Chapter in Software-Enabled Control: Information Technologies for Dynamical Systems, T. Samad and G. Balas, Eds., *Wiley-IEEE Press*, 2003.

[29]. Alessandro Astolfi, Lorenzo Marconi, Analysis and Design of Nonlinear Control Systems: In Honor of Alberto Isidori, *Springer-Verlag*, 2008.

[30]. Claudio Melchiorri, Non Linear Systems: Analysis and Control, *Automatic Control & System Theory.* http://www-lar.deis.unibo.it/people/cmelchiorri/Files_ACST/100_Non_ Linear_Systems.pdf

[31]. Oleg Makarenkov, Jeroen S. W. Lamb, Dynamics and bifurcations of non-smooth systems: a survey, *Physica D,* 13 April, 2012,

http://www.utdallas.edu/~makarenkov/preprints/NonsmoothSurveyDraft_colleagues.pdf

[32]. Alfio Quarteroni, A brief survey of partial differential equations, in Numerical Models for Differential Problems by A. Quarteroni, *Springer*, 2017.

[33]. Xiaobing Feng, Roland Glowinski, Michael Neilan, Recent Developments in Numerical Methods for Fully Nonlinear Second Order Partial Differential Equations, *SIAM Rev.*, Vol. 55, No. 2, 2013, pp. 205–267.

[34]. Christian Albrecht, Meike Reimann, Roadmaps and Recommendations for Strategic Action in the field of Systems of Systems in Europe, *Steinbeis-Edition*, Stuttgart, 2015. https://www.ifm.eng.cam.ac.uk/uploads/Roadmapping/Road2SoS-Ebook.pdf

[35]. Michael Henshaw, Carys Siemieniuch, M. Sinclair, D. DeLaurentis, T-AREA-SOS Project: The Systems of Systems Engineering Strategic Research Agenda, TAREA-PU-WP5-R-LU-26, 2013. https://www.researchgate.net/publication/316688269

[36]. S. A. Henson, M. J. D. Henshaw, V. Barot, C. E. Siemieniuch, M. A. Sinclair, H. Dogan, S. L. Lim, C. Ncube, M. Jamshidi, D. DeLaurentis, Towards a Systems of Systems Engineering EU Strategic Research Agenda, in *Proceedings of the 8th International Conference on System of Systems Engineering*, Maui, Hawaii, USA, 2-6 June 2013. https://www.researchgate.net/profile/Huseyin_Dogan/publication/261110218_Towards_a_Systems_of_Systems_Engineering_EU_Strategic_Research_Agenda/links/59086f44458515ebb4907cf4/Towards-a-Systems-of-Systems-Engineering-EU-Strategic-Research-Agenda.pdf

[37]. Hadi Farhangi, Dincer Konura, Cihan H. Daglia, Combining Max-Min and Max-Max Approaches for Robust SoS Architecting, *Procedia Computer Science*, Vol. 95, 2016, pp. 103–110.

[38]. Robert Abbott, Mirsad Hadzikadic, Complex Adaptive Systems, Systems Thinking, and Agent-Based Modeling, in Advanced Technologies, Systems, and Applications, *Springer*, Nov. 2017, pp. 1-8.

[39]. Onix Mohammad F. A., Fielt Erwin, Gable Guy G., Towards a Complex Adaptive Systems Roadmap for Information Systems Research, in *Proceedings of the PACIS'2017,* Vol. 106, 2017. http://aisel.aisnet.org/pacis2017/106

[40]. Lachlan Birdsey, A Framework for Large Scale Complex Adaptive Systems Modeling, Simulation, and Analysis, in *Proceedings of the 16th International Conference on Autonomous Agents and Multiagent Systems (AAMAS'2017)*, São Paulo, Brazil, 2017. http://www.ifaamas.org/Proceedings/aamas2017/pdfs/p1824.pdf

[40a]. William P. Fisher, A Practical Approach to Modelling Complex Adaptive Flows in Psychology and Social Science, *Procedia Computer Science*, Chicago, Illinois, USA, Vol. 114, 2017, pp. 165–174.

[40b]. Siddiqa A., Niazi M., A Novel Formal Agent-Based Simulation Modeling Framework of an AIDS Complex Adaptive System (CAS),

Chapter 7. An Overview of Systems, Control & Optimisation (SCO) in Recent
European R&D Programmes and Projects (2013-2017) under the Emergence of New
Concepts and Broad Industrial Initiatives

International Journal of Agent Technologies and Systems (IJATS), Vol. 5, No. 3, 2013, pp. 33-53.

[40c]. Peter Dobias, Cheryl Eisler, Implications of the Complex Adaptive Systems Paradigm, *Defence Research and Development Canada – Centre for Operational Research and Analysis*, Canada, 2017, pp. 1-32. http://cradpdf.drdc-rddc.gc.ca/PDFS/unc266/p805157_A1b.pdf

[40d]. Muaz A. Niazi, Complex Adaptive Systems Modeling: A multidisciplinary Roadmap, *Complex Adaptive Systems Modeling*, 2013. http://www.casmodeling.com/content/1/1/1, doi:10.1186/2194-3206-1-1

[41]. Dmitriy Khodyrev, Galina Antonova, Using adaptive Kalman filtering by bitrate control module of sound signal compression in real-time systems, in *Proceedings of the 13ᵗʰ IFAC Symposium on Information Control Problems in Manufacturing*, Moscow, Russia, 3-5 June 2009. https://ac.els-cdn.com/S1474667016340423/1-s2.0-S1474667016340423-main.pdf?_tid=bcced4f2-be6e-11e7-87ff-00000aab0f6c&acdnat=1509477028_879914898b90291d9c65ac5730eb7a0d

[42]. Zongze Wu, Shengli Xie, Kexin Zhang, Rong Wu, Rate Control in Video Coding, in Recent Advances in Video Coding. http://cdn.intechopen.com/pdfs/16267/InTech-Rate_control_in_video_coding.pdf

[43]. M. Kobayashi, M. Johnson, Y. Wada, et al., A Low Noise and High Sensitivity Image Sensor with Imaging and Phase-Difference Detection AF in All Pixels, *Canon Inc. Japan, IISW*, 2015, http://www.imagesensors.org/Past%20Workshops/2015%20Workshop/2015%20Papers/Sessions/Session_1/1-03-M.Kobayashi_Canon_IISW2015.pdf

[44]. European Commission Joint Research Center, (JRC) https://ec.europa.eu/jrc/sites/jrcsh/files/jrc_detailed_work_programme_2017-18.pdf and Nano Forum http://euronanoforum2017.eu/

[45]. Strategic Research Agenda of EPoSS - The European Technology Platform on Smart Systems Integration, *EPoSS Pre-Print September*, 2013, https://www.smart-systems-integration.org/public/documents/publications/EPoSS%20SRA%20Pre-Print%20September%202013

[46]. Sergey Y. Yurish, Digital and Intelligent Sensors & Sensor Systems: Practical Design, 2012, Rome, https://www.iaria.org/conferences2012/filesSENSORDEVICES12/Digital_and_Intelligent_Sensors_Tutorial_Yurish.pdf

[47]. Cristina Gaspar, Juuso Olkkonen, Soile Passoja, Maria Smolander, Paper as Active Layer in Inkjet-Printed Capacitive Humidity Sensors, *Sensors*, 2017. http://europepmc.org/backend/ptpmcrender.fcgi?accid=PMC5539845&blobtype=pdf

[48]. Utility of the Future, An MIT Energy Initiative response to an industry in transition, *Massachusetts Institute of Technology*, 2016.

https://energy.mit.edu/wp-content/uploads/2016/12/Utility-of-the-Future-Full-Report.pdf

[49]. Federal Ministry for Economic Affairs and Energy (BMWi). Report of the Federal Government on Energy Research 2017. Research Funding for the Energy Transition, *Energie-Wende*, March 2017. https://www.bmwi.de/Redaktion/EN/Publikationen/Energie/bundesbericht-energieforschung-2017.pdf?__blob=publicationFile&v=2

[50]. Directorate General for Internal Policies Policy Department A: Economic and Scientific Policy, European Energy Industry Investments, IP/A/ITRE/2013-046, February 2017 PE 595.356 EN. http://www.europarl.europa.eu/studies, http://www.europarl.europa.eu/RegData/etudes/STUD/2017/595356/IPOL_STU(2017)595356_EN.pdf

[51]. Innovate UK and Department for Business, Energy & Industrial Strategy: Industrial Strategy Challenge Fund: Joint Research and Innovation, May 2017.

[52]. Deloitte: Technology, Media and Telecommunications - Predictions 2017. https://www.deloitte.co.uk/tmtpredictions/assets/pdf/Deloitte-TMT-Predictions-2017.pdf

[53]. European Commission 2015: The Knowledge Future: Intelligent policy choices for Europe 2050, A report to the European Commission, Report by an expert group on Foresight on Key Long-term Transformations of European systems: Research, Innovation and Higher Education (KT2050). https://ec.europa.eu/research/foresight/pdf/knowledge_future_2050.pdf

[54]. European Parliamentary Research Service Scientific Foresight Unit: Ethical Aspects of Cyber-Physical Systems Scientific Foresight study, (STOA) PE 563.501, June, 2016. http://www.europarl.europa.eu/RegData/etudes/STUD/2016/563501/EPRS_STU%282016%29563501_EN.pdf

[55]. Federal Ministry of Science, Research and Economy (BMWFW) and the Federal Ministry for Transport, Innovation, and Technology (BMVIT): Austrian Research and Technology, Report 2017 Report under Section 8(1) of the Research Organisation, Act on federally subsidised research, technology and innovation in Austria, Vienna, 2017. https://wissenschaft.bmwfw.gv.at/fileadmin/user_upload/forschung/publikationen/FTB_2017_en_WEB.PDF and https://investinaustria.at/en/downloads/brochures/research-development-austria-2014.pdf

[56]. Ministry of Public Administration – December 2015: e-CROATIA 2020 STRATEGY (Proposal). https://uprava.gov.hr/UserDocsImages/e-Hrvatska/e-Croatia%202020%20Strategy%20(20.01.2016.).pdf

[57]. Ministry for Economic Affairs and the Interior: Denmark's National Reform Programme 2017, April 2017. http://www.oim.dk, http://english.oim.dk/media/18649/danmarks-nationale-reformprogram-2017-uk-150617.pdf

Chapter 7. An Overview of Systems, Control & Optimisation (SCO) in Recent
European R&D Programmes and Projects (2013-2017) under the Emergence of New
Concepts and Broad Industrial Initiatives

[58]. National Council for Research & Technology GSRT: National Strategic
Framework for Research & Innovation 2014-2020.
http://www.gsrt.gr/Financing/Files/ProPeFiles81/ESPEK%202014-
2020%20by%20ESET%282010-2013%29.FIN.v.3.pdf

[59]. Spanish National Plan for Scientific & Technical Research and Innovation
2013-2016.
http://www.idi.mineco.gob.es/stfls/MICINN/Investigacion/FICHEROS/S
panish_RDTI_Plan_2013-2016.pdf

[60]. Forskning och utveckling i Sverige 2015, Research and development in
Sweden: overview, international comparisons, Stockholm 2016-08-08.
https://www.scb.se/Statistik/UF/UF0301/2015A01/Forskning-och-
Utveckling-i-Sverige-2015-Preliminara-uppgifter.pdf and
https://publikationer.vr.se/en/product/the-swedish-research-barometer-
2016/

[61]. Sabine Hafner-Zimmermann, Michael J. de C. Henshaw, The future of
trans-Atlantic collaboration in modelling and simulation of Cyber-
Physical Systems: A Strategic Research Agenda for Collaboration,
1st edition, *Steinbeis-Edition*, Stuttgart, 2017.

[62]. Romain Jacob, Marco Zimmerling, Pengcheng Huang, Jan Beutel, Lothar
Thiele, Towards Real-time Wireless Cyber-physical Systems.
http://people.ee.ethz.ch/~jacobr/publications/16_ECRTS_WiP.pdf

[63]. Ovidiu Vermesan, Joël Bacquet (Editors), Cognitive Hyperconnected
Digital Transformation Internet of Things Intelligence Evolution, *River
Publishers, Gistrup Denmark and Delft the Netherlands,* 2017.
https://www.riverpublishers.com/pdf/ebook/RP_9788793609105.pdf

[64]. Ovidiu Vermesan and Peter Friess (Editors), Internet of Things
Applications - From Research and Innovation to Market Deployment,
River Publishers, 2014.
http://www.internet-of-things-research.eu/pdf/IoT-
From%20Research%20and%20Innovation%20to%20Market%20Deploy
ment_IERC_Cluster_eBook_978-87-93102-95-8_P.pdf

[65]. European Parliament: Briefing May 2015 EPRS | European Parliamentary
Research Service Author: Ron Davies Members' Research Service PE
557.012 EN The Internet of Things Opportunities.

[66]. Eva Geisberger, Manfred Broy (Eds.), Living in a networked world
Integrated research agenda Cyber-Physical Systems (agenda CPS) EU
Additional reports, *Acatech STUDY*, March 2015.
http://www.acatech.de/fileadmin/user_upload/Baumstruktur_nach_Websi
te/Acatech/root/de/Publikationen/Projektberichte/acaetch_STUDIE_agen
daCPS_eng_WEB.pdf

[67]. NIST: Special Publication 1500-202 Framework for Cyber-Physical
Systems: Volume 2, Working Group Reports Version 1.0, June 2017
http://nvlpubs.nist.gov/nistpubs/SpecialPublications/NIST.SP.1500-
202.pdf

[68]. NIST Smart Grid and CPS Newsletter, July 2017. https://www.nist.gov/engineering-laboratory/smart-grid/nist-smart-grid-and-cps-newsletter-july-2017

[69]. NSF (USA): Cyber-Physical Systems (CPS), PROGRAM SOLICITATION NSF 17-529. https://www.nsf.gov/pubs/2017/nsf17529/nsf17529.htm

[70]. Dmitrii Legatiuk, Kosmas Dragos, Kay Smarsly, Modeling and evaluation of cyber-physical systems in civil engineering, in *Proceedings of the Applied Mathematics and Mechanics*, 4 May 2017. https://www.researchgate.net/profile/Kay_Smarsly/publication/31684687 3_Modeling_and_evaluation_of_cyber-physical_systems_in_civil_engineering/links/59139b27aca27200fe4b459 7/Modeling-and-evaluation-of-cyber-physical-systems-in-civil-engineering.pdf

[71]. Evgeny Kusmenko, Alexander Roth, Bernhard Rumpe, Michael von Wenckstern, Modeling Architectures of Cyber-Physical Systems, in Modelling Foundations and Applications (ECMFA'17), *Springer International Publishing*, 2017. www.se-rwth.de/publications/

[72]. Smirnov A., Sandkuhl K., Shilov N., Multilevel self-organisation of cyber-physical networks: synergic approach, *Int. J. Integrated Supply Management*, Vol. 8, Nos. 1/2/3, 2013, pp. 90–106. https://www.researchgate.net/profile/Alexander_Smirnov5/publication/26 4837518_Multilevel_self-organisation_of_cyber-physical_networks_Synergic_approach/links/5698e97a08ae1c427906800 5/Multilevel-self-organisation-of-cyber-physical-networks-Synergic-approach.pdf

[73]. 1) The Impact of Control Technology, T. Samad and A. M. Annaswamy (Eds.), *IEEE Control Systems Society*, 2011. http://www.ieeecss.org, http://ieeecss.org/general/impact-control-technology and 2) The Impact of Control Technology, 2nd ed., T. Samad and A. M. Annaswamy (eds.), *IEEE Control Systems Society*, 2014. http://ieeecss.org/general/IoCT2-report

[74]. Lamnabhi-Lagarrigue F., Annaswamy A., Isaksson A., Systems and control for the future of humanity, research agenda: current and future roles, impact and grand challenges, *Ann. Rev. Control*, Vol. 3, 2017, pp. 1-64. (IFAC Report 2017)

[75]. Karl J. Åström, P. R. Kumar, Control: A perspective, *Automatica*, Vol. 50, 2014, pp. 3–43. http://dx.doi.org/10.1016/j.automatica.2013.10.012

[76]. Stephen Kahne, Sergio Bittanti, Dongil "Dan" Cho, Janos Gertler, Shinji Hara, Vladimir Kucera, Lennart Ljung, Sarah Spurgeon, The IFAC Story. The International Federation of Automatic Control, *1st Web Edition*, 2017. https://www.ifac-control.org/about/the-ifac-story

[77]. Sabato Manfredi, Multilayer control of networked cyber-physical systems, Application to Monitoring, Autonomous and Robot Systems, *Springer*, 2017.

[78]. Ying Tan, Handbook of Research on Design, Control, and Modeling of Swarm Robotics, *IGI Global*, 2015.

Chapter 7. An Overview of Systems, Control & Optimisation (SCO) in Recent
European R&D Programmes and Projects (2013-2017) under the Emergence of New
Concepts and Broad Industrial Initiatives

[79]. Theory and Modeling of Complex Systems in Life Sciences, 18-22 September, 2017. http://inadilic.fr/wp-content/uploads/conference-2017-program-final.pdf

[80]. FET- Quantum Systems Work Programme, Fetflag-03-2018: FET Flagship on Quantum Technologies, High-Level Steering Committee, 28 June 2017.

[81]. Alfons Crespo, Alejandro Alonso, Marga Marcos, Juan A. de la Puente, Patricia Balbastre, Mixed Criticality in Control Systems, in *Proceedings of the 19th World Congress the International Federation of Automatic Control Cape Town*, South Africa, 24-29 August 2014. https://www.sciencedirect.com/science/article/pii/S1474667016435664

[82]. GTAI (German Trade and Invest): Industrie 4.0, Smart Manufacturing for the Future, July 2014: https://www.gtai.de/GTAI/Content/EN/Invest/_SharedDocs/Downloads/GTAI/Brochures/Industries/industrie4.0-smart-manufacturing-for-the-future-en.pdf

[83]. Filos E., Adding Value to Manufacturing: Thirty Years of European Framework Program Activity, in *Proceedings of the 15th IFIP WG 5.5 Working Conference on Virtual Enterprises, PRO-VE'2014*, Amsterdam, The Netherlands, October 6-8, 2014, pp. 24-36. https://link.springer.com/chapter/10.1007/978-3-662-44745-1_3

[84]. Filos E., Four years of 'Factories of the Future' in Europe: achievements and outlook, *International Journal of Computer Integrated Manufacturing*, Vol. 30, Issue 1, 2017, pp. 15-22. http://www.tandfonline.com/doi/abs/10.1080/0951192X.2015.1044759?journalCode=tcim20

[85]. E. W. Reutzel, A. R. Nassar, A survey of sensing and control systems for machine and process monitoring of directed-energy, metal-based additive manufacturing, in *Proceedings of the SFF Symposium*, 2014. https://sffsymposium.engr.utexas.edu/sites/default/files/2014-027-Reutzel.pdf

[86]. Valentijn de Leeuw, Industrie 4.0 in the Chemical Industry, *ARC Insights*, March 2017.

[87]. Adrian Wells, Metacognitive Therapy for-Anxiety and Depression, The Guilford Press, 2008, pp.1-9.

[88]. Adrian Wells, Gerald Matthews, Modelling cognition in emotional disorder, The S-REF model, 1994.

[89]. T. O. Nelson, L. Narens, Metamemory: A theoretical framework and new findings, in Psychology of Learning and Motivation, G. H. Bower (Ed.), Vol. 26, 1990, pp. 125–173.

[90]. Henrik Nordahl, Adrian Wells, Testing the metacognitive model against the benchmark CBT model of social anxiety disorder: Is it time to move beyond cognition ?, PLOS ONE, 12(5), 2017. https://doi.org/10.1371/journal.pone.0177109

[91]. Modelling cognition in emotional disorder: The S-REF model. Available from: https://www.researchgate.net/publication/222615343_Modelling_cogniti on_in_emotional_disorder_The_S-REF_model [accessed Oct. 18 2017].

[92]. Joëlle Proust, The Philosophy of Metacognition, Mental Agency and Self-Awareness, *Oxford University Press*, 2013, Paperback 2015.

[93]. Manfred Morari, Predicting the Future of Model Predictive Control, Lecture in Honor of Professor David Clarke Oxford, January 9, 2009. http://www.eng.ox.ac.uk/control/events/slides/Morari.pdf

[94]. Marta Kwiatkowska, Gethin Norman, David Parker, A Framework for Verification of Software with Time and Probabilities, Springer, Vol. 6246 of LNCS, September 2010, pp. 25-45. http://www.prismmodelchecker.org/papers/formats10inv.pdf

[95]. Sofie Andersson, Alexandros Nikou, Dimos V. Dimarogonas, Control Synthesis for Multi-Agent Systems under Metric Interval Temporal Logic Specifications, 2017. https://arxiv.org/pdf/1703.02780.pdf

[96]. Marta Kwiatkowska, David Parker, Clemens Wiltsche: PRISM-games: Verification and Strategy Synthesis for Stochastic Multi-player Games with Multiple Objectives, *International Journal on Software Tools for Technology Transfer, Springer,* 2017. http://qav.comlab.ox.ac.uk/papers/kpw17.pdf

[97]. Češka Milan, Dannenberg Frits, Kwiatkowska Marta, Paoletti Nicola, Precise Parameter Synthesis for Stochastic Biochemical Systems, *Lecture Notes in Computer Science*, 2014.

[98]. Fabio Cremona, Marten Lohstroh, David Broman, Stavros Tripakis, Edward A. Lee, Hybrid Co-simulation: It's About Time, *Technical Report*, No. UCB/EECS-2017-6, 6 April, 2017. http://www2.eecs. berkeley.edu/Pubs/TechRpts/2017/EECS-2017-6.html

[99]. E. A. Lee, Living Digital Things, in *Proceedings of the 3rd IEEE International Conference on Collaboration and Internet Computing*, 15-17 October 2017. http://platoAndTheNerd.org https://chess.eecs.berkeley.edu/pubs/1199/LivingDigitalThings_CIC.pdf

[100]. Life 3.0: Being Human in the Age of Artificial Intelligence MAX TEGMARK, *Knopf Doubleday Publishing Group*, 2017. https://www.nature.com/nature/journal/v548/n7669/pdf/548520a.pdf

[101]. Nick Bostrom's: Superintelligence, 2014 (Oxford University Press)

[102]. Stuart Russell, Artificial intelligence: The future is super-intelligent, *Nature*, Vol. 548, 31 August 2017, pp. 520–521. (Critique of [100])

Chapter 7. An Overview of Systems, Control & Optimisation (SCO) in Recent
European R&D Programmes and Projects (2013-2017) under the Emergence of New
Concepts and Broad Industrial Initiatives

Appendix

7.A1. Additional Information on Selective Project Groups in SCO Topics

In Section 7.4.2 of this chapter, we mentioned examples of titles of projects in Systems, Control and Optimisation (SCO). In this Annex we provide additional information on selective groups of projects, background references, links and publications. Although projects have a typical duration of 1-5 years, (except for a limited number, receiving a follow up grant), they maintain their web-sites beyond the end of the EC contract, but publications may not be updated. Certain thematic groups, to some extent, overlap and some topics such as CPS and Optimisation are currently evolving towards unifying themes. Therefore, apart from the standard disciplines, the reader may expect some topics to be re-located under new headings. Furthermore, projects dealing with multiple topics are mentioned in this annex only once. The references are added in-situ and are not numbered like in the main chapter sections. The key, first level, source about EC projects since 1990 is the Cordis portal and search engine: http://cordis.europa.eu/projects/home_en.html.

7.A2. Project Groups

7.A2.1. Algebraic and Geometric Methods (See Also under PDEs Group Below)

Project A-DAP Project ID: 329084, (Approximate Solutions of the Determinantal Assignment Problem and Distance Problems). *Note*: The idea has emerged as the abstract problem formulation of poles and zeros assignment of linear systems. Developments allow the transformation of the DAP framework from a synthesis methodology (exact problems) to a design approach that can handle model uncertainty and capable to develop approximate solutions. Varieties: Linear and Grassmann.

Project: CONNECT, ID-734922, Combinatorics of Networks and Computation, cordis.europa.eu/project/rcn/207071_en.html. *Note*: The project contains the work packages: "Geometric networks",

"Stochastic Geometry and Networks", "Restricted orientation geometry", "Graph-based algorithms for UAVs and for MIR[1]"

- Werner M. Seiler, Eva Zerz Algebraic Theory of Linear Systems: A Survey. http://www.mathematik.uni-kassel.de/~Seiler/Papers/PDF/LinSys.pdf

7.A2.2. Automata-based System Design

Project: RSCS, ID: 268310, (Automata Based Interfaces for Dynamic Resource Scheduling in Control Systems). *Note*: Dynamic network resource scheduling techniques (bus scheduling) are based on Automata interfaces and hybrid systems. Verification techniques for distributed algorithms and resource utilization in embedded control software, combining control and dynamic scheduling mechanisms on hard deadlines. Applications cover self-configuring wireless control networks and quadrotor control:

http://cordis.europa.eu/docs/results/268/268310/final1-rscs-final-report.pdf

Project: SENTIENT, ID: 755953, (ERC), Scheduling of Event-Triggered Control Tasks, (2018-2023). *Note*: Research aims at the improvement of Event-triggered control (ETC) techniques, modelled as timed-priced-game-automata (TPGA) to prevent data communication collisions and ensure prescribed performances for the control approach.

Project: PACOMANEDIA, ID: 308253 (Partially Coherent Many-Body Nonequilibrium Dynamics for Information Applications), ERC and EPSRC (UK). *Note*: study of the nonequilibrium dynamics of many strongly coupled quantum systems and the high control demanded on every quantum bit and interactions with other quantum bits. Investigation of networks of interacting spins as a possible Automaton for running an entire quantum algorithm, whether magnon wave-packets can be used for linear optics-type computation.

Project: MCUNLEASH, ID-259267, Model Checking Unleashed, 2010-2015, http://cordis.europa.eu/result/rcn/178218_en.html

---

[1]MIR: Musical Information Retrieval

Chapter 7. An Overview of Systems, Control & Optimisation (SCO) in Recent
European R&D Programmes and Projects (2013-2017) under the Emergence of New
Concepts and Broad Industrial Initiatives

- Background: Rajeev Alur and David L. Dill, A theory of timed automata, *Theoretical Computer Science*, Vol. 126, 1994, pp. 183-235, *Elsevier.*   https://ac.els-cdn.com/0304397594900108/1-s2.0-03043 97594900108-main.pdf?_tid=5436e58e-bd73-11e7-a36d-0000aab0f27&acdnat=1509369049_27a1f063a4b9d476f0d7aa4bc31cd 7ac

- Background: D. Kaynar, N. Lynch, R. Segala, F. Vandrager, The Theory of Timed I/O Automata, 2nd Edition Morgan & Claypool Publishers, 2010 (link to earlier EC project AMETIST)

- Loïc Paulevé, Morgan Magnin, Olivier Roux, Sufficient Conditions for Reachability in Automata Networks with Priorities, *Journal of Theoretical Computer Science (TCS), Elsevier,* 2015. http://www.sciencedirect.com/science/article/pii/S0304397515007872.

<10.1016/j.tcs.2015.08.040>. <hal-01202671> and https://hal.archives-ouvertes.fr/hal-01202671/document, 2015, Final report: http://cordis.europa.eu/result/rcn/178218_en.html

- Leonardo Banchi, Nicola Pancotti and Sougato Bose, Quantum gate learning in qubit networks: Toffoli gate without time-dependent control, *Quantum Information*, Vol. 2, 2016. http://discovery.ucl.ac.uk/1502564/1/npjqi201619.pdf

- Stefano Pirandola, Riccardo Laurenza, Carlo Ottaviani1 & Leonardo Banchi, Fundamental limits of repeaterless quantum communications, *Nature Communications*, 26 Apr 2017.

- Milka Hutagalung, Martin Lange and Etienne Lozes: Buffered Simulation Games for Büchi Automata, in Automata and Formal Languages Z. Esik and Z. Fulop (Eds.),, (AFL 2014) EPTCS 151, 2014, pp. 286–300.

- Maxime Folschette, Loïc Paulevé, Morgan Magnin, Olivier Roux, Sufficient Conditions for Reachability in Automata Networks with Priorities, *Journal of Theoretical Computer Science (TCS), Elsevier*, 2015: http://www.sciencedirect.com/science/article/pii/S0304397515007872

7.A2.3. Dynamical Systems

Project: SFSYSCELLBIO, ID-661650 (H2020), (Slow-Fast Systems in Cellular Biology), 2016-18. *Note*: Mathematical modelling to handle the structural complexity of cellular processes and their dynamics using ODE models in cell biology i.e. cell division cycle, NF-kB signalling pathway, and the p53 system. Questions of biological interest are: existence and stability of equilibria, periodic oscillations, switching phenomena, and bifurcations. Slow-fast dynamical systems, i.e. systems with solutions varying on very different timescales are abundant in biology in general and in cellular biology. Methods for systems with multiple time scale dynamics, known as geometric singular perturbation theory (GSPT).

Project: ILDS, ID-655209, H2020, Integrability and Linearization of Dynamical Systems, 2016-2018. *Note*: R&D directions: (1) Methods and tools of algebraic geometry and computational algebra, to study the integrability of the nonlinear and high dimension dynamical systems, combining symbolic computations with methods of the theory of integrability of dynamical systems. (2) Bifurcations of limit cycles and critical periods arising after perturbations of systems of differential equations, including isochronicity (which is equivalent to the problem of linearization). (3) Global topological linearization.

- General Background: X. D. Koutsoukos and P. J. Antsaklis, Hybrid Dynamical Systems: Review and Recent Progress, Chapter in Software-Enabled Control: Information Technologies for Dynamical Systems, T. Samad and G. Balas (Eds.), *Wiley-IEEE*, 2003.

- Background: R. Alur, C. Courcoubetis, N. Halbwachs, T. A. Henzinger, P.-H. Ho, X. Nicollin, A. Olivero, J. Sifakis, and S. Yovine, The algorithmic analysis of hybrid systems, *Theor. Comput. Sci.*, Vol. 138, Issue 1, 1995, pp. 3–34.

- Background: B. De Schutter, W. Heemels, J. Lunze, and C. Prieur, Survey of modeling, analysis, and control of hybrid systems, in Handbook of Hybrid Systems Control – Theory, Tools, Applications, J. Lunze and F. Lamnabhi-Lagarrigue (Eds.), Chapter 2, pp. 31–55. *Cambridge University Press*, Cambridge, UK, 2009.

- Nicola Guglielmi and Ernst Hairer, Classification of Hidden Dynamics in Discontinuous Dynamical Systems, *Siam j. Applied Dynamical Systems*, 14 June 2015:
http://www.unige.ch/~hairer/preprints/discsys.pdf

- Munters W., Meyers J., An optimal control framework for dynamic induction control of wind farms and their interaction with the atmospheric boundary layer, *Philos Trans A Math Phys Eng Sci.*, 2017.
http://dx.doi.org/10.1098/rsta.2016.0100

- ETH projects: for example: http://www.nano-tera.ch/projects/360.php#desc

7.A2.4. Non-linear Systems and Bifurcations

Project: MUDIBI, ID: 300281, Multiple-Discontinuity Induced Bifurcations in Theory and Applications, 2012-2014. Note: Deals with switching manifold in non-linear systems.

Project: PINQUAR, ID-657042, H2020, (Polaritons in the Quantum Regime), 2015-2017. *Note*: Semiconductor optical microcavities host hybrid light-matter quasi-particles known as polaritons, formed by strong coupling between cavity photons and quantum well excitons. Enhancing the polariton non-linearity is a pre-requisite for accessing the quantum regime. The project analyses the extreme non-linear optical phenomena within the mean-field limits, bifurcations, exceptional points, and spin-squeezing, non-classical photon statistics can be engineered in time and energy, and quantum simulations can be performed in a controlled solid-state platform suitable for all-optical integrated circuits.

Project: CDSANAB, ID-750865, H2020, (Complex Dynamics and Strange Attractors Through Non-Autonomous Bifurcations) 2018-2020. *Note*: focus on common Saddle-node and Hopf bifurcations, through multiscale analysis techniques and tools used in life sciences to reliably track complex behaviour near saddle-node bifurcations. Concerning randomly driven systems, it is aimed at shear-induced chaos.

Project: UBPDS, ID-655212, H2020, (Unfoldings and Bifurcations of Polynomial Differential Systems). *Note*: focus on planar polynomial differential systems having some topological structures, which have centres or degenerate equilibria. Rigorous bio-mathematical models. *http://cordis.europa.eu/project/rcn/198654_en.html*

Project: SOLIRING, ID- 691011, H2020, (Solitons and Frequency Combs in Micro-resonators), 2016-2018. *Note*: It is a multi-lateral transfer of knowledge among leading research groups in the UK, Sweden, South Africa and Russia in the fields of nonlinear fibre optics, optical micro-resonators, nano-photonics, optical microcavities, lasers and nonlinear dynamics, specifically with the second-order and Raman nonlinearities.

- Background: Lennart Ljung, Identification of Nonlinear Systems, Report LiTH-ISY-R-2784, *Linköpings*, Sweden, 2007. http://www.diva-portal.org/smash/get/diva2:316888/FULLTEXT01.pdf

- Arianna Dal Forno, Ugo Merlone, Viktor Avrutin, Dynamics in Braess Paradox with Non-Impulsive Commuters, *Hindawi Publishing Corporation, Discrete Dynamics in Nature and Society*, Vol. 2015, 2015.

- Quan Kai-Yuan Cai, Reviews·A Survey of Repetitive Control for Nonlinear Systems. http://pub.nsfc.gov.cn/sficen/ch/reader/create_pdf.aspx?file_no=20100 2045&fl ag=1&journal_id=sficen&year_id=2010

- Ali Baharev, Ferenc Domes, Arnold Neumaier, A robust approach for finding all well-separated solutions of sparse systems of nonlinear equations, *Springer*, 2016. http://www.mat.univie.ac.at/~neum/ms/maniSol.pdf

- Oleg Makarenkov, Jeroen S. W. Lamb, Dynamics and bifurcations of non-smooth systems: a survey, *Physica D*, April 13, 2012, Imperial College London. http://www.utdallas.edu/~makarenkov/preprints/NonsmoothSurveyDraf t_colleagues.pdf

- Luisa Faella1 and Carmen Perugia, Optimal control for evolutionary imperfect transmission problems, Boundary Value Problems, 2015.

7.A2.5. Complex Systems

Project: **CRITICS, ID-643073, H2020, (Critical Transitions in Complex Systems), 2015-19.** *Note*: Training network addressing mathematical theories for the existence of early-warning signals for sudden changes in dynamical behaviour, so-called critical transitions, which have been reported by applied scientists in various contexts (e.g. epileptic seizures, stock market collapses, earthquakes, and climatic phenomena). From the SCO point of view, topics bifurcation theory, primarily for low-dimensional deterministic dynamical systems.

- Background: Rebecca Dodder and Robert Dare, Complex Adaptive Systems and Complexity Theory, Inter-related Knowledge Domains, *MIT*, October 2000.

- Background: D. Sumpter & S. C. Nicolis, Survey of self-organisation phenomena and complex non-linear phenomena, Lecture presentation, 2008. http://www2.math.uu.se/~david/web/AppDyn/Lecture1a.pdf

- Background: W. H. Sandholm, Population Games and Evolutionary Dynamics, *The MIT Press*, Cambridge, Mass, USA, 2010.

- http://mp.ipme.ru/ipme/labs/ccs/index.html

- Complex Systems: Control And Modeling Problems (CSCMP), September 2017. http://www.iccs.ru/cscmp/cscmp_topics_e.html#

- Martí Rosas-Casals, Sergi Valverde and Ricard v. Solé, Topological Vulnerability of European Power Grid under errors and attacks, *Int. J. Bifurcation Chaos*, Vol. 17, No. 2465, 2007. https://doi.org/10.1142/S0218127407018531

- Theory and Modeling of Complex Systems in Life Sciences, 18-22 Sept. 2017. http://inadilic.fr/wp-content/uploads/conference-2017-program-final.pdf

- Robert J. Glass, Arlo L. Ames, Theresa J. Brown, S. Louise Maffitt, *et al.*, Complex Adaptive Systems of Systems (CASoS) Engineering: Mapping Aspirations to Problem Solutions, Sandia National Laboratories, *in Proceedings of the 8th International Conference on Complex Systems,* At Quincy, MA, USA, Vol. 26, June 2011.

- Robert Abbott and Mirsad Hadzikadic, Complex Adaptive Systems, Systems Thinking, and Agent-Based Modeling, *Advanced Technologies, Systems, and Applications, 2017.*

- Ian Sommerville, Dave Cliff, Radu Calinescu, Justin Keen, Tim Kelly, Marta Kwiatkowska, John McDermid and Richard Paige, Large-scale Complex IT Systems, *Communications of the ACM*, September 2011.

Complex Adaptive Systems (CAS) in Non-engineering Disciplines:

- William P. Fisher, Jr.: A Practical Approach to Modelling Complex Adaptive Flows in Psychology and Social Science, *in Proceedings of the Complex Adaptive Systems Conference with Theme: Engineering Cyber Physical Systems, CAS*, 2017, Chicago, Illinois, USA, *Procedia Computer Science*, Vol. 114, 2017, pp. 165–174.

- Siddiqa A., & Niazi M., A Novel Formal Agent-Based Simulation Modeling Framework of an AIDS Complex Adaptive System (CAS), *International Journal of Agent Technologies and Systems (IJATS)*, Vol. 5, No. 3, 2013, pp. 33-53.

- Muaz A Niazi, Complex Adaptive Systems Modeling, A multidisciplinary Roadmap, *Complex Adaptive Systems Modeling, Springer*, 2013. http://www.casmodeling.com/content/1/1/1

- Bonnie Johnson and Alejandro Hernandez, Exploring Engineered Complex Adaptive Systems of Systems, Complex Adaptive Systems, Publication 6 Cihan H. Dagli, Editor in Chief Conference Organized by Missouri University of Science and Technology 2016, Los Angeles, CA, *Procedia Computer Science*, Vol. 95, 2016, pp. 58-65. *Note*: Complex Adaptive Systems of Systems (CASoS): Focus on a specific area of CAS e.g. ecology, social sciences, large communication networks, biological sciences etc., using agent-based simulation models or else a complex network model based on data from CAS e.g. Gene regulatory Networks, Social Networks, Ecological Networks etc.

7.A2.6. Formal Methods

Background and Examples of Recent Activities:

- Classic survey paper: EDMUND M. CLARKE, JEANNETTE M. WING, ET AL: Formal Methods: State of the Art and Future Directions,

Chapter 7. An Overview of Systems, Control & Optimisation (SCO) in Recent
European R&D Programmes and Projects (2013-2017) under the Emergence of New
Concepts and Broad Industrial Initiatives

ACM Computing Surveys, Vol. 28, No. 4, December 1996. http://www.site.uottawa.ca/~bochmann/ELG7187C/Course*Notes*/Litera ture/Clarke%20-%20FM%20State%20of%20the%20art%20-%2096.pdf

- Broad Background (Book): HOSSAM A. GABBAR: Modern Formal Methods and Applications, *Springer*, 2006.

- International SPIN Symposium on Model Checking of Software, and the RERS Challenge 2017 at the 7th International Challenge on the Rigorous Examination of Reactive Systems. http://spinroot.com/spin/symposia/ws17/index.html and http://www.rers-challenge.org/2017/

7.A2.6.1. Verification

Project: UnCoVerCPS "Unifying Control and Verification of Cyber-Physical Systems", Background Related to UnCoVerCPS: http://cordis.europa.eu/project/rcn/194114_en.html

- M. Burger, G. Notarstefano and F. Allgöwer, A polyhedral approximation framework for convex and robust distributed optimization, *IEEE Transactions on Automatic Control*, Vol. 59, No. 2, 2014, pp. 3740–3754.

- Mathias Bürger, Giuseppe Notarstefano, Francesco Bullo, Frank Allgöwer, A distributed simplex algorithm for degenerate linear programs and multi-agent assignments, *Automatica*, Vol. 48, No. 9, 2012, pp. 2298–2304. https://fenix.tecnico.ulisboa.pt/downloadFile/ 3779579952578/distributed-simplex.pdf

- German Research Foundation within the Cluster of Excellence in Simulation Technology (EXC 310/1) and the Priority Program SPP 1305. The research of G. Notarstefano has received funding from the European Community's Seventh Framework Programme (FP7/2007–2013) under grant agreement no. 224428 (CHAT)

Project: MoVeS: "Modelling, Verification and Control of Complex Systems: From Foundations to Power Network Applications"

- Background: General report on R&D in aerospace: Verification of the control laws in aerospace, Laminar wing: On the right path, Clean Sky Magazine, Skyline, No. 20, November 2016, http://www.cleansky.eu/sites/default/files/inline-files/Skyline%2020%20-%20final_0.pdf. *Note*: In particular, the validation of the simulation performed to define the GLA[1] concepts • the verification of the control laws capability when coupled to an aeroelastic system during gust occurrence (simulated in wind tunnel) • the verification of the interaction and signal chain functionality between control laws and feed forward/wing accelerometer sensors. To achieve these aims an aero-servo-elastic A/C half-model has been manufactured, with flexible wing reproduced by stiffness scaling the real-size wing dynamic response under gust excitation loads, equipped with active control movables (aileron and elevator), relevant actuation system and sensors (accelerometers, alpha flow sensor); and integrating control laws engineering model.

- Stefan Mitsch, Grant Olney Passmore and André Platzer: Collaborative Verification-Driven Engineering of Hybrid Systems, *Math. Comput. Sci.*, 2014. https://arxiv.org/pdf/1403.6085.pdf //[EU-US]

- Jade Alglave, Alastair F. Donaldson, Daniel Kroening and Michael Tautschnig, Making Software Verification Tools Really Work, http://www0.cs.ucl.ac.uk/staff/j.alglave/papers/atva11.pdf *Note*: Supported by the Engineering and Physical Sciences Research Council (EPSRC) under grants no. EP/G051100/1 and EP/H017585/1. http://www0.cs.ucl.ac.uk/staff/j.alglave/papers/atva11.pdf

Project: VERIWARE, ID-246967, From Software Verification to Everyware Verification, 2010-2016

- Češka Milan, Dannenberg Frits, Kwiatkowska Marta, Paoletti Nicola, Precise Parameter Synthesis for Stochastic Biochemical Systems, *Lecture Notes in Computer Science*, 2014.

- Lu Feng, Clemens Wiltsche, Laura Humphrey, Ufuk Topcu: Controller Synthesis for Autonomous Systems Interacting with Human Operators, 2015. http://www.prismmodelchecker.org/papers/iccps15.pdf *Note*: Joint funding: EC ERC Advanced Grant AdG-246967 VERIWARE, and

---

[1] GLA: Gust Load Alleviation, control system (control laws, sensors and devices)

e.g. AFOSR grant # FA9440-12-1-0302, and ONR grant # N000141310778.

Project: DataVerif, ID-301166, Temporal Reasoning with Data for Verification, 2012-2015

- Kshitij Bansal and Stephane Demri, A Note on the Complexity of Model-Checking Bounded Multi-Pushdown Systems, 2012, pp. 1-28. https://arxiv.org/pdf/1212.1485.pdf

Project: i-FLY, ID-37180, FP6-AEROSPACE, Safety, Complexity and Responsibility Based Design and Validation of Highly Automated Air Traffic Management

Project: HARMONICS, ID-269851, Assessment of Reliability of Modern Nuclear I&C Software, 2011-2015

- Alessandro Abate: Systems Verification (Lecture slides), http://aims.robots.ox.ac.uk/wp-content/uploads/2016/01/2016_aims_sv_day1.pdf

Project: MOGENTES, ID-216679, Model-based Generation of Tests for Dependable Embedded Systems, http://www.mogentes.eu/, 2008-2012

- General background: NASA: Robust Software Engineering, 2012. https://ti.arc.nasa.gov/tech/rse/publications/vnv/

- Vijay D'Silva, Daniel Kroening, Georg Weissenbacher, A Survey of Automated Techniques for Formal Software Verification, *Transactions on CAD*, 2008, pp. 1-14. http://www.kroening.com/papers/tcad-sw-2008.pdf

- He N., Ruemmer P., Kroening D., Test-Case Generation for Embedded Simulink via Formal Concept Analysis, 2011. http://www.kroening.com/publications/view-publications-hrk2011-dac.html

7.A2.6.2. Verification-Refactoring Methods (e.g. Model Checking, Hybrid Systems)

Project: SPHINX, ID-328378, EC, MSCA, A Co-Evolution Framework for Model Refactoring and Proof Adaptation in Cyber-Physical Systems, 2014-2016

- Stefan Mitsch, Jan-David Quesel and Andre Platzer, Refactoring, Refinement, and Reasoning. A Logical Characterization for Hybrid Systems, 2014.
http://www.cis.jku.at/images/userUploads/pdfs/d807661852e0b493121 5b1a729bbb464-Mitsch2014.pdf

- Andreas Muller, Stefan Mitsch, Werner Retschitzegger and Wieland Schwinger, A Conceptual Reference Model of Modeling and Verification Concepts for Hybrid Systems, 2014.

- Albert Rizaldi, Cyber-Physical Systems (CPS) Seminar, 2017.
http://www.i6.in.tum.de/Main/TeachingSs2017SeminarCyberPhysicalS ystems.

- A Co-Evolution Framework for Model Refactoring and Proof Adaptation in Cyber-Physical Systems.
http://www.cis.jku.at/projects/4?view=project

- Marcello M. Bersani, Francesco Marconi, Matteo Rossi, Madalina Erascu, Formal Verification of Data-Intensive Applications through Model Checking Modulo Theories, 2017.
http://spinroot.com/spin/symposia/ws17/SPIN_2017_paper_43.pdf

Project: EQUALIS, ID-308087, ERC, Enhancing the Quality of Interacting Systems 2013-2019,
http://cordis.europa.eu/project/rcn/104739_en.html

Project: CASSTING, ID- 601148, ICT, Collective Adaptive System Synthesis with non-zero-sum games,
http://cordis.europa.eu/docs/projects/cnect/8/601148/080/deliverabl es/001-D51SITEWEBCASSTING.pdf

Project: BUCOPHSYS, Project ID: 639365 (ERC), Bottom-up hybrid control and planning synthesis with application to multi-robot multi-human coordination (2015-2020). *Note:* Decentralized

Chapter 7. An Overview of Systems, Control & Optimisation (SCO) in Recent
European R&D Programmes and Projects (2013-2017) under the Emergence of New
Concepts and Broad Industrial Initiatives

control at the continuous level, including both continuous state and discrete plan/abstraction information.

- Ocan Sankur, Patricia Bouyer, Nicolas Markey, Pierre-Alain Reynier, Robust Controller Synthesis in Timed Automata, http://www.lsv.fr/Publis/PAPERS/PDF/SBMR-concur13.pdf

Joint funding: by ANR projects ImpRo (ANR-2010-BLAN-0317) and ECSPER (ANR-2009-JCJC-0069), by EC Starting grant EQualIS (308087), Cassting (FP7-ICT-601148).

- Marta Kwiatkowska, Gethin Norman and David Parker, A Framework for Verification of Software with Time and Probabilities, *Springer,* Vol. 6246 of LNCS, September 2010, pp. 25-45. http://www.prismmodelchecker.org/papers/formats10inv.pdf

- Sofie Andersson, Alexandros Nikou, Dimos V. Dimarogonas, Control Synthesis for Multi-Agent Systems under Metric Interval Temporal Logic Specifications, 2017. https://arxiv.org/pdf/1703.02780.pdf *Note*: H2020 ERC Starting Grand BUCOPHSYS, the Swedish Research Council (VR), the Swedish Foundation for Strategic Research (SSF) and the Knut & Alice Wallenberg Foundation).

- Marta Kwiatkowska, David Parker, Clemens Wiltsche, PRISM-games: Verification and Strategy Synthesis for Stochastic Multi-player Games with Multiple Objectives, 2017. http://qav.comlab.ox.ac.uk/papers/kpw17.pdf

7.A2.6.3. Model-checking and Run Time Verification

Project: MISSA, ID-212088, Model Checking of Hybrid Systems using Shallow Synchronization,
http://cordis.europa.eu/result/rcn/196174_en.html

Project: DICE, ID-644869, H2020. *Note*: Intensive applications based on the Storm technology, a framework for developing streaming applications, on array-based systems formalism, introduced by Ghilardi et al., "a suitable abstraction of infinite state systems that we used to model the runtime behaviour of Storm-based applications. Data-intensive applications; Storm technology".

- Martin Leucker, Christian Schallhart: A brief account of runtime verification, *The Journal of Logic and Algebraic Programming*, Vol. 78, 2009, pp. 293–303.

- Vaiapury Karthikeyan, Aksay Anil, Lin Xinyu, Izquierdo Ebroul, Papadopoulos Christopher, A new cost-effective 3D measurement audit and model comparison system for verification tasks, *Springer*, 2013.

7.A2.7. Consensus Methods (Including Non-ICT / Non-engineering Systems)

Project: DEMAND, ID-334098, MSCA, DECENTRALIZED MONITORING and ADAPTIVE CONTROL for NETWORKED DYNAMICAL SYSTEMS, 2013-2017, http://cordis.europa.eu/result/rcn/197821_en.html

- General Background: Schenato Sandro Zampieri, et al., (Univ. Padova), Applications of Consensus Algorithms to Wireless Sensor Networks, 2008. http://automatica.dei.unipd.it/tl_files/utenti/ lucaschenato/Papers/Presentations/Schenato_Berkeley08.pdf

- Background: Srdjan S. Stankovic, Nemanja Ilic, Milos S. Stankovic and Karl Henrik Johansson, Distributed Change Detection Based on a Consensus Algorithm, *IEEE Transactions on Signal Processing*, Vol. 59, No. 12, December 2011.
https://people.kth.se/~kallej/papers/wsn_tsp11.pdf

- Qingling Wang, Huijun Gao, Fuad Alsaadi, Tasawar Hayat, An overview of consensus problems in constrained multi-agent coordination, *Systems Science & Control Engineering*, Vol. 2, Issue 1, 2014.
http://www.tandfonline.com/doi/full/10.1080/21642583.2014.897658

- Theodor S. Borsche and Florian Dorfler, On Placement of Synthetic Inertia with Explicit Time-Domain Constraints.
https://arxiv.org/pdf/1705.03244.pdf

- Theodor S. Borsche, Impact of demand and storage control on power system operation and dynamics, Ph.D. dissertation, ETH Zurich, Feb 2016.

Chapter 7. An Overview of Systems, Control & Optimisation (SCO) in Recent
European R&D Programmes and Projects (2013-2017) under the Emergence of New
Concepts and Broad Industrial Initiatives

- Marcello Colombino, Dominic Groß, Jean-Sebastien Brouillon and Florian Dorfler, Global phase and magnitude synchronization of coupled oscillators with application to the control of grid-forming power inverters. https://arxiv.org/pdf/1710.00694.pdf

- Srdjan S. Stankovic, Milos S. Stankovic, Dusan M. Stipanovic, Consensus Based Overlapping Decentralized Estimation With Missing Observations and Communication Faults, in *Proceedings of the 17th World Congress The International Federation of Automatic Control*, Seoul, Korea, 6-11 July 2008. https://pdfs.semanticscholar.org/2cd4/cf36e90cbdf6c8330525efd88781 05edad75.pdf *Note*: Recent work on stochastic approximations supported by MSCA -2012-334098.

- Miloš S. Stanković, Srdjan S. Stanković, Dušan M. Stipanović, Consensus-based decentralized real-time identification of large-scale systems, *Automatica*, Vol. 60, 2015, pp. 219–226. This work was supported by the EU Marie Curie CIG (PCIG12-GA-2012-334098). http://dx.doi.org/10.1016/j.automatica.2015.07.018. *Note*: Related to EC project PRODI.

- Bo Shen, Zidong Wang, Y. S. Hung, Distributed H∞-Consensus Filtering in Sensor Networks with Multiple Missing Measurements: The Finite-Horizon Case. http://bura.brunel.ac.uk/bitstream/2438/4703/3/ Fulltext.pdf. *Note*: supported in part by the Engineering and Physical Sciences Research Council (EPSRC), U.K. under Grant GR/S27658/01, the Royal Society of the U.K., and the Alexander von Humboldt Foundation of Germany.

- Stanković Miloš S., Ilić Nemanja, Stanković Srdjan S., Distributed Stochastic Approximation: Weak Convergence and Network Design, 2014.

Project: POSTO, ID-286536, Methodology and SW libraries for the design and development of GALILEO/EGNOS-based Positioning applications for smartphones. *Note*: Research on Range Consensus (RANCO) algorithm dealing with the problem of multiple faulty measurements. It has a low computational complexity compared to the MSS RAIM if more than one measurement is faulty. It calculates a position solution based on four satellites and compares this estimate with the pseudo ranges of all the satellites that did not contribute to this solution. The residuals of this comparison are then used as a measure of statistical consensus. www.posto-project.eu

Project BIOLEDGE, ID-289126, FP7-KBBE, Bio- knowledge Extractor and Modeller for Protein Production,
http://cordis.europa.eu/project/rcn/100449_en.html

- Isık Barıs Fidaner, Ayca Cankorur-Cetinkaya, Duygu Dikicioglu, *et al.*, CLUSTERnGO: a user-defined modelling platform for two-stage clustering of time-series data, *Bioinformatics*, Vol. 32, No. 3, 2016, pp. 388–397.

- Maciej Rybinski and José Francisco Aldana-Montes, tESA: a distributional measure for calculating semantic relatedness, *Journal of Biomedical Semantics*, 2016, 7, 67.

Project: MONFISPOL (Modeling and Implementation of Optimal Fiscal and Monetary Policy Algorithms in Multi-Country Econometric Models)

Project: DOVSA (Development of Virtual Screening Algorithms: Exploring Multiple Ligand Binding Modes Using Spherical Harmonic Consensus Clustering), *Note:* development of virtual screening algorithms to help deal with cases where multiple ligands may be associated with multiple pocket sub-sites or which may bind multiple targets, using a spherical harmonic surface shape-based approach.

- Background: Ajay D. Kshemkalyani, Mukesh Singhal, Distributed Computing Principles, Algorithms, and Systems, *Cambridge University Press*, Cambridge, New York, Melbourne, Madrid, Cape Town, Singapore, São Paulo, ISBN-13 978-0-521-87634-6, Chapter 14: Consensus and agreement algorithms, related to project FP7-ICT-223866-FeedNetBack. *Note*: Related recent publication to FeedNetBack project see next reference:

- Lara Briñon Arranz, Cooperative Control Design of Multi-Agent Systems, Application to Underwater Missions, Padova, 24th July 2012. http://automatica.dei.unipd.it/tl_files/events/Seminar_Brinon.pdf

- Lara Briñon Arranz, Alexandre Seuret, Carlos Canudas de Wit. Cooperative Control Design for Time Varying Formations of Multi-Agent Systems, *IEEE Transactions on Automatic Control*,

Chapter 7. An Overview of Systems, Control & Optimisation (SCO) in Recent
European R&D Programmes and Projects (2013-2017) under the Emergence of New
Concepts and Broad Industrial Initiatives

Institute of Electrical and Electronics Engineers, Vol. 59, No. 8, 2014, pp. 2283-2288.

- Zhongkui Li, Guanghui Wen, Zhisheng Duan, Wei Ren: Designing Fully Distributed Consensus Protocols for Linear Multi-agent Systems with Directed Graphs (2014), https://arxiv.org/pdf/1312.7377.pdf

Consensus & Optimisation

- Kostas Margellos, Alessandro Falsone, Simone Garatti and Maria Prandini, Distributed constrained optimization and consensus in uncertain networks via proximal minimization. https://arxiv.org/pdf/1603.02239.pdf

- K. Margellos, A. Falsone, S. Garatti and M. Prandini, Proximal minimization based distributed convex optimization, in *Proceedings of the American Control Conference*, 2016, pp. 2466–2471.

Project: HEARING MINDS, ID-324401, Hearing Minds: Optimizing Hearing Performance in Deaf Cochlear Implanted Individuals, 2013-2017. *Note*: of fitting models based on Bayesian networks to reinforcement learning models such as partially-observable Markov decision processes (POMDPs); (ii) of evaluation tools to measure functional hearing capacities in 'difficult' listeners such as young children or elderly adults. Interdisciplinary work with speech perception such as linguistics, biomedical physics, mathematics and audiology.

7.A2.8. PDEs, System Modelling (e.g. Control for Wave, -HD, -MHD Equations)

Project: CEMCAST, ID: 245479 (Centre of Excellence for Modern Composites Applied in Aerospace and Surface Transport Infrastructure), www.cemcast.pollub.pl *Note*: addresses the cooperation and dynamics of auto-parametric L-shaped mechanical beam structures: (a) providing new equations of motion which took into account rotary inertia in the 3D space dynamics and (b) solving a linear eigenvalue problem of the structure. The differential equations of motion were based on the Hamilton principle of least actions.

- General background, Alfio Quarteroni: Survey of partial differential equations, in Numerical Models for Differential Problems 2017, https://www.springer.com/gb/book/9788847058835, https://doi.org/10.1137/110825960

- Fotios Georgiades, Jerzy Warminskia, Matthew P. Cartmell, Linear modal analysis of L-shaped beam structures, *Mechanical Systems and Signal Processing*, Vol. 38, Issue 2, 20 July 2013, pp. 312-332. https://doi.org/10.1016/j.ymssp.2012.12.006 *Note*: Theoretical linear modal analysis of Euler–Bernoulli L-shaped beam structures is performed by solving two sets of coupled partial differential equations of motion within the Nonlinear Normal Mode theory.

Project: CHANGE, ID: 694515, (ERC): New Challenges for (adaptive) PDE Solvers: the Interplay of Analysis and Geometry, 2016-2021. *Note:* Simulation of Partial Differential Equations (PDEs-geometric modeling and processing, discretisation of PDEs: isogeometric methods and variational methods on polyhedral partitions, as extensions of standard finite elements.

Project: HAPDEGMT, Project ID: 615112, ERC, Harmonic Analysis, Partial Differential Equations and Geometric Measure Theory. *Note:* Harmonic Analysis, Geometric Measure Theory and Calderón-Zygmund theory

Project: HYDRON, Project ID: 332136, MSCA-IEF, Quasi Linear PDEs. *Note:* First-order quasi-linear PDEs -critical phenomena (shock waves). Hydrodynamic reductions, bidifferential calculus, and numerical analysis in the critical regime. Space-time singularities. Possible applications on biofilm biogenesis and the regulation and mode of action of the ubiquitously biofilm-promoting c-di-GMP. Perspectives for developing anti-biofilm drugs.

- O. Chvartatskyia, F. Muller-Hoissen and N. Stoilov: 'Riemann Equations' in Bidifferential Calculus, 2017. https://arxiv.org/pdf/1409.1154.pdf

Project: INVARIANT, Project ID: 335079, (Invariant manifolds in dynamical systems and PDE), *Note:* Mathematical Physics challenges on invariant manifolds in the context of fluid mechanics and elliptic PDEs, where many important problems and conjectures due to Arnold,

Chapter 7. An Overview of Systems, Control & Optimisation (SCO) in Recent
European R&D Programmes and Projects (2013-2017) under the Emergence of New
Concepts and Broad Industrial Initiatives

Kelvin, Ulam, Yau. Investigation of high-energy eigenfunction of the harmonic oscillator or the hydrogen atom.

- Margalef-Bentabol Juan, Peralta-Salas D., Realization problems for limit cycles of planar polynomial vector fields. http://digital.csic.es/handle/10261/152100, 11-Nov-2016, Publisher: Academic Press, Journal of Differential Equations Supported by ERC grant and CAST, *Note*: Research Network Program of the European Science Foundation (ESF) http://doi.org/10.1016/j.jde.2015.10.044

Project: Vort3DEuler, Project ID: 616797, ERC, (3D Euler, Vortex Dynamics and PDE), 2014-2019. *Note:* PDE arising from fluid mechanics. De-singularization procedures are carried out (including a time renormalization) to obtain an evolution equation (the binormal equation). Proof of existence of solutions to the viscous, non-resistive magnetohydrodynamics (MHD) equation.

- Fefferman Charles L., McCormick David S., Robinson James C. and Rodrigo Jose L., Local existence for the non-resistive MHD equations in nearly optimal Sobolev spaces, *Archive for Rational Mechanics and Analysis*, Vol. 223, No. 2, 2017, pp. 677-691. This version is available from Sussex Research Online: http://sro.sussex.ac.uk/66123/

- Charles L. Fefferman, David S. McCormick, James C. Robinson, Jose L. Rodrigo, Local Existence for the Non-Resistive MHD Equations in Nearly Optimal Sobolev Spaces

Project: ExaHyPE, (An Exascale Hyperbolic PDE Engine), 2015-2019. *Note:* Development of an Exascale-ready engine to solve hyperbolic systems of partial differential equations (PDE) on next-generation supercomputers. Applications in Seismology (seismic wave propagation, as for example for earthquake simulation) and Astrophysics (relativistic equations to model systems of neutron stars collapsing to a black hole and generating gravitational waves. hyperbolic simulation engine based on high-order communication-avoiding Finite-Volume/Discontinuous-Galerkin schemes yielding high computational efficiency.

Project: GEOPARDI, ID-279389, ERC, Numerical Integration of Geometric Partial Differential Equations 2011-2016. *Note:* reproduce qualitative behavior of differential equations over long time.

Deterministic and stochastic equations, with a general aim at deriving hybrid methods.

Project: SCAPDE, ID-320845, ERC. *Note:* Sato and Hormander micro-local analysis, including spectral theory, scattering theory, control theory, and some aspects in non-linear equations, using dispersive estimates and paraproduct techniques. Applications to the Cauchy problem for non-linear waves in domains and optimal control operator in control theory. Markov Chain Monte Carlo algorithm of Metropolis type via PDE's tools. Generalization of the classical pseudo-differential calculus.

- Bony Jean-François, Hérau Frédéric, Michel Laurent, Tunnel effect for semiclassical random walks, Mathematical Sciences Publishers Analysis & PDE, *Mathematical Sciences Publishers*, Vol. 8, No. 2, 2015, pp. 289-332. <10.2140/apde.2015.8.289> 2015

Project: DYCON, ID-694126, (ERC), Dynamic Control and Numerics of Partial Differential Equations, 2016-2021. *Note:* Challenges in the management of natural resources, meteorology, aeronautics, oil industry, biomedicine, human and animal collective behaviour. Control of Partial Differential Equations (PDE) and their numerical approximation methods.

Project: NUMERIWAVES, ID-246775 (New analytical and numerical methods in wave propagation). *Note:* Control problems for continuous wave equations, investigates non-uniform meshes preserving the control theoretical properties of continuous waves. Numerical approximation for the Kolmogorov model, as hypoelliptic Partial Differential Equation (PDE), Augmented Lyapunov exhibited through the corresponding Lie brackets. Application in sonic-boom minimization for supersonic aircrafts and the management of hydraulic resources.

- Xiaobing Feng, Roland Glowinski and Michael Neilan, Recent Developments in Numerical Methods for Fully Nonlinear Second Order Partial Differential Equations, *SIAM Rev.*, Vol. 55, No. 2, 2013, pp. 205–267, Annex

7.A2.9. Symbolic Control

Project: SPEEDD, Scalable Data Analytics, Scalable Algorithms, Software Frameworks and Visualization – Big Data, http://speedd-

Chapter 7. An Overview of Systems, Control & Optimisation (SCO) in Recent
European R&D Programmes and Projects (2013-2017) under the Emergence of New
Concepts and Broad Industrial Initiatives

project.eu   and   http://speedd-project.eu/sites/default/files/D2.11-Dissemination%20Report_v_1.0_final.pdf

Project: PROCSYS, Project ID-725144, H2020 (ERC): Towards Programmable Cyber-physical Systems: a Symbolic Control Approach, 2017-2022. *Note:* Develop advanced functionalities using a high-level programming language for CPS. Correctness of the controllers will be guaranteed by following the correct by construction synthesis paradigm using symbolic control techniques. Continuous physical dynamics is abstracted by a symbolic model, which is a purely discrete dynamical system; a high-level symbolic controller is then synthesized automatically from the high-level program and the symbolic model.

- Pietro Grandinetti, Carlos Canudas de Wit, Federica Garin, An efficient one-step-ahead optimal control for urban signalized traffic networks based on an averaged Cell-Transmission model, in *Proceedings of the 14th Annual European Control Conference (ECC15)*, Jul 2015, Linz, Austria. https://hal.archives-ouvertes.fr/hal-01188535/file/final.pdf

- Majid Zamani, Peyman Mohajerin Esfahani, Rupak Majumdar, Alessandro Abate and John Lygeros, Symbolic control of stochastic systems via approximately bisimilar finite abstractions. http://www.dcsc.tudelft.nl/~mohajerin/Publications/journal/2014/Bisimulation.pdf This work is supported by the European Commission MoVeS project FP7- 257005

- Zamani I. Tkachev and A. Abate, Bisimilar symbolic models for stochastic control systems without state-space discretization, in *Proceedings of the 17th International Conference on Hybrid Systems: Computation and Control*, ACM New York, NY, April 2014, pp. 41–50.

- Elias Alevizos, Alexander Artikis, George Paliouras, (Event Forecasting with Markov Chains), 2017, http://cer.iit.demokritos.gr/papers/Forecasting-DEBS17.pdf

Project: datACRON, ID 687591 Big Data Analytics for Time Critical Mobility Forecasting, 2016-2018. http://cordis.europa.eu/project/rcn/199835_en.html

- Peyman Mohajerin Esfahani, Tobias Sutter, Daniel Kuhn and John Lygeros, From Infinite to Finite Programs: Explicit Error Bounds with Applications to Approximate Dynamic Programming, 2017. https://arxiv.org/pdf/1701.06379.pdf

- Elias Alevizos, Anastasios Skarlatidis, Alexander Artikis, Georgios Paliouras, Complex Event Recognition under
Uncertainty: A Short Survey. http://speedd-
project.eu/sites/default/files/
Alevizos_et_al_15_cer_survey.pdf

7.A2.10. Robotics

UAVs: SESAR JU: European Drones Outlook Study-Unlocking the value for Europe, Nov. 2016
http://www.sesarju.eu/sites/default/files/documents/reports/European_
Drones_Outlook_Study_2016.pdf

Project: EUROBOTICS, 2014-2020 Robotics 2020 Strategic Research Agenda for Robotics in Europe Produced by euRobotics, aisbl Draft 0v42 11/10/2013,
https://ec.europa.eu/research/industrial_technologies/pdf/robotics-ppp-roadmap_en.pdf

- Christoph Klein: ROBOTICS IN H2020, 2014-03-27, Presentation, Leuven, Belgium. Presentation, https://www.cascade-fp7.eu/Public/public-presentations/ROBOTICS%20IN%20H2020.pdf

- European Commission, Digital Single Market POLICIES, Robotics. https://ec.europa.eu/digital-single-market/en/policies/robotics

- European Commission: Robotics Public-Private Partnership in Horizon 2020. https://ec.europa.eu/digital-single-market/en/robotics-public-private-partnership-horizon-2020

Project: SPARC (PPP): Robotics 2020, Multi-Annual Roadmap For Robotics in Europe Horizon 2020 Call ICT-2017 (ICT-25, ICT-27 & ICT-28) Release B 02/12/2016 https://www.eu-robotics.net/cms/upload/topic_groups/H2020_Robotics_Multi-Annual_Roadmap_ICT-2017B.pdf, and http://www.internet-of-things-research.eu/pdf/IoT-From%20Research%20and%20Innovation%20to%20Market%20Deployment_IERC_Cluster_eBook_978-87-93102-95-8_P.pdf

- Sabato Manfredi, Multilayer control of networked cyber-physical systems: Application to Monitoring, Autonomous & Robot Systems, *Springer*, 2017.

- Ying Tan, Handbook of Research on Design, Control, and Modeling of Swarm Robotics, *IGI Global*, 2015.

Project: ERRIC, ID-264207, (Empowering Romanian Research on Intelligent Information Technologies), http://www.erric.eu/

- ERRIC Workshop on Service Orientation in Holonic and Multi-Agent Manufacturing and Robotics, organized by the University Politehnica of Bucharest, the Faculty of Automatic Control and Computer Science and the University of Valenciennes and Hainaut-Cambresis. See https://sohoma17.sciencesconf.org/ and http://www.sohoma16.cimr. pub.ro/documents_SOHOMA/cfp_SOHOMA16.pdf

Project: NANOBIOTOUCH, ID-228844, Nano-resolved Multi-scale Investigations of Human Tactile Sensations and Tissue Engineered Nano-bio Sensors, 2010-2013

Project: DECORO, ID-628629, Developmental Context-Driven Robot Learning (2014-2016). *Note:* The principal scientific objectives of this project are to study how the full sensorial context of an embodied robot platform can be used to drive behaviour; how such behaviour can be self-organized in real-time to enable learning in dynamic (and learning) contexts; and how to also adapt the structure of the neural substrate to enable a development of the set of symbols used.

Project: SARAFun, ID-644938, H2020 (Smart Assembly Robot with Advanced Functionalities), http://media.h2020sarafun.eu/2016/05/ LeafletFront-708x1024.png, 2015-2018. *Note:* Addresses the improvement of robots functional programming in terms of the time required to swiftly obtain ready to use software, not by smart architecting and coding but mainly by enhancing the actual sensory and cognitive abilities of the robot. Learning "best practices" of a skilled human in progressive stages though smart vision and other tools, initially supervised, with an ABB FRIDA robot.

Project: AUTORECON, ID-285189, FP7-NMP, 2011-2014, Autonomous Co-operative Machines for Highly Reconfigurable

Assembly Operations of the Future, http://cordis.europa.eu/result/rcn/164248_en.html

Project: VERSATILE: Flexible Robotic Cells with Dual Robot Arms that Can Adapt Automatically to the High Number of Different Products in Industries Such as Automotive, Aerospace and Handling and Packaging, https://ec.europa.eu/programmes/horizon2020/en/h2020-sections-projects.

Project: ESROCOS, ID-730080 European Space Robot Control Operating System, 2016-2019, http://cordis.europa.eu/project/rcn/206157_en.html and wrt ROS: http://rosin-project.eu/wp-content/uploads/2017/03/ROSIN-press-release.pdf

Project: RECONFIG, ID-600825 (Cognitive, Decentralized Coordination of Heterogeneous Multi-Robot Systems via Reconfigurable Task Planning)

- General reference: Open Source Robotics Foundation. ROS.org | Powering the world's robots. url: http://www.ros.org, (visited on 11/25/2016)

- Adrian Leva, Reactive Controller Synthesis for Mobile Robotics, Master Thesis, Technical Report MPI-SWS-2017-001, University of Kaiserslautern, 2017. https://www.mpi-sws.org/tr/2017-001.pdf

- Karayiannidis Y., Doulgeri Z., Robot Force/Position Tracking on a Surface of Unknown Orientation. In: Bruyninckx H., Přeučil L., Kulich M. (eds.) European Robotics Symposium 2008. Springer Tracts in Advanced Robotics, *Springer*, Vol. 44, 2008, Berlin, Heidelberg. https://doi.org/10.1007/978-3-540-78317-6_26

- Abdelrahem Atawnih, Zoe Doulgeri and George A. Rovithakis, Operational Space Prescribed Tracking Performance and Compliance in Flexible Joint Robots, *Journal of Dynamic Systems Measurement and Control*, July 2015. https://www.researchgate.net/publication/272368706 *Note*: research co-financed by the EU-ESF and Greek national funds by the program 'Education and Lifelong Learning' of the National Strategic Reference Framework (NSRF) - Research Funding Program ARISTEIA I./II

- Dimitrios Papageorgiou, Abdelrahem Atawnih, Zoe Doulgeri, A Passivity Based Control Signal Guaranteeing Joint Limit Avoidance in

Redundant Robots, in *Proceedings of the 24th Mediterranean Conference on Control and Automation (MED 2016)*, Athens, Greece.

- Achilles Theodorakopoulos, George A. Rovithakis and Zoe Doulgeri, An Impedance Control Modification Guaranteeing Compliance Strictly Within Preselected Spatial Limits, in *Proceedings of the IEEE/RSJ International Conference on Intelligent Robots and Systems (IROS) Congress Center Hamburg*, Sept 28 - Oct 2, 2015. Hamburg, Germany. https://www.researchgate.net/profile/Achilles_Theodorakopoulos/publi cation/282349096_An_Impedance_Control_Modification_Guaranteein g_Compliance_Strictly_Within_Preselected_Spatial_Limits/links/5610 f7bd08ae0fc513f1712b/An-Impedance-Control-Modification-Guaranteeing-Compliance-Strictly-Within-Preselected-Spatial-Limits.pdf

- Filippo Arrichiello, Marino Pierri, Distributed Fault-Tolerant Control for Networked Robots in the Presence of Recoverable/Unrecoverable Faults and Reactive Behaviors, *Front. Robot. AI*, 21 February 2017. https://doi.org/10.3389/frobt.2017.00002

- Abdelrahem Atawnih, Zoe Doulgeri and George A. Rovithakis, Operational Space Prescribed Tracking Performance and Compliance in Flexible Joint Robots, J. Dyn. Sys., Meas., Control 137 (7), 074503 (Jul 01, 2015) (6 pages) Paper No: DS-14-1332. History: Received August 14, 2014; Revised December 22, 2014; Online February 09, 2015.

- Emmanouil Kourtikakis, Emmanouil Kapellakis, John Fasoulas and Michael Sfakiotakis, An Embedded Controller for the Pendubot, *in Proceedings of the 15th International Symposium on Ambient Intelligence and Embedded Systems (AmiEs'16)*, 2016.

- Kaklamani Georgia, Cheneler David, Grover Liam M., Adams Michael J., Bowen James, Mechanical properties of alginate hydrogels manufactured using external gelation, *Journal of the Mechanical Behavior of Biomedical Materials*, Vol. 36, 2014, pp. 135–142. 1878-0180 2014.

- Li Zhijun, *et al.*, Decentralised adaptive fuzzy control of coordinated multiple mobile manipulators interacting with non-rigid environments, *IET Control Theory & Applications*, 2013. http://dx.doi.org/10.1049/iet-cta.2011.0334

- Dickerhof Markus, Kimmig Daniel, Adamietz Raphael, Iseringhausen Tobias, Segal Joel, Vladov Nikola, Pfleging Wilhelm, Torge Maika, A generative manufacturing-based concept and equipment for flexible, scalable manufacturing of microsystems, IPAS 2014, *in Proceedings of the 7ᵗʰ International Precision Assembly Seminar*, Chamonix, 16-18 February 2014. http://publica.fraunhofer.de/documents/N-283498.html

- Fabio Cremona, Marten Lohstroh, David Broman, Stavros Tripakis, Edward A. Lee, Hybrid Co-simulation: It's About Time, Technical Report No. UCB/EECS-2017-6, April 2017. http://www2.eecs. berkeley.edu/Pubs/TechRpts/2017/EECS-2017-6.html

7.A2.11. Self-Organisation & Self-assembling Systems

Project: NLL, ID: 617521, (ERC) and TUBITAK, Nonlinear Laser Lithography 2014-2019. *Note:* New method for regulating self-organised formation of metal-oxide nanostructures at high speed via non-local feedback, thereby achieving unprecedented levels of uniformity over indefinitely large areas by simply scanning the laser beam over the surface. Control the self-organised pattern through the laser field using, e.g., a spatial light modulator

- Serim Ilday, Ghaith Makey, Gursoy B. Akguc, *et al.:* Rich complex behaviour of self-assembled nanoparticles far from equilibrium, *Nature Communications,* Vol. 8, 2017. http://europepmc.org/backend/ ptpmcrender.fcgi?accid=PMC5414064&blobtype=pdf

https://www.nature.com/articles/srep38674.pdf

7.A2.12. Decision Making/Processes, Markov DP, POMDP, Multi-agents Systems

Project: CompLACS, ID: 270327, Composing Learning for Artificial Cognitive Systems (2011-2015), www.CompLACS.org ERC and French ANR project ExTra-Learn. *Note:* AI application to multiple bandit problems, Markov Decision Processes (MDPs), Partially Observable MDPs(POMDPs), continuous stochastic control, and multi-agent systems. Multi-task Reinforcement learning (MTRL) and approximate dynamic programming (ADP).

Chapter 7. An Overview of Systems, Control & Optimisation (SCO) in Recent
European R&D Programmes and Projects (2013-2017) under the Emergence of New
Concepts and Broad Industrial Initiatives

- Daniele Calandriello, Alessandro Lazaric, Marcello Restelli, Sparse Multi-task Reinforcement Learning, *NIPS - Advances in Neural Information Processing Systems 26*, December 2014, Montreal, Canada. https://hal.inria.fr/hal-01073513/document

- A. Castelletti, S. Galelli, M. Restelli and R. Soncini-Sessa, Tree-based feature selection for dimensionality reduction of large-scale control systems, in *Proceedings of the IEEE Symposium on Adaptive Dynamic Programming & Reinforcement Learning*, 2011.

Project: PURe-MaS, ID-275217, Planning under Uncertainty for Real-world Multiagent Systems, (2011-2013). *Note:* Artificial Intelligence designing scalable flexible multi-agents. Self-interested agents, relevant in domains such as smart grids or cars driving on a highway. Non-cooperative techniques by exploiting local interactions between agents. Beyond POMDP for intelligent transportation and smart grids.

- Joris Scharpff, Matthijs T. J. Spaan, Leentje Volker, Mathijs de Weerdt, Planning under Uncertainty for Coordinating Infrastructural Maintenance, in *Proceedings of the 23rd International Conference on Automated Planning and Scheduling (ICAPS 2013)*, Rome, Italy.

7.A2.13. Control of Embryonic Stem Cells Systems - Regulatory Systems

Project: EUROSYSTEM, ID-200720, FP7-HEALTH, European Consortium for Systematic Stem Cell Biology,

Project: NEUROSTEM, ID-250342, (ERC), 2010-2015, A Systems Level Approach to Proliferation and Differentiation Control in Neural Stem Cell Lineages. *Note:* Research contributes to the understanding how self-renewal is controlled in neural stem cell and how defects in this process can lead to the formation of brain tumours in model organisms. The approach we take is to determine the transcriptional network that acts in neuroblasts to control self-renewal using time-resolved transcriptional profiling to determine, how this network changes in the differentiating daughter cell and develop tools for medium-throughput functional analysis of the key network players and expand this analysis to other stem cell systems inside and outside the fly nervous system to determine how modifications of stem cell systems like transit amplifying pools or perpetual adult proliferation are reflected in network architecture. "Furthermore, regulatory circuits within individual ESCs undergoing fate computation could be fundamentally disorganized or chaotic in order to compute cell fate trajectories, a possibility explicitly captured by a recent theory of stem cell decision-making centred on 'critical-like dynamics' at the edge of chaos".

- Richard B. Greaves, Sabine Dietmann, Austin Smith, Susan Stepney, Julianne D. Halley, A conceptual and computational framework for modelling and understanding the nonequilibrium gene regulatory networks of mouse embryonic stem cells, *PLOS Computational Biology*, 1 September 2017. https://doi.org/10.1371/journal.pcbi.1005713

- D. Halley, K. Smith-Miles, D. A. Winkler, T. Kalkana, S. Huang, A. Smith, Self-organizing circuitry and emergent computation in mouse embryonic stem cells, *Stem Cell Research*, Vol. 8, 2012, pp. 324–333.

7.A2.14. Advanced Controller Synthesis - Novel Concepts and Methods

Multiple funding research project: Supported in part by EPSRC (UK) grant EP/N031962/1, FWF (Austria) S 11405-N23 (RiSE/SHiNE), AFOSR Grant FA9550-14-1-0261 and NSF Grants IIS-1447549, CNS-1446832, CNS-1445770, CNS-1445770, CNS-1553273, CNS-1536086, CNS 1463722, and IIS-1460370. *Note:* The design of PID controllers for complex, safety-critical cyber-physical systems is challenging due to the hybrid, stochastic, and nonlinear dynamics they exhibit. Motivated by the need for high-assurance design techniques a new method for the automated synthesis of PID controllers for stochastic hybrid systems from probabilistic reachability specifications is derived, providing rigorous guarantees of safety and robustness for the resulting closed-loop system, while ensuring prescribed performance levels for the controller. Approach applied on an artificial pancreas case study, for which safety and robustness guarantees are paramount.

Notes about COOL: The goal of the co-synthesis[1] approach realized in COOL is to refine the system specification by generating hardware

---

[1] Background on Co-Synthesis: Rolf Ernst, *et al.*: Hardware-software co-synthesis for microcontrollers - IEEE Design & Test of Computers, 2004,

specifications for synthesis and simulation in VHDL and software specifications for compilation in C. COOL is a hardware/software co-design tool which has been developed for dataflow dominated systems. COOL uses a homogeneous system modelling approach using a subset of VHDL for specification. A graphical user interface has been developed to specify these systems in a structural and hierarchical way. The main objective of COOL is heterogeneous implementation. Several algorithms for hardware/software partitioning have been developed allowing the designer.
https://ls12-www.cs.tu-dortmund.de/daes/en/research/hwsw-co-design/cool.html

- Schmuck A. K., Majumdar R., Leva A., Dynamic hierarchical reactive controller synthesis, *Discrete Event Dyn. Syst.*, Vol. 27, No. 2, 2017, pp. 261-299. https://doi.org/10.10 https://link.springer.com/article/10.1007/s10626-017-0239-8#citeas07/s10626-017-0239-8 *Note*: Large-scale reactive controller synthesis problems with intrinsic hierarchy and locality can be modelled as a hierarchical two player game over a set of local game graphs w.r.t. to a set of local strategies on multiple, interacting abstraction layers. Proposed is a reactive controller synthesis algorithm that allows for dynamic specification changes at each step. This re-calculation becomes computationally tractable by the proposed decomposition.

- Fedor Shmarov, Nicola Paoletti, Ezio Bartocci, Shan Lin, Scott A. Smolka and Paolo Zuliani, Automated Synthesis of Safe and Robust PID Controllers for Stochastic Hybrid Systems, 2017. https://arxiv.org/pdf/1707.05229.pdf

- Idress Husien, Nicolas Berthier, Sven Schewe, A Hot Method for Synthesising Cool Controllers, *SPIN*, 2017. http://spinroot.com/spin/symposia/ws17/SPIN_2017_paper_35.pdf

Support from EPSRC (UK)

- Matthias Rungger, Manuel Mazo Jr., Paulo Tabuada, Specification-Guided Controller Synthesis for Linear Systems and Safe Linear-Time

---

http://w3.ualg.pt/~jmcardo/ensino/ihs2004/Benner93.pdf and also: COSYMA, a platform for the investigation of hardware/software co-synthesis of small embedded systems.

Temporal Logic, in *Proceedings of the 16ᵗʰ International Conference on Hybrid Systems: Computation and Control*, 2013, pp. 333-342. http://www.mmazojr.net/Manuel_Mazo_Jr/Publications_files/hscc118f -rungger.pdf

Note: Related US initiative, ExCAPE: (Expeditions in Compute-Augmented Program Engineering), https://excape.cis.upenn.edu/ documents/2nd_year_report.pdf

- Rudiger Ehlers, Robert Konighofer and Roderick Bloem, Synthesizing Cooperative Reactive Mission Plans, https://online.tugraz.at/ tug_online/voe_main2.getvolltext?pCurrPk=90746

- EU-US cooperation: Temporal Logic Mission planning (LTLMoP) being developed in the. United states and European Union (PhD): MENG GUO Licentiate Thesis Stockholm, Sweden 2014: Cooperative Motion and Task Planning Under Temporal Tasks https://people.kth.se/~kallej/grad_students/guo_licthesis14.pdf

- Marco Faella, Controller Synthesis for Linear Hybrid Systems. Formal Methods for Cyber-Physical Systems, Verona, September 12-16, 2017. *Note*: advanced tools: Symbolic polyhedra manipulation with PPL, Controller synthesis with SpaceEx, Optimizing compilation (Graphite) – Stochastic games (PRISM) – Petri Nets (Romeo)

- M. Benerecetti, M. Faella, Automatic Synthesis of Switching Controllers for Linear Hybrid Systems: Reachability Control, *ACM Trans. on Embedded Computing Systems*, Vol. 16, No. 4, 2017.

- M. Benerecetti, M. Faella, Automatic Synthesis of Switching Controllers for Linear Hybrid Systems: Safety Control. Theoretical Computer Science, 493, *Elsevier*, 2013; M. Benerecetti, M. Faella, Tracking Differentiable Trajectories across Polyhedra Boundaries, in *Proceedings of the ACM International Conference on Hybrid Systems: Computation and Control (HSCC'13)*.

- M. Benerecetti, M. Faella, S. Minopoli, Reachability Games for Linear Hybrid Systems, in *Proceedings of the 15ᵗʰ International Conference on Hybrid Systems: Computation and Control (HSCC'2012)*, 2012; M. Benerecetti, M. Faella, S. Minopoli, Revisiting Synthesis of Switching Controllers for Linear Hybrid Systems, IEEE CDC 2011.

French National Projects:

- Nicolas Berthier, Herve Marchand. Discrete Controller Synthesis for Infinite State Systems with ReaX, in *Proceedings of the IEEE International Workshop on Discrete Event Systems*, Cachan, France, pp. 420-427, May 2014. https://hal.inria.fr/hal-00974553/document

Note: Research supported by the French ANR project Ctrl-Green (ANR-11-INFR 012 11), INFRA and MINALOGIC

- Nicolas Berthier, Hervé Marchand, Deadlock-free Discrete Controller Synthesis for Infinite State Systems, *in Proceedings of the 54th IEEE Conference on Decision and Control*, Osaka, Japan, December 2015.

Chapter 8

Model Detection Using Innovations Squared Mismatch Method: Application to Probe Based Data Storage System

Sayan Ghosal and Murti Salapaka

8.1. Introduction

Systems which switch from one model to another are common in many different applications [1]. In such applications, observer banks are widely employed to decipher system models [2-3] to enable closed loop control of the switched systems. In principle, the observer which generates the minimum tracking error is considered closest to the active system model. Here, switching indices are derived from tracking errors that decide the closest system. Various switching indices can be found in existing literature; for example, switching indices in [2-3] integrate the square of the tracking error with tunable forgetting factor. [4] uses a switching index based on a combination of instantaneous and integrated values of tracking errors. Modified switching indices are proposed in [5] to simplify computation and improve stability conditions. One of the main emphases of these studies include analyzing and improving tracking performance of closed-loop switching systems while ensuring stability. [6-7] and focus on speed of detection to ascertain which model is current for a system that switches between two models. Such need for bandwidth of detection is particularly relevant for data storage applications where system characteristics when interacting with the bit 1 is different than interacting with bit 0.

The maximum a posteriori probability (MAP) based methods are prevalent in bit patterned media recording and magnetic data storage [8-10] which are relatively mature technologies compared to probe based storage systems. Motivated by these examples, a recursive MAP based algorithm (see Chapter 10, [11]) is adopted in [7] and applied to online

Sayan Ghosal
Seagate Technology, Shakopee, MN, USA

system identification for probe based storage systems. Further, in [7] a new signal called the innovation squared mismatch (ISM) signal is developed. It is demonstrated that ISM offers considerable advantage over MAP while being much simpler to implement. For high bandwidth read operations, the transients caused by fast transition between bits lead to inter symbol interference (ISI). In order to mitigate ISI, dynamic programming based sequence detection schemes are also employed in [6-7]. It is demonstrated that sequence detection in conjunction with ISM can lead to significant advances in read bandwidth.

The data density for the hard disk drive technology is facing severe restrictions due to thermal limits [12]. Further, the minimum bit size for solid state drive memories is limited by the resolution of the lithography processes [13], where new technology for higher resolution lithography is hugely expensive. The probe based surface manipulation and reading methods offer an alternative route which can pack binary data at the nanometer or sub nanometer scale. Since the invention of atomic force microscope [14], probe based data storage systems are being explored for data storage with extremely high areal density [15]. In pioneering demonstrations of topographic probe based data storage [16], an indentation of sample surface is used to encode the bit 1 and absence of indentation represents the bit 0. Even though extremely high data density in the order of 3-4 Tb/in^2 are reported in literature (for example, in [17]), a big challenge is the mechanical wear and tear of the probe as well as of the media [18] which significantly reduces device lifetime.

In order to circumvent the mechanical wear, the dynamic mode of operation for probe storage is proposed in [7] which drastically eliminates mechanical wear and tear during the reading process. Here the flexure probe is forced sinusoidally and the probe contacts the media intermittently. The information on the media encoded topographically modulates the flexure oscillations which is used to decipher information on the media. It is demonstrated in [6-7] that the oscillating cantilever while interacting with raised topography assumes one model behavior (model 1) which is different than model behavior assumed by the probe when interacting with a lowered topography (model 0). Thus for topographic data storage where topography encode binary bits, the read task becomes detection of which model is active [7]. The conventional dynamic mode atomic force microscopy (AFM, see Fig. 8.1) utilizes cantilever oscillation amplitude and phase to interpret sample topography. However, application of MAP and ISM offer significant detection bandwidth gains over conventional dynamic mode signals.

Moreover, ISM offers similar detection speed, better detection accuracy and substantial computational gain over MAP. Further, inter symbol interference (ISI) arises during fast data detection which is addressed using the maximum likelihood sequence detection (MLSD) methods. In this chapter, optimal sequence detection based on the ISM signal (the ISM-MLSD method) is outlined where the time spent by the probe on single bit (bit duration) is known. ISM-MLSD consistently outperforms symbol-by-symbol ISM and MAP, however, at the expense of complex signal processing.

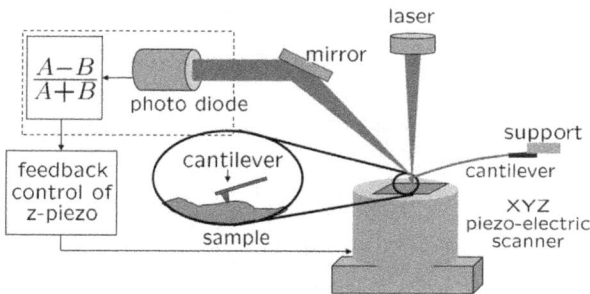

Fig. 8.1. The atomic force microscope (AFM). The sample sits on a positioning device which can move in all three spatial directions with respect to a flexible cantilever. The cantilever deflection which depends on inter atomic forces between the tip and the sample is recorded using a laser, photo-detector setup. In the dynamic mode, the cantilever is excited sinusoidally where its orbit depends on the topography of sample. Traditionally the oscillation amplitude A_{def} and phase ϕ_{def} of the cantilever deflection are utilized for dynamic mode imaging. In [7], equivalent linear models of the cantilever are obtained when the cantilever oscillates in free air (model 1) and when it interacts with the sample (model 2). Further, observer based architectures designed using models 1 and 2 are utilized in [7] for fast detection of topography.

8.2. Real Time Plant Detection with Innovations Squared Mismatch (ISM)

In this section, a system is assumed to be switching between two linear models. The individual models can be first identified during a slower testing phase using the identification methods described in [19]. Assuming that the switching system models are already identified, a fast detection methodology called the innovations squared mismatch (ISM) is presented and compared with an existing maximum a-posteriori probability (MAP) based approach for fast real-time detection of which of the system models is current. The technique of system parameter estimation followed by real-time detection is shown later to be a powerful tool for probe based storage system.

8.2.1. Preliminaries, Problem Formulation and MAP

Consider a continuous time system which is linear and time varying described by

$$
\begin{aligned}
\dot{x}(t) &= A(t)x(t) + B(t)g(t), \\
y(t) &= C(t)x(t) + D(t)g(t) + \zeta(t),
\end{aligned}
\tag{8.1}
$$

where $\zeta(t)$ denotes continuous time measurement noise assumed to be zero-mean Gaussian white processes with covariances described by $E[\zeta(t)\zeta^T(t_0)] = R_n \delta(t - t_0)$. It is assumed that the quadruple $(A(t), B(t), C(t), D(t)) \in (A_i, B_i, C_i, D_i)$ with $i \in \{1, 2\}$, that is the dynamics can switch between two known linear systems. The objective of this section is to present a simple yet fast real time detection of the active model. Two continuous time steady state Kalman observers with the first observer matched to the model (A_1, B_1, C_1, D_1) and the second matched to model (A_2, B_2, C_2, D_2) are realized. Let $\hat{x}^{(i)}$ denote the estimated state from the $i^{(th)}$ observer and $\hat{y}^{(i)}$ denotes the corresponding estimated output. The estimated states and outputs are governed by

$$
\begin{aligned}
\dot{\hat{x}}^{(i)}(t) &= A_i\hat{x}^{(i)}(t) + B_ig(t) + L_i(y(t) - \hat{y}^{(i)}(t)), \\
\hat{y}^{(i)}(t) &= C_i\hat{x}^{(i)}(t) + D_ig(t),
\end{aligned}
\tag{8.2}
$$

where $i = 1, 2$ represent the first and the second observer respectively. Let $\tilde{x}^{(i)}$ denote the error between the original and the estimated states, i.e. $\tilde{x}^{(i)} = x - \hat{x}^{(i)}$ and $e^{(i)}$ denote the error between the system output and the estimated output, i.e. $e^{(i)} = y - \hat{y}^{(i)}$. It follows from (8.1) and (8.2) that the error dynamics is described by

$$
\begin{aligned}
\dot{\tilde{x}}^{(i)}(t) = &\left((A(t) - A_i) - L_i(C(t) - C_i)\right)x(t) \\
&+ (A_i - L_iC_i)\tilde{x}^{(i)}(t) + \left(B(t) - B_i + L_iD_i - L_iD(t)\right)g(t) - L_i\zeta(t).
\end{aligned}
\tag{8.3}
$$

The output error (also referred in this chapter as innovation) is given by

$$e^{(i)}(t) = \left(C(t) - C_i\right)x(t) + C_i\tilde{x}^{(i)}(t) + \left(D(t) - D_i\right)g(t) + \zeta(t). \quad (8.4)$$

A prevalent approach for detecting which model is active is based on a maximum a posteriori probability (MAP) based approach [11].

Here, detection method based on MAP is summarized. Consider a discrete time system which stays in model i given by

$$
\begin{aligned}
x[m+1] &= A_{d,i}x[m] + B_{d,i}g[m], \\
y[m] &= C_{d,i}x[m] + D_{d,i}g[m] + \zeta[m],
\end{aligned}
\quad (8.5)
$$

with $\zeta \sim N(0, R_{d,i})$. A steady state discrete-time Kalman observer matched to model i with gain $L_{d,i}$ is governed by

$$\hat{x}^{(i)}[m+1\,|\,m] = A_{d,i}\hat{x}^{(i)}[m\,|\,m-1] + B_{d,i}g[m]$$

$$+ L_{d,i}\left(y[m] - C_{d,i}\hat{x}^{(i)}[m\,|\,m-1] - D_{d,i}g[m]\right), \quad (8.6)$$

$$\hat{y}^{(i)}[m] = C_{d,i}\hat{x}^{(i)}[m\,|\,m-1] + D_{d,i}g[m].$$

Let θ_i denote model *i* and let $\mathcal{Y}_m$ be the sequence of observations $\mathcal{Y}_m = \{y[0], y[1], \cdots, y[m]\}$. The a posteriori probability that the plant parameter is θ_i given the observation vector $\mathcal{Y}_m$ is denoted by $p(\theta_i \mid \mathcal{Y}_m)$. The decision θ_{MAP} is chosen from the candidates θ_i so that $p(\theta_i \mid \mathcal{Y}_m)$ is maximized. It can be shown that (see [7, 11]) $p(\theta_i \mid \mathcal{Y}_m)$ can be updated from $p(\theta_i \mid \mathcal{Y}_{m-1})$ using

$$p(\theta_i \mid \mathcal{Y}_m) = c_m \mid \Omega_{m/i} \mid^{-\frac{1}{2}} \exp\left\{-\frac{1}{2}\tilde{y}_{m,i}^T \Omega_{m/i}^{-1} \tilde{y}_{m,i}\right\} p(\theta_i \mid \mathcal{Y}_{m-1}), \quad (8.7)$$

where $\tilde{y}_{m,i} = y - \hat{y}_{m,i}$. c_m is a normalization constant so that $\sum_{i=1}^{2} p(\theta_i \mid \mathcal{Y}_m) = 1$. The output covariance $\Omega_{m/i} = E\left[\tilde{y}_{m,i}\tilde{y}_{m,i}^T\right]$ in (8.7) and covariance of state estimation error $\Sigma_{m+1/m,i}$ where $\Sigma_{m+1/m,i} = E\left[\tilde{x}_{m+1,i}\tilde{x}_{m+1,i}^T\right]$ with $\tilde{x}_{m+1,i} = (x[m+1] - \hat{x}^{(i)}[m+1\,|\,m])$ can be updated recursively as well (see [7, 11]). Equation (8.7) provides the way to recursively estimate the probability of the system being either in model 1 or 2. The recursion typically starts with the assumption that both the candidates θ_i are equally probable at the beginning, that is $p(\theta_i | \mathcal{Y}_{-1}) = 1/2$ for $i \in \{1,2\}$.

8.2.2. Innovations Squared Mismatch (ISM)

In this part, a method based on a new signal reported in [6-7] is presented. We term this method as the innovations squared mismatch (ISM) technique. Consider the continuous time varying system described in (8.1). A new set of state space variables will be defined such that the observer outputs in terms of the new variables follow a sign changing behavior depending on the present system model. This sign changing phenomenon in turn will be utilized for system model detection. The state space equations describing the original time varying system together with two observers matched to models 1 and 2 contain the dynamical equations of $x(t)$, $\tilde{x}^{(1)}(t)$, which are given by

$$
\begin{bmatrix} \dot{x}(t) \\ \dot{\tilde{x}}^{(1)}(t) \\ \dot{\tilde{x}}^{(2)}(t) \end{bmatrix} = \begin{bmatrix} A(t) \\ ((A(t)-A_1)-L_1(C(t)-C_1)) \\ ((A(t)-A_2)-L_2(C(t)-C_2)) \end{bmatrix} x(t)
$$
$$
+ \begin{bmatrix} 0 & 0 \\ (A_1-L_1C_1) & 0 \\ 0 & (A_2-L_2C_2) \end{bmatrix} \begin{bmatrix} \tilde{x}^{(1)}(t) \\ \tilde{x}^{(2)}(t) \end{bmatrix} \qquad (8.8)
$$
$$
+ \begin{bmatrix} B(t) \\ (B(t)-B_1+L_1(D_1-D(t))) \\ (B(t)-B_2+L_2(D_2-D(t))) \end{bmatrix} g(t) + \begin{bmatrix} 0 \\ -L_1 \\ -L_2 \end{bmatrix} \zeta(t).
$$

From (8.4), the output errors can be represented as

$$
\begin{bmatrix} e^{(1)}(t) \\ e^{(2)}(t) \end{bmatrix} = \begin{bmatrix} (C(t)-C_1) \\ (C(t)-C_2) \end{bmatrix} x(t) + \begin{bmatrix} C_1 & 0 \\ 0 & C_2 \end{bmatrix} \begin{bmatrix} \tilde{x}^{(1)}(t) \\ \tilde{x}^{(2)}(t) \end{bmatrix}
$$
$$
+ \begin{bmatrix} D(t)-D_1 \\ D(t)-D_2 \end{bmatrix} g(t) + \begin{bmatrix} 1 \\ 1 \end{bmatrix} \zeta(t). \qquad (8.9)
$$

A new set of variables z_1 and z_2 are derived from the state space vectors $\tilde{x}^{(1)}(t)$ and $\tilde{x}^{(2)}(t)$ via $z_1(t) := (\tilde{x}^{(1)}(t) - \tilde{x}^{(2)}(t))$ and $z_2(t) := (\tilde{x}^{(1)}(t) + \tilde{x}^{(2)}(t))$. The dynamics of the new variables is given by

$$\begin{bmatrix} \dot{z}_1(t) \\ \dot{z}_2(t) \end{bmatrix} = \begin{bmatrix} (L_2 - L_1)C(t) - 2M_2 \\ 2A(t) - C(t)(L_1 + L_2) - 2M_1 \end{bmatrix} x(t) + \begin{bmatrix} M_1 & M_2 \\ M_2 & M_1 \end{bmatrix} \begin{bmatrix} z_1(t) \\ z_2(t) \end{bmatrix}$$

$$+ \begin{bmatrix} B_2 - B_1 + L_1 D_1 - L_2 D_2 & +(L_2 - L_1)D(t) \\ 2B(t) - (B_1 + B_2) + L_1 D_1 + L_2 D_2 & -(L_1 + L_2)D(t) \end{bmatrix} g(t) \qquad (8.10)$$

$$+ \begin{bmatrix} (L_2 - L_1) \\ -(L_1 + L_2) \end{bmatrix} \zeta(t),$$

where matrices M_1 and M_2 are defined by $M_1 = \frac{1}{2}[(A_1 + A_2) - (L_1 C_1 + L_2 C_2)]$ and $M_2 = \frac{1}{2}[(A_1 - A_2) - (L_1 C_1 - L_2 C_2)]$. The output vectors e_{dif} and e_{sum} are defined by $e_{dif}(t) := e^{(1)}(t) - e^{(2)}(t)$ and $e_{sum}(t) := e^{(1)}(t) + e^{(2)}(t)$ where it follows from (8.9) that

$$\begin{bmatrix} e_{dif}(t) \\ e_{sum}(t) \end{bmatrix} = \begin{bmatrix} C_2 - C_1 \\ 2C(t) - C_1 - C_2 \end{bmatrix} x(t)$$

$$+ \begin{bmatrix} \dfrac{(C_1 + C_2)}{2} & \dfrac{(C_1 - C_2)}{2} \\ \dfrac{(C_1 - C_2)}{2} & \dfrac{(C_1 + C_2)}{2} \end{bmatrix} \begin{bmatrix} z_1(t) \\ z_2(t) \end{bmatrix} \qquad (8.11)$$

$$+ \begin{bmatrix} D_2 - D_1 \\ D(t) - D_1 - D_2 \end{bmatrix} g(t) + \begin{bmatrix} 0 \\ 2 \end{bmatrix} \zeta(t).$$

Define the following:

$$N_1 := \frac{1}{2}(C_1 + C_2), \quad A_{m1} := \left((A_2 - A_1) + L_2(C_1 - C_2)\right),$$

$$N_2 := \frac{1}{2}(C_1 - C_2), \quad A_{m2} := \left((A_2 - A_1) + L_1(C_1 - C_2)\right),$$

$$B_{m1} := \left(B_2 - B_1 + L_2(D_1 - D_2)\right), \quad C_{m1} = C_{m2} := (C_2 - C_1),$$

$$B_{m2} := \left(B_2 - B_1 + L_1(D_1 - D_2)\right), \quad D_{m1} = D_{m2} := (D_2 - D_1).$$

After some algebra it can be shown that (see [7]) in the Laplace domain e_{dif} and e_{sum} are given by

$$e_{dif}(s) = e_{mi}(s) + \zeta_{dif}(s), \text{ and}$$
$$e_{sum}(s) = (-1)^i e_{mi}(s) + \zeta_{sum}(s),$$

(8.12)

where $e_{mi} = \mathcal{G}_{mi}^{(1)} x + \mathcal{G}_{mi}^{(2)} g$ represents the signal component. $\zeta_{dif} = \mathcal{G}_{(dif,\zeta)} \zeta$ and $\zeta_{sum} = \mathcal{G}_{(sum,\zeta)} \zeta$ denote the noise components in e_{dif} and e_{sum} respectively ($i \in \{1,2\}$). Further, the matrices $\mathcal{G}_{mi}^{(1)}$, $\mathcal{G}_{mi}^{(2)}$, $\mathcal{G}_{(dif,\zeta)}$ and $\mathcal{G}_{(sum,\zeta)}$ are defined by

$$\mathcal{G}_{m1}^{(1)}(s) = (N_1 \mathcal{S}_1 + N_2 \mathcal{S}_2 - (N_1 \mathcal{S}_2 + N_2 \mathcal{S}_1)) A_{m1} + C_{m1},$$
$$\mathcal{G}_{m1}^{(2)}(s) = (N_1 \mathcal{S}_1 + N_2 \mathcal{S}_2 - (N_1 \mathcal{S}_2 + N_2 \mathcal{S}_1)) B_{m1} + D_{m1},$$
$$\mathcal{G}_{m2}^{(1)}(s) = (N_1 \mathcal{S}_1 + N_2 \mathcal{S}_2 + N_1 \mathcal{S}_2 + N_2 \mathcal{S}_1) A_{m2} + C_{m2},$$
$$\mathcal{G}_{m2}^{(2)}(s) = (N_1 \mathcal{S}_1 + N_2 \mathcal{S}_2 + N_1 \mathcal{S}_2 + N_2 \mathcal{S}_1) B_{m2} + D_{m2},$$
$$\mathcal{G}_{(dif,\zeta)} = (N_1 \mathcal{S}_1 + N_2 \mathcal{S}_2)(L_2 - L_1) + (N_1 \mathcal{S}_2 + N_2 \mathcal{S}_1)(L_1 + L_2), \text{ and}$$
$$\mathcal{G}_{(sum,\zeta)} = (N_1 \mathcal{S}_2 + N_2 \mathcal{S}_1)(L_2 - L_1) - (N_1 \mathcal{S}_1 + N_2 \mathcal{S}_2)(L_1 + L_2) + 2.$$

(8.13)

Further $\mathcal{S}_1$ and $\mathcal{S}_2$ are defined as (see [7]),

$$\mathcal{S}_1 = \left((sI - M_1)M_2^{-1}(sI - M_1) - M_2\right)^{-1}(sI - M_1)M_2^{-1}, \text{ and}$$
$$\mathcal{S}_2 = \left((sI - M_1)M_2^{-1}(sI - M_1) - M_2\right)^{-1}.$$

(8.14)

Notice that e_{mi} is multiplied with $(-1)^i$ in (8.12), which suggests that when the system is in model 2, the transfer functions from $[x^T(s), g^T(s)]^T$ to $e_{dif}(s)$ and $e_{sum}(s)$ have the same value and the same sign, opposite to the case when the system is in model 1. Thus, it is expected that when the signal components in $e_{dif}(t)$ and $e_{sum}(t)$ reach steady-state after system switching, the relative sign between them can be employed to infer the active model.

A simple discrete time implementation for checking relative sign between $e_{dif}(t)$ and $e_{sum}(t)$ can be achieved by sampling the signals over a time window of size M and then correlating the sampled signals. Let the test signal corresponding to sample index m be denoted by T[m] described by

$$T[m] = e_{dif}[m]e_{sum}[m] + e_{dif}[m-1]e_{sum}[m-1] + \cdots + e_{dif}[m-M+1]e_{sum}[m-M+1].$$

(8.15)

If $T[m] > 0$ decision is made that plant is following the dynamics of model 2, otherwise model 1. The window size M strikes a balance between reducing the effects of noise and the detection speed of the test signal $T[m]$. Typically, increasing M tends to average out (thus reduce) the effect of noise present in $T[m]$. However, increasing M also reduces the detection speed.

Here the case when $g(t)$ being sinusoidal is further analyzed. As a consequence of (8.12), when the system is in model 1, signal components in $e_{dif}(t)$ and $e_{sum}(t)$ will be exactly out of phase in steady state. Conversely, when the system follows dynamics described by model 2, signal components in $e_{dif}(t)$ and $e_{sum}(t)$ will be in phase in steady state. An analog phase demodulator can be used to extract the phase of the sinusoids $e_{dif}(t)$ and $e_{sum}(t)$ and the resulting phase signals differ by 180 degree when the system is in model 1 and 0 degree when the system is in model 2. Hence, the difference of the phase signals, denoted by ϕ_m, can be used for real time detection of system model.

Another simple way to extract the model information is to use the envelope of the signal $(e^{(1)}(t))^2 - (e^{(2)}(t))^2$ (see [7]). Define the signal $ISM_{lpf} := \left((e^{(1)}(t))^2 - (e^{(2)}(t))^2 \right)_{lpf}$ which is the output of a suitably chosen low pass filter with input $(e^{(1)}(t))^2 - (e^{(2)}(t))^2$. ISM_{lpf} retains negative sign in model 1 and positive sign in model 2 and this sign inversion can be used in real time to interpret the active model [7]. In the context of probe based detection, the phase difference between $e_{dif}(t)$ and $e_{sum}(t)$ is called the ISM based phase signal: ϕ_m. Similarly, the test signal ISM_{lpf} is referred to as the ISM based amplitude signal: ISM_{lpf}.

A comparison of the computational effort required by each of the two methods MAP and ISM is discussed in [7]. It turns out that ISM is computationally much simpler than MAP.

8.3. Plant Detection with Known Dwell Interval; Sequence Detection

In many applications including the probe based data storage system detailed later in this chapter, each symbol is encoded in the media for an equivalent duration of T seconds for the read operation. Thus the probe encounters one kind of interaction during entire T seconds and the interaction behavior can change only after every T seconds. Here during the dwell-time of T seconds the probe interacts with the media for deciphering the same bit of information.

The test signals for each of the schemes, ISM and MAP, have finite response time; $T[m]$ in (8.15) and $p(\theta_i \mid \mathcal{Y}_m)$, $i \in \{1, 2\}$ from (8.7) take time to change to new steady state values when the system changes from model 1 to model 2 and vice versa. The effect of contributions of past interactions on the present interaction causes inter symbol interference (ISI). ISI is more severe when the system switches fast, where the dwell interval T is small. A significant gain over symbol-by-symbol schemes can be achieved by deciphering the entire sequence of system models from the observation data. A widely accepted sequence detection technique in the communication systems and data storage community is the Maximum Likelihood Sequence Detection (see [20]) method. As demonstrated here, Maximum Likelihood Sequence Detection (MLSD) can be applied in the context of binary system model detection as well. The detailed explanation of MLSD method applied for probe based data storage is discussed in [7]. Here, only the key results and highlights will be presented.

In [7] MLSD is applied to process sequence of samples from the test signal $s_m(t)$ generated by two ways; firstly $s_m(t)$ is generated from suitably low pass filtered $(e^{(1)})^2$ signal which is termed as regular MLSD and secondly $s_m(t)$ is generated using low pass filtered $(e^{(1)})^2 - (e^{(2)})^2$ signal which is termed as ISM-MLSD. It is demonstrated that
ISM-MLSD provides significant performance benefits over symbol-by-symbol ISM, MAP based detections and also regular MLSD.

8.4. Application of ISM and ISM-MLSD to Probe Based Data Storage

Probe based data storage has the potential to revolutionize data storage where greater than 5Tb/in^2 densities can be achieved. In the dynamic mode of operation for probe based data storage, the cantilever flexure is sinusoidally oscillated near its first resonant frequency. The flexure has a sharp tip near its end that intermittently interacts with the storage media

[7]. The tip-sample interaction occurring in every oscillation cycle alters the cantilever orbit. In conventional dynamic mode method the cantilever oscillation amplitude and phase of the fundamental harmonic (referred here as A_{def} and ϕ_{def} respectively) are utilized for signal detection purposes. It is demonstrated that the cantilever-sample system, which is nonlinear, can be mapped to an equivalent linear-time-invariant model [19]. Parameters of the equivalent model, if identified, can be utilized to infer changes in the sample topography [6]. In probe based data storage applications, high and low topography are used to encode different bits [21]. Observer based concepts have played an important role in improving imaging with AFM [6, 22-25]. Binary topography detection technique, where the detection of bits is the main task, is presented here.

The first mode model of the cantilever is specified by the parameters ω_0 (first resonant frequency) and Q_0 (quality factor). Dynamics of the cantilever flexure is governed by the following differential equation (see [26]).

$$\ddot{p} + \frac{\omega_0}{Q_0}\dot{p} + \omega_0^2 p = F(t) = g(t) + h(t),$$

$$h(t) = \varphi(p,\dot{p}), \quad y = p + \zeta,$$

(8.16)

where p is the cantilever tip deflection and $\dot{p}$ is the tip velocity. The force per unit mass on the cantilever F consists of the excitation g and nonlinear tip-sample interaction force h . Output y is corrupted with the measurement noise ζ . Sinusoidal excitation g is normally chosen at or very near ω_0 . The resulting system consists of the system G specified by (ω_0, Q_0) and the nonlinear feedback term h as shown in Fig. 8.1. It is established that a cantilever steadily interacting with the sample in the dynamic mode can be modelled as an equivalent cantilever specified by (ω_{eq}, Q_{eq}) [19]. The equivalent dynamics is described by

$$\ddot{p} + \frac{\omega_{eq}}{Q_{eq}}\dot{p} + \omega_{eq}^2 p = g(t).$$

(8.17)

(ω_{eq}, Q_{eq}) depend on the slowly varying cantilever oscillation amplitude which can be identified using the steady-state or recursive techniques described in [19]. A continuous time state-space model for the free air cantilever dynamics as specified in (8.16) consists of the matrices (A_1, B_1, C_1) that can be considered as model 1 for switched detection framework. The recursive method, namely the bias-compensated exponentially weighted recursive least square (BCEWRLS) method [19, 27], is used to estimate the discrete-time parameters of the equivalent cantilever dynamics (8.17) when the cantilever interacts with the sample for a specified set-point amplitude. Once these equivalent parameters are known they can be easily converted to continuous-time domain to construct another set of state space matrices (A_2, B_2, C_2) (see [6]) that represent the model 2 in the model detection strategy.

Here it is assumed that the system dynamics instantaneously changes to the equivalent cantilever (nominal cantilever) model when it starts (stops) interacting with the sample. It also holds true that the equivalent cantilever model differs from the free air model only in the A matrix. So that $B_1 = B_2 (:= B)$ and $C_1 = C_2 (:= C)$. Hence from (8.10) and (8.11), it follows that

$$\begin{bmatrix} \dot{z}_1(t) \\ \dot{z}_2(t) \end{bmatrix} = \underbrace{\begin{bmatrix} M_1 & M_2 \\ M_2 & M_1 \end{bmatrix}}_{A_{ISM}} \begin{bmatrix} z_1(t) \\ z_2(t) \end{bmatrix}$$

$$+ \underbrace{\begin{bmatrix} (A_2 - A_1) & 0 & (L_2 - L_1) \\ 0 & (A_2 - A_1) & -(L_1 + L_2) \end{bmatrix}}_{B_{ISM}} \begin{bmatrix} x(t) \\ b_m x(t) \\ \zeta(t) \end{bmatrix}, \qquad (8.18)$$

where the variable $b_m = -1$ when the cantilever oscillates in air and $b_m = +1$ when the cantilever follows the equivalent dynamics. The output equations for e_{dif} and e_{sum} are governed by

$$\begin{bmatrix} e_{dif}(t) \\ e_{sum}(t) \end{bmatrix} = \underbrace{\begin{bmatrix} C & 0 \\ 0 & C \end{bmatrix}}_{C_{ISM}} \begin{bmatrix} z_1 \\ z_2 \end{bmatrix} + \underbrace{\begin{bmatrix} 0 & 0 & 0 \\ 0 & 0 & 2 \end{bmatrix}}_{D_{ISM}} \begin{bmatrix} x(t) \\ b_m x(t) \\ \zeta(t) \end{bmatrix}. \qquad (8.19)$$

The system described by (8.18) and (8.19) is used for verifying the switching model for the probe based storage system. Discrete time equivalent systems of (A_i, B_i, C_i), $i \in \{1, 2\}$ are used for MAP based detection as shown in (8.7). It should be noted that although the BCEWRLS scheme is extremely effective for measurement of equivalent parameters, the speed of parameter convergence is slower than what can be achieved by a detection scheme like ISM or MAP. Thus typically, we estimate the nominal and equivalent plant dynamics from a slower learning phase using BCEWRLS algorithm, and subsequently utilize the detection strategies.

8.4.1. Validation of Equivalent Model and Detection Framework

Here we demonstrate the efficacy of a detection framework over an estimation framework for the purpose of fast identification of plant models in real time. The transfer function of a cantilever with resonant frequency of 63.147 kHz ($f_0 = \omega_0 / 2\pi$) and quality factor 227.849 (Q_0) is chosen as the model 1 for switching system. A discrete time system is easily obtained from the continuous time model corresponding to sampling frequency of 2 MHz. The discrete time model 1 is described by parameters $\{b_1, b_0, a_2, a_1, a_0\}$ which are found to be {-1.9599, 0.9991, 3.972e-4, -2.036e-4, -4.990e-4}. Another cantilever transfer function model with resonant frequency of $f_0 + 0.5$ kHz and quality factor $0.8 * Q_0$ is chosen to represent the model 2. The parameters $\{a_2, a_1, a_0\}$ are the same in both models while $\{b_1, b_0\}$ for model 2 are found to be {-1.9591, 0.9989}.

In Fig. 8.2 the plant is made to switch randomly between model 1 and 2 at a bit duration of 2 ms. The BCEWRLS algorithm (see [19]) shown in Fig. 8.2 works with $\lambda = 0.995$. ISM based detection algorithm as described in Section 8.2.2 is also employed. The decision variable T[m] for discrete time implementation of ISM (8.15) is calculated and decision is made to be model 2 for $T[m] > 0$, otherwise model 1. Fig. 8.2 shows the significant speed improvement of implementing ISM algorithm over BCEWRLS scheme for system detection purposes. Near the time 0.9 μs, system switches from model 1 to model 2. At that time both the estimates of b_0 and b_1 are in transient states in Fig. 8.2 (a) and (b).

However, during the time scales shown in Fig. 8.2 (a) and (b), ISM based detection algorithm is observed to catch up with plant changes at much faster rates than the estimation scheme. For example, in Fig. 8.2 (b), near time 0.9 μs, b_1 changes from low to high value. The ISM decision changes within 0.0175 μs to the correct decision value. However the BCEWRLS algorithm takes almost 0.350 μs to reach steady value.

Hence a speed improvement order of 20 can be achieved using ISM based detection over BCEWRLS method alone. Even though the exact convergence time may vary from one example to another, ISM based detection is consistently observed to be faster than BCEWRLS based estimation. In Fig. 8.2 (b), near time 0.9 μs, b_1 changes from low to high value. The ISM decision changes within 0.0175 μs to the correct decision value.

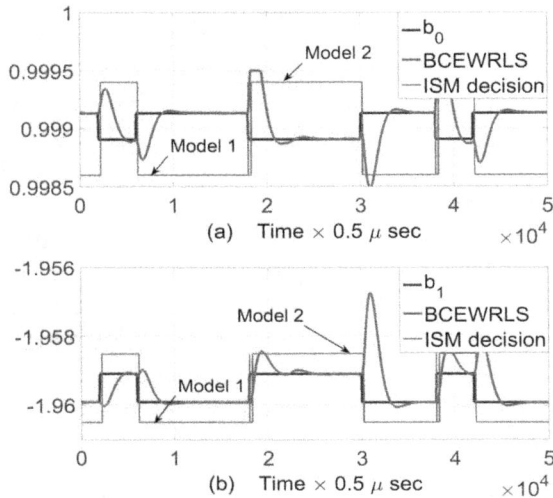

Fig. 8.2. Simulation results: (a) Estimation of b_0 using BCEWRLS ($\lambda = 0.995$) and decision with ISM from simulation; (b) Estimation of b_1 with ISM based decision from simulation. ISM decision signal is much faster and hence justifies its effectiveness for real time deciphering of the active system model.

A 'close to reality' tip-sample interconnected system is employed for verification of the effectiveness of BCEWRLS in estimating the equivalent cantilever parameters. A piecewise linear model developed in [28] is utilized to model the tip-sample interaction which works reasonably well for capturing interactions with hard substrates like mica. The model parameters are chosen so that the cantilever tip-sample interaction is predominantly repulsive. A cantilever with resonant frequency of 63.15 kHz and quality factor 227.85 is chosen for simulation. The cantilever is oscillated with free air oscillation amplitude of 24 nm. The oscillating cantilever is gradually moved towards and away from the sample surface to increase the interaction length l_{int} (see [6] for detailed definition of l_{int}) followed by reducing it. As shown in Fig. 8.3 when l_{int} increases, the amplitude of oscillation A_{def} reduces and vice versa. Equivalent cantilever parameters (A_{def} and f_{eq}) corresponding to each cantilever oscillation amplitude are obtained from analytical expressions derived using averaging theory techniques [29] and compared with estimated values using BCEWRLS with λ=0.995. It is evident from Fig. 8.3 that the equivalent cantilever model is indeed a very useful tool to model oscillating cantilever interacting with a sample. Further, BCEWRLS estimated values are close match with theoretical predictions.

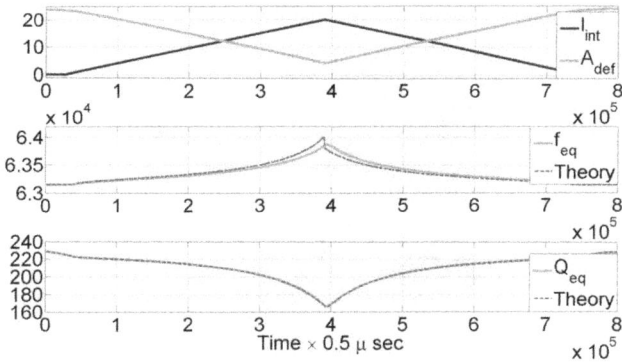

Fig. 8.3. (Simulation Results) Verification of equivalent cantilever model using averaging theory technique. The oscillating cantilever is made to interact with sample for gradually ascending interaction (approach) followed by descending interaction (retract). With increasing (reducing) l_{int}, the amplitude of oscillation A_{def} reduces (increases). BCEWRLS scheme is applied to calculate equivalent cantilever parameters; equivalent resonant frequency f_{eq} and equivalent quality factor Q_{eq}. Using averaging theory f_{eq} and Q_{eq} are analytically calculated. Theoretically predicted and BCEWRLS estimated values are decent match.

The validity of the switching model discussed before is explained now. The separation between the cantilever base and the sample (called sep) is raised and lowered with a pseudo random bit pattern of bit duration 200 μs. A high value of sep in Fig. 8.2 indicates model 1 (free air oscillations) and a low value indicates model 2 (interaction state). A frequency sweep method is applied to obtain the frequency response of the cantilever around the first resonance frequency and from the

311

frequency response the state space matrices (A_1, B, C) are obtained. In order to get the model 2, the cantilever tip is made to interact with the sample with an interaction length (l_{int}) of 2 nm. The equivalent cantilever model corresponding to l_{int} of 2 nm is obtained using the BCEWRLS scheme with forgetting factor $\lambda = 0.995$. From the estimated equivalent parameters, A_2 matrix for the continuous time equivalent system is obtained. From the state space models with (A_1, B, C) and (A_2, B, C), the system $(A_{ISM}, B_{ISM}, C_{ISM}, D_{ISM})$ as described in (8.18) and (8.19) is constructed. Now, the outputs e_{dif} and e_{sum} from the ISM system (see Fig. 8.4) are compared to the outputs e_{dif} and e_{sum} obtained from the 'close to reality' cantilever-sample interconnected system together with the two observers.

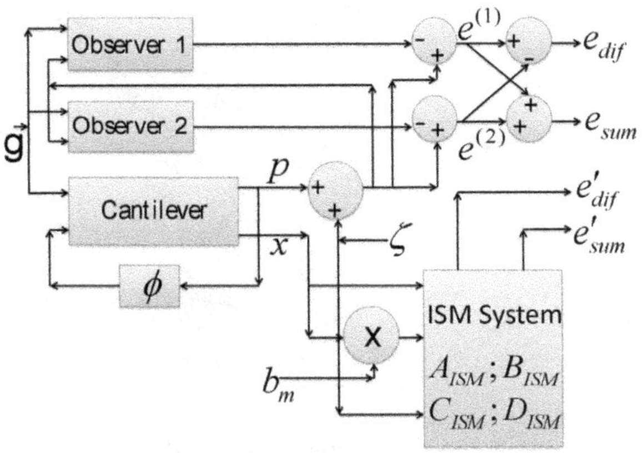

Fig. 8.4. ISM system: Simulation model for verification of the switching model.

From Fig. 8.5 it can be observed that whether the model is 1 (sep high) or 2 (sep low), e_{dif} and e'_{dif} are always near perfect match with each other. From Fig. 8.5 (b), it is clear that e_{sum} and e'_{sum} match correctly in model 1. In model 2, the phase behaviors of e_{sum} and e'_{sum} are very close match. However, the amplitude behaviors are not the same. A better model than a simple switching model is required for predicting the behavior of amplitude of e_{sum}. Nevertheless, the switching model is effective since it allows a nice balance between simplicity and a reasonable ground for analytical understanding.

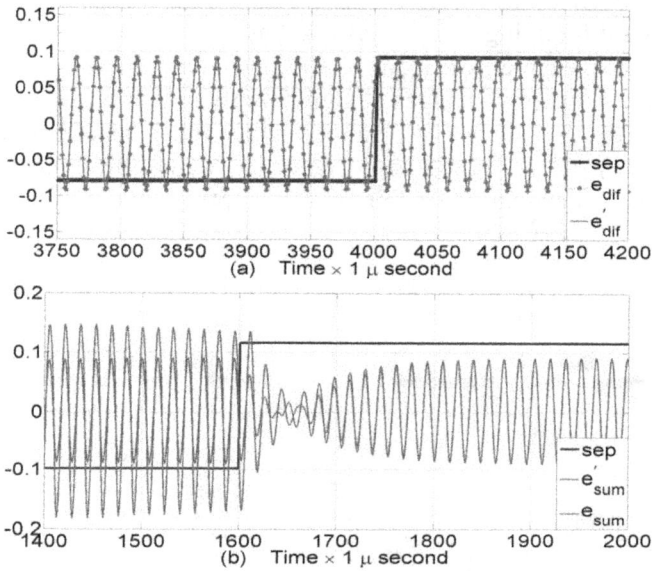

Fig. 8.5. Simulation results: (a) e_{dif} and with sep from simulation. These signals have no detectable difference; (b) e_{sum} and e'_{sum} from simulation. These signals match perfectly for no interaction from sample (model 1). The phase behaviors of e_{sum} and e'_{sum} match during interaction with sample (model 2). e'_{dif}

8.4.2. Detection Performance from Experiments

In the following the detection method is experimentally assessed. A field-programmable gate array (FPGA) based circuit actuates the piezoelectric nano-positioning system according to a pseudo random bit pattern of bit duration 350 µs. In Fig. 8.6 (a) and (b), the signal hgt denotes the height of the sample. High hgt implies interaction and low hgt implies free air cantilever oscillations where there is no interaction with the sample. The innovation signal $e^{(1)}$ is obtained from observer matched to the free air cantilever model (f_0 =71.73 kHz, Q_0 =130.21). The second observer is designed to match the equivalent cantilever model where f_{eq} is determined to be 100 Hz less than f_0. Q_{eq} is determined to be same as Q_0.

An interesting observation is that ϕ_m exhibits better steady behavior than ISM_{lpf} in experiment. In Fig. 8.6, value of A_{def} shows a downward drift from left to right. However, the magnitude levels of ISM_{lpf} are seen to be consistent implying that a threshold based detection is difficult from A_{def} but it is possible from ISM_{lpf}. The ISM phase signal ϕ_m is found to detect the topography changes at significantly higher bandwidth. When hgt changes, ϕ_m switches value almost seven times faster than ϕ_{def} (see Fig. 8.6 (b)). Clearly ISM based detection outperforms conventional AFM signals for binary detection of topography.

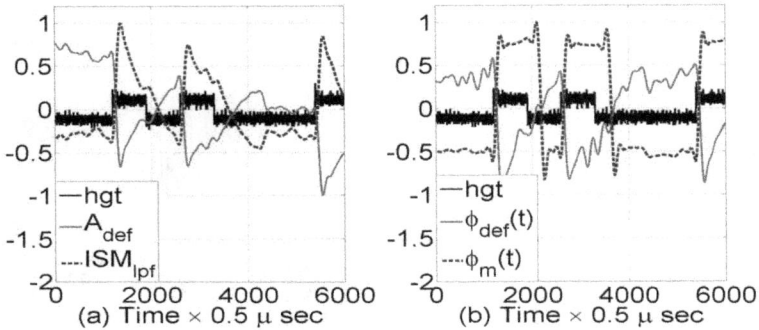

Fig. 8.6. Experimental data: (a) Conventional amplitude A_{def} and ISM_{lpf} from experiment. A_{def} shows an overall drift from left to right. However, ISM_{lpf} has more consistent levels; (b) Regular phase ϕ_{def} and ISM based phase ϕ_m signals from experiment. ϕ_m has a very consistent binary behavior and better steady state magnitudes than ϕ_{def}.

Fig. 8.7 (a) and (b) show a comparison of experimentally obtained decision images by recursive MAP and ISM. It is evident that the decision image from MAP results in considerable fraction of incorrect decisions (see Fig. 8.7 (a)). However, from Fig. 8.7 (b) it is observed that ISM yields high fidelity results.

8.4.3. Symbol by Symbol and Sequence Detection

A comparison of detection of plant models with finite dwell time as discussed in Section 8.3 is presented here. For simulations, sep is varied with a dwell time T (or equivalently bit duration) of 250 µs. Test signals for ISM in (8.15) and MAP in (8.7) are averaged over this time interval T before making decisions for symbol-by-symbol detection of plant models.

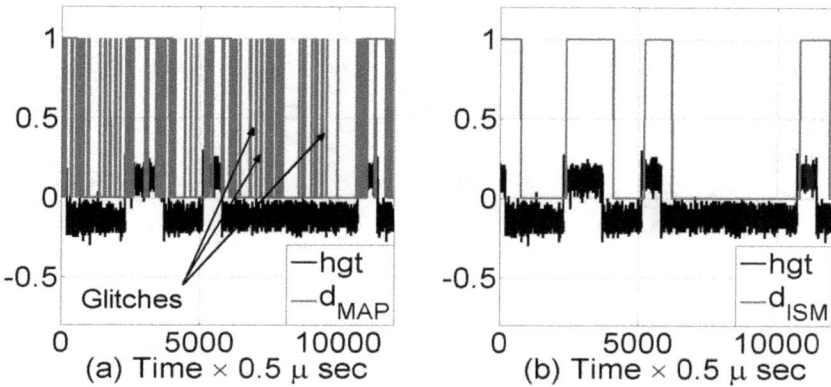

Fig. 8.7. Experimental data: (a) MAP, and (b) ISM based decision images d_{MAP} and d_{ISM}. d_{MAP} is severely effected by decision glitches whereas d_{ISM} has correct decision making capability.

In the previous work on MLSD for probe based data storage (see [30]), the test signal $s_m(t)$ is chosen by applying a matched filter on the innovation signal $e^{(1)}$. Apart from the assumption of known dwell time T, in [30] it is also assumed that the exact time instant when the oscillating cantilever tip impulsively interacts with the sample is known. In [6] it is shown that ISM performs as good as MLSD applied on test signal generated through processing $e^{(1)}$ as recommended in [30]. However, from the perspective of implementation, having accurate knowledge of time instants where the cantilever interacts with the sample during each dwell time T is a difficult task.

Hence, regular MLSD and ISM-MLSD are applied on test signals $s_m(t)$ chosen as appropriately low pass filtered $(e^{(1)})^2$ and $(e^{(1)})^2 - (e^{(2)})^2$ signals respectively. This method is simpler to implement and the samples during each interval T can be collected simply by uniform sampling of the observation data without knowing the time instants of cantilever tip-sample interaction (see [7] for details). The bit error rates of each method are observed where the bits are generated pseudo randomly. Average BERs are obtained for fixed on-sample interaction length l_{int} which is further varied from 1 nm to 2 nm. The resulting BER vs l_{int} plots are compared in Fig. 8.8. In Fig. 8.8, ISM-MLSD performs better than other methods. Further experimental results can be found in [7].

315

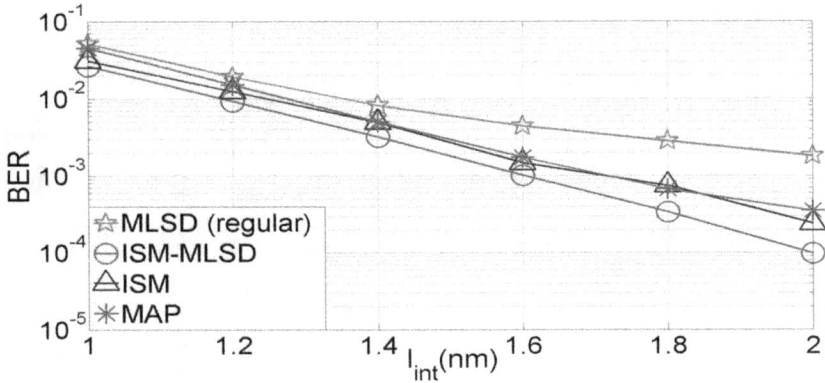

Fig. 8.8. (Simulation results) Bit error rate (BER) performance of symbol-by-symbol ISM and MAP with regular MLSD and ISM-MLSD as a function of interaction length l_{int}. ISM-MLSD consistently outperforms all the other methods. Dwell time T is chosen as 250 μs.

8.5. Conclusions

In this chapter a real time switching-model detection strategy ISM is presented for a system that switches between two models. The concept of utilizing the known system models and sign-inverting test signals for ISM is developed. A MAP based detection algorithm from existing literature is briefly outlined and compared with ISM. Detection schemes are modified for the situation when a dwell time with which the plant model switches is known a priori. Sequence detection method namely MLSD on innovation squared mismatch signal is proposed to further improve detection performance in the presence of inter symbol interference. In the application part, the dynamic mode probe based fast topography detection system is cast as a binary switched system. The results indicate the superior speed of operation and the steady-state value advantages of model detection approach over the conventional AFM signals. The results also suggest that ISM is simpler to implement yet performs better than recursive MAP for real time detection. Performance of sequence based detections are compared with ISM and MAP based symbol-by-symbol detections when the bit duration is known. It is verified that the proposed combination of ISM-MLSD offers significant gain in detection accuracy compared to other methods through simulation and experiments.

References

[1]. Z. Sun, S. S. Ge, Analysis and synthesis of switched linear control systems, *Automatica,* Vol. 41, 2005, pp. 181-195.

[2]. A. S. Morse, Supervisory control of families of linear set-point controllers - Part I. Exact matching, *IEEE Transactions on Automatic Control,* Vol. 41, No. 10, 1996, pp. 1413-1431.

[3]. A. S. Morse, Supervisory control of families of linear set-point controllers. 2. robustness, *IEEE Transactions on Automatic Control,* Vol. 42, No. 11, 1997, pp. 1500-1515.

[4]. K. S. Narendra, J. Balakrishnan, Adaptive control using multiple models, *IEEE Transactions on Automatic Control,* Vol. 42, No. 2, 1997, pp. 171-187.

[5]. F. Gao, S. E. Li, D. Kum, H. Zhang, Synthesis of multiple model switching controllers using H∞ theory for systems with large uncertainties, *Neurocomputing,* Vol. 157, 2015, pp. 118-124.

[6]. S. Ghosal, G. Saraswat, A. Ramamoorthy, M. Salapaka, Topography detection using innovations mismatch method for high speed and high density dynamic mode AFM, in *Proceedings of the American Control Conference (ACC),* Washington DC, USA, 2013, pp. 5500-5505.

[7]. S. Ghosal, M. Salapaka, Model detection with application to probe based data storage, *Automatica,* Vol. 74, 2016, pp. 171-182.

[8]. M. Carosino, Y. Chen, B. J. Belzer, K. Sivakumar, J. Murray, P. Wettin, Iterative detection and decoding for the four-rectangular-grain TDMR model, in *Proceedings of the IEEE 51st Annual Allerton Conference on Communication, Control, and Computing (Allerton),* 2013, pp. 653-659.

[9]. S. M. Khatami, B. Vasic, Detection for two-dimensional magnetic recording systems, in *Proceedings of the International Conference on Computing, Networking and Communications (ICNC),* 2013, pp. 535-539.

[10]. K. Cai, Z. Qin, S. Zhang, Y. Ng, K. Chai, R. Radhakrishnan, Modeling, detection, and LDPC codes for bit-patterned media recording, in *Proceedings of the IEEE GLOBECOM Workshops,* 2010, pp. 1910-1914.

[11]. B. D. Anderson, J. B. Moore, Optimal filtering, *Englewood Cliffs,* Vol. 21, 1979, pp. 22-95.

[12]. G. Campardo, F. Tiziani, M. Iaculo, Memory mass storage, *Springer Science & Business Media,* 2011.

[13]. S. K. Lai, Flash memories: Successes and challenges, *IBM Journal of Research and Development,* Vol. 52, No. 4.5, 2008, pp. 529-535.

[14]. G. Binnig, C. F. Quate, C. Gerber, Atomic force microscopez, *Physical Review Letters,* Vol. 56, No. 9, 1986, pp. 930-933.

[15]. C. D. Wright, M. M. Aziz, P. Shah, L. Wang, Scanning probe memories-- Technology and applications, *Current Applied Physics,* Vol. 11, No. 2, 2011, pp. e104-e109.

[16]. H. J. Mamin, R. P. Ried, B. D. Terris, D. Rugar, High-density data storage based on the atomic force microscope, in *Proceedings of the IEEE,* Vol. 87, No. 6, 1999, pp. 1014-1027.

[17]. K. Tanaka, Y. Cho, Actual information storage with a recording density of 4 Tbit/in. 2 in a ferroelectric recording medium, *Applied Physics Letters,* Vol. 97, No. 9, 2010, p. 092901.

[18]. M. A. Lantz, B. Gotsmann, P. Jaroenapibal, T. D. Jacobs, S. D. O'Connor, K. Sridharan, R. W. Carpick, Wear-Resistant Nanoscale Silicon Carbide Tips for Scanning Probe Applications, *Advanced Functional Materials,* Vol. 22, No. 8, 2012, pp. 1639-1645.

[19]. S. Ghosal, A. Gannepalli, M. Salapaka, Toward quantitative estimation of material properties with dynamic mode atomic force microscopy: a comparative study, *Nanotechnology,* Vol. 28, No. 32, 2017, p. 325703.

[20]. G. D. J. R. Forney, Maximum-likelihood sequence estimation of digital sequences in the presence of intersymbol interference, *IEEE Transactions on Information Theory,* Vol. 18, 1972, pp. 363-378.

[21]. C. D. Wright, M. M. Aziz, P. Shah, L. Wang, Scanning probe memories-- Technology and applications, *Current Applied Physics,* Vol. 11, No. 2, 2011, pp. e104-e109.

[22]. S. Ghosal, M. Salapaka, Fidelity imaging for atomic force microscopy, *Applied Physics Letters,* Vol. 106, No. 1, 2015, p. 013113.

[23]. D. R. Sahoo, A. Sebastian, M. V. Salapaka, Harnessing the transient signals in atomic force microscopy, *International Journal of Robust and Nonlinear Control,* Vol. 15, No. 16, 2005, pp. 805-820.

[24]. T. De, P. Agarwal, D. R. Sahoo, M. V. Salapaka, Real-time detection of probe loss in atomic force microscopy, *Applied Physics Letters,* Vol. 89, No. 13, 2006, pp. 133119-1 – 133119-3.

[25]. D. R. Sahoo, T. De Murti, V. Salapaka, Observer based imaging methods for atomic force microscopy, in *Proceedings of the IEEE 44th Conference on Decision and Control, European Control Conference (CDC-ECC'05),* 2005, pp. 1185-1190.

[26]. A. Sebastian, A. Gannepalli, M. V. Salapaka, A review of the systems approach to the analysis of dynamic-mode atomic force microscopy, *IEEE Transactions on Control Systems Technology,* Vol. 15, No. 5, 2007, pp. 952-959.

[27]. P. Agarwal, M. V. Salapaka, Real time estimation of equivalent cantilever parameters in tapping mode atomic force microscopy, *Applied Physics Letters,* Vol. 95, No. 8, 2009, pp. 083113-083113.

[28]. A. Sebastian, M. Salapaka, D. J. Chen, J. Cleveland, Harmonic and power balance tools for tapping-mode atomic force microscope, *Journal of Applied Physics,* Vol. 89, No. 11, 2001, pp. 6473-6480.

[29]. J. A. Sanders, F. Verhulst, J. A. Murdock, Averaging methods in nonlinear dynamical systems, *Springer,* 2007.

[30]. N. Kumar, P. Agarwal, A. Ramamoorthy, M. V. Salapaka, Maximum-likelihood sequence detector for dynamic mode high density probe storage, *IEEE Transactions on Communications,* Vol. 58, No. 6, 2010, pp. 1686-1694.

Chapter 9

H∞ Tracking Adaptive Fuzzy Sliding Mode Design Controller for a Class of Non Square Nonlinear Systems

S. Aloui, M. Elloumi and Y. Koubaa

9.1. Introduction

Output tracking and regulation problems of nonlinear systems with parametric uncertainties and unmodelled dynamics have been widely studied in recent time. In the last decade, differential geometry has proven to be successful to the study of nonlinear system control [2]. This approach involves coming up with a transformation of the nonlinear system into an equivalent linear system through a change of variables and a suitable control input, so that the conventional linear control techniques can be applied. However, this control design can only be applied to a nonlinear system whose dynamics are exactly known [1]. In order to relax some exact dynamic restrictions, several adaptive schemes have been introduced [3]. Moreover, several works are focused on the combination of classical adaptive techniques and universal approximators such as fuzzy systems, neural networks [31, 33, 35]. In fact, many important adaptive fuzzy control schemes have been developed to systematically incorporate the expert information [10, 16, 32, 34]. Nonetheless, a prescribed tracking performance can not be guaranteed in the conventional fuzzy adaptive control systems owing to the fact that the influence of the matching error and external disturbances on tracking error cannot be efficiently eliminated. Other studies which combine the backstepping design techniques with the fuzzy adaptive output feedback control are constructed [37-40]. Since the control

S. Aloui

National Engineering School of Sfax, Laboratory of Sciences and Techniques of Automatic Control & Computer Engineering (Lab-STA), University of Sfax, Sfax, Tunisia strategy based on sliding mode technique can offer many good properties [23, 28], such as the low sensitivity to the matching parameter variations and its robustness against to a large class of perturbations or modelled uncertainties, many works proceed to integrate this technique into the adaptive fuzzy controller in order to improve the stability and the robustness of closed loop system [17, 24, 28-30]. These studies are based on the use of two adaptive fuzzy systems to approximate the unknown system dynamics to design the control system. The adaptation laws of adjustable parameters are derived based on Lyapunov stability analysis. However, the chattering phenomenon represents a limitation for this technique. In order to overcome this drawback, several techniques have been proposed. In [18], the main idea is to change the discontinuous switching action by a continuous saturation function. In [19], an asymptotic observer in the control loop to eliminate chattering despite the discontinuous control law. Both the boundary layer approach and the observer solution assume that the unmodelled dynamics are completely unknown. In [20], the authors have used two fuzzy systems to approximate the equivalent control term and the switching control term, for Single Input Single Output (SISO) nonlinear systems. To attain the same objective, other approaches based on the higher order Sliding Mode Control (SMC) have been developed [21, 9]. Instead of influencing the first sliding variable time derivative, the signum function appears in its higher order time derivative. In [22], the authors propose a method to mitigate the chattering phenomenon by using an integral sliding mode. However most of the recent research works considered Multi Inputs Multi Outputs (MIMO) nonlinear system in the following form $y^{(r)} = F(x) + G(x)U$, where $x \in \mathbb{R}^n$ is the state vector; $U \in \mathbb{R}^p$ is the input vector; $Y \in \mathbb{R}^m$ is the output vector where the number of inputs and outputs is equal (i.e $p = m$). Then, the controller synthesis developed for square systems are applied directly. Recently, there has been growing attention is non-square systems characterized by the fact that the number of inputs is not equal to the number of outputs (i.e $p \neq m$). The usual solution when designing feedback control scheme for nonlinear non-square systems is to square (i.e $p = m$) the system by eliminating or adding variables. In the litterature, several works have proposed different techniques to study the difficult control problem of this class of systems in particular the underactuated systems [4-5]. In [11], an adaptive fuzzy sliding mode controller for a class of underactuated systems was proposed. In fact, the underactuated system is decoupled into subsystems, and respectively a sliding surface is defined for each subsystem. In [12], a stable hierarchical sliding-mode control method for a class of second-order underactuated systems is presented. The underactuated system is divided into two subsystems. For each part, a first-level sliding surface is defined. For these two first-level sliding surfaces, a second-level sliding surface is introduced. The sliding mode control is derived using Lyapunov approach. The control law can drive the subsystems toward their sliding surfaces and attain their desired values. The asymptotic stability of all the sliding surface is ensured. In [13], a cascade sliding mode controller for a class of large-scale underactuated systems is proposed. Firstly, two states are chosen to construct the first-layer sliding surface. Secondly, the first-layer sliding surface and one of the left states are used to construct the second-layer sliding surface. This procedure goes on till the last-layer sliding surface is obtained. The cascade sliding-mode controller is proved to be globally stable in the sense that all signals involved are bounded. In [14], a robust controller using sliding mode control method for a class of underactuated mechanical systems with mismatched uncertainties is proposed. A distributed compensator is added to the sliding mode surfaces. For an underactuated system, which consists of $2n$ state variables, the controller has the $(2n-1)$-layer structure. Using Lyapunov approach, the stability of all the sliding surfaces is proven. In [15], two sliding-mode controllers based on the incremental hierarchical structure and the aggregated hierarchical structure for a class of underactuated systems are presented. Their design steps and the choice of parameters are given. Differing from the other approaches we intend, in this chapter, to study a fuzzy adaptive control law for a large class of MIMO nonlinear systems with external disturbances in their original non-square form by relaxing the assumption $p \neq m$ [6-9]. A robust control algorithm is developed by combining the adaptive fuzzy sliding mode approach, and an adaptive PI term in order to simultaneously to reduce the chattering phenomenon and to ensure the good tracking performances in presence of the external disturbances. The chapter is organized as follows. The problem under investigation is first formulated in Section 9.2. The controller design and the main results are described in Section 9.3. The simulation studies are given in Section 9.4. Section 9.5 gives some conclusions on the main works developed in this chapter.

9.2. Generalized Conventional Sliding Mode

Consider the following MIMO nonlinear perturbed non-square system given by:

$$
\begin{cases}
y_1^{(r_1)} = f_1(x) + g_{11}(x)u_1 + \cdots + g_{1p}(x)u_p + d_{1_{r_1}} \\
\cdots \\
y_m^{(r_m)} = f_m(x) + g_{m1}(x)u_1 + \cdots + g_{mp}(x)u_p + d_{m_{r_m}},
\end{cases}
\tag{9.1}
$$

where $x \in \mathbb{R}^n$ is the state vector, $U = [u_1 \ldots u_p] \in \mathbb{R}^p$ is the control input vector, $y = [y_1 \ldots y_m] \in \mathbb{R}^m$ is the output vector and $p \neq m$. f_i, $g_{ij} \in \mathbb{R}^n \to \mathbb{R}$, $i = 1 \cdots m \ j = 1, 2 \cdots p$ are smooth unknown function vectors. $D = [d_1 \ \cdots \ d_m]^T$ is the perturbation vector, where $d_i = [0 \ \ 0 \ \ d_{i_{r_i}}] \in \mathbb{R}^{r_i}$ and $d_{i_{r_i}}$ is unknown.

Assumption 1. The external disturbance d_i is assumed to belong to $L_2[0,t]$, $\forall t \in [0,\infty)$ and $|d_i| < D_i \ \forall i$.

Denote

$$
Y = \begin{bmatrix} y_1^{(r_1)} \\ \cdots \\ y_m^{(r_m)} \end{bmatrix}, \ F(x) = \begin{bmatrix} f_1(x) \\ \cdots \\ f_m(x) \end{bmatrix}, \ U = \begin{bmatrix} u_1 \\ \cdots \\ u_p \end{bmatrix} \ d = \begin{bmatrix} d_{1_{r_1}} \\ \cdots \\ d_{m_{r_m}} \end{bmatrix}
$$

$$
G(x) = [G_1 \ \ G_2 \ \ \cdots \ \ G_m]^T,
$$

where $G_i = [g_{i1} \ \ g_{i2} \ \ \cdots \ \ g_{ip}]^T, i = 1, \cdots, m$, G is of full rank.

Then (9.1) can be written as:

$$
Y = F(x) + G(x)U + d \tag{9.2}
$$

Definition 2.1. Let $G(x)$ be a $m \times p$ matrix, and $G^+(x)$ be the pseudo-inverse of $G(x)$. If $G(x)$ is of full rank, then $G(x)^+$ can be computed as follows:

$$G^+(x) = \begin{cases} G^T(x)[G(x)G^T(x)]^{-1}, & m < p \\ G^{-1}(x), & m = p \\ [G(x)^T G(x)]^{-1} G^T(x), & m > p \end{cases} \tag{9.3}$$

The output tracking error e_i is:

$$e_i = y_i - y_{d_i} \tag{9.4}$$

y_{d_i} is the desired reference trajectory of y_i.

Assumption 2. Let us consider desired trajectories y_{d_i}, $i = 1 \; \cdots \; m$ that are known bounded functions of time with bounded known derivatives and are assumed to be *ri-time* differentiable.

The sliding mode surfaces are defined in the space of the tracking errors by the following equation:

$$s_i = \sum_{j=1}^{r_i} \alpha^i_{(r_i-j)} e_i^{(r_i-j)} \tag{9.5}$$

with $\alpha^i_{(r_i-1)} = 1$.

$$S = [s_1 \; \cdots \; s_m]^T \tag{9.6}$$

The parameters $\alpha^i_{(r_i-2)}, \cdots, \alpha^i_0$ are chosen such that all roots $h_i(p) = p^{(r_i-1)} + \alpha^i_{(r_i-2)} p^{(r_i-2)} + \cdots + \alpha^i_1 p + \alpha^i_0$ are in the left half of the complexe plane. (here p denotes the complex Laplace transform variable).

To achieve the control objective, it is sufficient to find a control law U so that all initial states lying off a sliding hyperplane $H / H = \{e \mid S(e) = 0\}$ would hit H in finite time and then remain on it.

The control law is designed as:

$$U = u_{eq} + u_{sw} \tag{9.7}$$

The equivalent control law u_{eq} is determined by $\dot{S} = 0$.

The time derivative of S can be obtained as

$$\dot{S} = F(x) + G(x)U + d - Y_d^{(n_r)} + \sum_{i=2}^{n_r} \Lambda_{(i-2)} E^{(i-1)}, \qquad (9.8)$$

where:

$$n_r = \max_i \{r_i\}$$

$$Y_d^{(n_r)} = [y_{d_i}^{(r_1)} \quad y_{d_i}^{(r_2)} \quad \cdots \quad y_{d_i}^{(r_m)}]^T \quad i = 1, 2, \cdots m$$

$$\Lambda_i = diag[\alpha_i^1 \text{ or } \varnothing, \alpha_i^2 \text{ or } \varnothing, \cdots, \alpha_i^m \text{ or } \varnothing] \quad i = 0, 1, \cdots, n-2$$

$$E^{(i)} = [e_1^{(i)} \quad e_2^{(i)} \quad \cdots \quad e_m^{(i)}]^T \quad i = 1, 2, \cdots n_r - 1$$

Three cases are determined

Case 1: $m = p$

We may use the equivalent control law:

$$u_{eq} = G^{-1}(x)\left(-F(x) + Y_d^{(n_r)} - \sum_{i=2}^{n_r} \Lambda_{(i-2)} E^{(i-1)} \right) \qquad (9.9)$$

The switching control term is defined by:

$$u_{sw} = -G^{-1}(x)\eta sgn(S) \text{ with } \eta > D_i \qquad (9.10)$$

sgn is the signum function.

The resulting sliding mode control law is:

$$U = G^{-1}(x)\left(-F(x) + Y_d^{(n_r)} - \sum_{i=2}^{n_r} \Lambda_{(i-2)} E^{(i-1)} - \eta sgn(S) \right) \qquad (9.11)$$

Case 2: $m < p$

In the same way, the sliding mode control law is:

$$U = G^+(x)\left(-F(x) + Y_d^{(n_r)} - \sum_{i=2}^{n_r}\Lambda_{(i-2)}E^{(i-1)} - \eta sgn(S)\right) \quad (9.12)$$

Case 3: $m > p$, in this case, the regularized inverse of $\tilde{G}(x) = (G(x)G^+(x))$ is defined by [23]:

$$\tilde{G}^T(x)[\varepsilon I + \tilde{G}(x)\tilde{G}^T(x)]^{-1} \quad (9.13)$$

ε is small positive constant and I is the identity matrix of appropriate dimension.

Such that:

$$[\varepsilon I + \tilde{G}(x)\tilde{G}^T(x)][\varepsilon I + \tilde{G}(x)\tilde{G}^T(x)]^{-1} = I$$

This implies that:

$$\tilde{G}(x)\tilde{G}^T(x)[\varepsilon I + \tilde{G}(x)\tilde{G}^T(x)]^{-1} = I - \varepsilon[\varepsilon I + \tilde{G}(x)\tilde{G}^T(x)]^{-1} \quad (9.14)$$

In order to avoid the singularity problem of $(G(x)G^+(x))$, the equivalent control term is defined by the following expression:

$$u_{eq} = G^+(x)\tilde{G}^T(x)\left(\varepsilon I + \tilde{G}(x)\tilde{G}(x)^T\right)^{-1}\left(-F(x) + Y_d^{(n_r)} - \sum_{i=2}^{n_r}\Lambda_{(i-2)}E^{(i-1)} + u_r\right)$$

We introduce another term u_r in the control law in order to cancel the approximation error.

By substituting (9.15) in (9.8), the time derivative of S can be rewritten as follows:

$$\dot{S} = F(x) + \tilde{G}(x)\tilde{G}^T(x)\left(\varepsilon I + \tilde{G}(x)\tilde{G}^T(x)\right)^{-1}(-F(x) + Y_d^{(n_r)}$$
$$- \sum_{i=2}^{n_r}\Lambda_{(i-2)}E^{(i-1)} + u_r) - Y_d^{(n_r)} + \sum_{i=2}^{n_r}\Lambda_{(i-2)}E^{(i-1)} \quad (9.15)$$

From (9.14), (9.15) can be rewritten as follows:

$$\dot{S} = F(x) + (I - \varepsilon[\varepsilon I + \tilde{G}(x)\tilde{G}^T(x)]^{-1})(-F(x) + Y_d^{(n_r)}$$

$$-\sum_{i=2}^{n_r}\Lambda_{(i-2)}E^{(i-1)} + u_r) - Y_d^{(n_r)} + \sum_{i=2}^{n_r}\Lambda_{(i-2)}E^{(i-1)}$$

$$\dot{S} = F(x) - F(x) + Y_d^{(n_r)} - \sum_{i=2}^{n_r}\Lambda_{(i-2)}E^{(i-1)} + (I - \varepsilon[\varepsilon I$$

$$+ \tilde{G}(x)\tilde{G}^T(x)]^{-1})u_r - \varepsilon[\varepsilon I + \tilde{G}(x)\tilde{G}^T(x)]^{-1}(F(x)$$

$$+ Y_d^{(n_r)} - \sum_{i=2}^{n_r}\Lambda_{(i-2)}E^{(i-1)}) - Y_d^{(n_r)} + \sum_{i=2}^{n_r}\Lambda_{(i-2)}E^{(i-1)}$$

$$\dot{S} = (I - \varepsilon[\varepsilon I + \tilde{G}(x)\tilde{G}^T(x)]^{-1})u_r - \varepsilon[\varepsilon I + \tilde{G}(x)\tilde{G}^T(x)]^{-1}$$

$$(F(x) + Y_d^{(n_r)} - \sum_{i=2}^{n_r}\Lambda_{(i-2)}E^{(i-1)}) \tag{9.16}$$

From (9.16), in order to avoid the singularity problem of $(I - \varepsilon[\varepsilon I + \tilde{G}(x)\tilde{G}^T(x)]^{-1})$, we consider the following expression of u_r:

$$u_r = [I + \tilde{G}(x)\tilde{G}^T(x)]\tilde{G}(x)[\varepsilon I + \tilde{G}^T(x)\tilde{G}(x)]^{-1}\tilde{G}^T(x)[\varepsilon I + \tilde{G}(x)\tilde{G}^T(x)]^{-1}$$

$$\varepsilon[\varepsilon I + \tilde{G}(x)\tilde{G}^T(x)]^{-1}(-\hat{F}(\hat{x} \mid \theta_1) + Y_d^{(n_r)} - \sum_{i=2}^{n_r}\Lambda_{(i-2)}E^{(i-1)}) \tag{9.17}$$

Remark 1. (9.17) is a approach solution of $\dot{S} = 0$.

By using (9.17), (9.16) can be rewritten as follows:

$$\dot{S} = \Delta_1(x, \varepsilon) \tag{9.18}$$

The sliding mode control law is given by:

$$U = G^+(x)\tilde{G}^T(x)(\varepsilon I + \tilde{G}(x)\tilde{G}^T(x))^{-1}(-F(x) + Y_d^{(n_r)} - \sum_{i=2}^{n_r}\Lambda_{(i-2)}E^{(i-1)} + u_r - \eta sgn(S)) \tag{9.19}$$

In order to obtain the sliding mode control, it is obvious that the system functions $F(x)$ and $G(x)$ and the switching parameters η have to be known.

In the next section, we will use adaptive fuzzy systems to approximate unknown nonlinear functions $F(x)$ and $G(x)$. Then, in order to eliminate the chattering phenomenon without deteriorating the tracking performances, the discontinuous term in the conventional sliding mode technique is replaced by an adaptive PI term [34, 9].

9.3. Design of a Robust Adaptive Fuzzy Controller

Control objectives: Determine a robust adaptive control law $U(x)$ for the nonlinear system given by (9.1) such that the vector y can follow a given desired reference signal vector y_d, under that the following conditions are met:

1) The closed-loop system is stable, i.e., all the signals involved are uniformly bounded.

2) For a given disturbance attenuation level $0 < \rho < 1$, the following H_∞ tracking performances is achieved as:

$$\int_0^t e^T Q e \, dt \le e(0)^T P e(0) + \frac{1}{\gamma_f} \tilde{\theta}_1^T(0)\tilde{\theta}_1(0) + \frac{1}{\gamma_g} \tilde{\theta}_2^T(0)\tilde{\theta}_2(0) + \rho^2 \int_0^t W^T W \, dt$$

$$(9.20)$$

for all $W \in L_2[0,t], \forall t \in [0,\infty)$, where $Q = Q^T \ge 0$ (A^T denotes the transpose of A) and $P = P^T \ge 0$ are symmetric positive definite weighting matrices, $\tilde{\theta}_1 = \theta_1^* - \theta_1$ and $\tilde{\theta}_2 = \theta_2^* - \theta_2$ are parameter approximation errors defined next, γ_f, γ_g are the adaptation parameters. Let $W = [w_1 \quad \cdots \quad w_m]^T$ and W gathers all the terms relating to the approximation errors and the external disturbances.

We introduce the notation $W \in L_2[0 \quad t] \, t \in \mathbb{R}^+ \Leftrightarrow$

$$\int_0^t W^T W \, dt < \infty \quad \forall t \in \mathbb{R}^+ \qquad (9.21)$$

Step 1: Approximation of $F(x)$ and $G(x)$

Both functions $F(x)$ and $G(x)$ in (9.2) are unknown, that is why we design two fuzzy systems $\hat{F}(x|\theta_1)$ and $\hat{G}(x|\theta_2)$ to approximate them.

Let $A_i^k, k = 1, \cdots, n_i$ be the fuzzy sets defined on the universe of discourse of the *ith* input i.e. x_i. The fuzzy system is characterized by a set of if-then rules in the following form:

Rule k: If x_1 is A_1^k and $\cdots$ and x_n is A_n^k then z is G^k (k = 1, \cdots N),

(9.22)

where $A_i^k \in \{A_i^1, \cdots, A_i^{n_i}\}$ $i = 1, \cdots, n$, G^k are fuzzy sets defined respectively for x_i $i = 1, \cdots, n$ and z and $N = \prod_{i=1}^n n_i$ is the total number of rules.

z is given by:

$$z = \theta^T \xi(x),$$ (9.23)

where $\theta^T = \theta_1, \theta_2, \cdots, \theta_N$ is a vector grouping all consequence parameters and $\xi = \xi_1, \xi_2, \cdots, \xi_N^T$ is a set of fuzzy basis functions defined as

$$\xi_k(x) = \frac{\prod_{i=1}^n \mu_{A_i^k}(x_i)}{\Sigma_{i=1}^n \left(\prod_{i=1}^n \mu_{A_i^k}(x_i) \right)},$$ (9.24)

where $\mu_{A_i^k}(x_i)$ is the membership function and represents the fuzzy meaning of the symbol A_i^k. θ_k is the value of the singleton associated with G^k.

The output of $\hat{F}(x \mid \theta_1)$ and $\hat{G}(x \mid \theta_2)$ can be respectively obtained by:

$$\hat{F}(x \mid \theta_1) = \left[\hat{f}_1(x \mid \theta_{f_1}) \cdots \hat{f}_m(x \mid \theta_{f_m}) \right]^T = \theta_1^T \xi_f(x),$$ (9.25)

where $\hat{f}_j(x \mid \theta_{f_j}) = \theta_{f_j}^T \xi_f(x)$ $j = 1 \cdots m$ and $\theta_1^T = \begin{bmatrix} \theta_{f_1}^T \\ \cdots \\ \theta_{f_m}^T \end{bmatrix}$

$$\hat{G}(x|\theta_2) = \begin{bmatrix} \hat{g}_{11}(x|\theta_{g_{11}}) & \cdots & \hat{g}_{1m}(x|\theta_{g_{1p}}) \\ \cdots & \cdots & \cdots \\ \hat{g}_{p1}(x|\theta_{g_{m1}}) & \cdots & \hat{g}_{pm}(x|\theta_{g_{mp}}) \end{bmatrix} = \theta_2^T \Phi(x) \qquad (9.26)$$

$$\hat{g}_{ij}(x|\theta_{g_{ij}}) = \theta_{g_{ij}}^T \xi_g(x) \ i=1\cdots p, \ j=1\cdots m.$$

$$\theta_2^T = \begin{bmatrix} \theta_{g_{11}}^T & \cdots & \theta_{g_{1p}}^T \\ \vdots & \ddots & \vdots \\ \theta_{g_{m1}}^T & \cdots & \theta_{g_{mp}}^T \end{bmatrix}; \ \Phi(x) = \begin{bmatrix} \xi_g(x) & 0 & 0 \\ 0 & \ddots & 0 \\ 0 & 0 & \xi_g(x) \end{bmatrix}$$

Step 2: Reduction of chattering phenomenon by adding a PI term.

The presence of the signum function in the term u_{sw} in the classical sliding mode technique leads to the chattering phenomenon, which can excite the high frequency dynamics. To avoid this problem and to achieve the previous control objectives, an adaptive Proportional Integral (PI) controller is introduced. In fact, the integral term when added to the proportional term accelerates the convergence of the process towards set point and eliminates the residual steady-state error that occurs with a proportional controller. Hence, the inputs and outputs of the continuous time PI controller are of the following form:

$$u_{PI} = \begin{bmatrix} u_{PI_1} \\ \cdots \\ u_{PI_m} \end{bmatrix} = \begin{bmatrix} k_{P_1} s_1 + k_{i_1} \int_{t_0}^{t} s_1(\tau)d\tau \\ \cdots \\ k_{P_m} s_m + k_{i_m} \int_{t_0}^{t} s_m(\tau)d\tau \end{bmatrix}, \qquad (9.27)$$

where k_{P_j} and k_{i_j} $j=1\cdots m$ are the control gains to be computed.

(27) can be rewritten as:

$$u_{PI} = \hat{\rho}(S|\theta_p) = [(\hat{\rho}_1(s_1|\theta_{\rho_1}) \ \cdots \ (\hat{\rho}_m(s_m|\theta_{\rho_m})]^T$$
$$= [\theta_{\rho_1}^T \Theta(s_1) \ \cdots \ \theta_{\rho_m}^T \Theta(s_m)]^T = \Theta(S)\theta_\rho \qquad (9.28)$$

where adjustable parameters vector θ_{ρ_j} are given by $\theta_{\rho_j} = [k_{p_j}, k_{i_j}]^T$

and $\Theta^T(s_j) = [s_j, \int_{t_0}^{t} s_j(\tau)d\tau]$ are regressive vectors, $j = 1 \cdots m$,

$\theta_\rho = [\theta_{\rho_1}^T \cdots \theta_{\rho_m}^T]^T$, $\Theta(S) = diag[\Theta^T(s_1) \cdots \Theta^T(s_m)]$.

Step 3: Control law design

The control law is obtained:

$$U = \hat{G}^+(x \mid \theta_2)\left(\hat{G}(x \mid \theta_2)\hat{G}^+(x \mid \theta_2)\right)^T \left(\varepsilon I + \left(\hat{G}(x \mid \theta_2)\hat{G}^+(x \mid \theta_2)\right)\right)$$

$$\left(\hat{G}(x \mid \theta_2)\hat{G}^+(x \mid \theta_2)\right)^T)^{-1}(-\hat{F}(x \mid \theta_1) + Y_d^{(n_r)} - \sum_{i=2}^{n_r}\Lambda_{(i-2)}E^{(i-1)}) \quad (9.29)$$

$$+ u_r - u_{PI} - \hat{u}_{rob})$$

The term $\hat{u}_{rob}$ in (9.29) developed in the next is the robust compensator defined as:

$$\hat{u}_{rob} = \frac{2}{\rho}S \quad\quad\quad (9.30)$$

The term u_r is also to be designed next.

Theorem 1. Consider the class of MIMO nonlinear non-square systems (9.1), the control law (9.29) where the functions $\hat{F}(x \mid \theta_1)$, $\hat{G}(x \mid \theta_2)$ and u_{PI} are given by (9.25), (9.26) and (9.28) the parameter vectors θ_1, θ_2, θ_ρ are adjusted by the adaptive laws (9.31). Suppose that the assumptions 2, 3, 4 and 5 are satisfied, then (9.29) ensures that all the closed-loop signals are bounded and the tracking errors decrease asymptotically to zero.

Adaptive laws:

$$\dot{\theta}_1 = \gamma_f \xi_f(x)S^T$$

$$\dot{\theta}_2 = \gamma_g \Phi(x)US^T \quad\quad\quad (9.31)$$

$$\dot{\theta}_\rho = \gamma_\rho \Theta(S)^T S$$

Proof

First, let us define the following variables:

$$\theta^*_{p_i} = arg\ min_{\theta_{p_i} \in \Omega_{p_i}} (\ sup_{S \in R^m} \| \hat{\rho}_i(S|\theta_{p_i}) - \eta sgn(s_i) \|),\qquad (9.32)$$

where $\eta sgn(s_i)$ is the discontinuous term of the conventional sliding mode control with the signum function "sign" such $\eta > D_i$ and Ω_{p_i} denotes the set of the suitable bounds on θ_{p_i}.

$$and\ \Omega_{p_i} = \{\theta_{p_i} \in \Re^2 : \| \theta_{p_i} \| \leq M_{p_i}\}\qquad (9.33)$$

Denote:

$$w_{i_{PI}} = -\hat{\rho}(S|\theta^*_{p_i}) + \eta sgn(s_i)\qquad (9.34)$$

$$W_{PI} = [w_{1_{PI}}\quad \cdots\quad w_{m_{PI}}]^T\qquad (9.35)$$

The approximation error W_{PI} belongs to $L_2[0,t]$, $\forall t \in [0,\infty)$.

Define also the optimal parameter vector θ^*_1, θ^*_2 of the previous fuzzy systems:

$$\theta^*_1 = arg\ min_{\theta_1 \in \Omega_{\theta_1}} (\ sup_{x \in \Omega_x} \| F(x) - \hat{F}(x|\theta_1) \|),\qquad (9.36)$$

$$\theta^*_2 = arg\ min_{\theta_2 \in \Omega_{\theta_2}} (\ sup_{x \in \Omega_x} \| G(x) - \hat{G}(x|\theta_2) \|),\qquad (9.37)$$

where $\Omega_{\theta_1}, \Omega_{\theta_2}$ denote the sets of suitable bounds on θ_1, θ_2 respectively and Ω_x denotes the set of the suitable bounds on x.

Assume that the constraint sets $\Omega_{\theta_1}, \Omega_{\theta_2}$ are specified as:

$$\Omega_{\theta_1} = \{\theta_1 : \| \theta_1 \| \leq M_f\}\qquad (9.38)$$

$$and\ \Omega_{\theta_2} = \{\theta_2 : \| \theta_2 \| \leq M_g\}\qquad (9.39)$$

Note while assuming that the fuzzy parameter vectors θ_1, θ_2 and x never reach the boundary of Ω_{θ_1}, Ω_{θ_2} and Ω_x respectively, we can define the vector of the minimum approximation errors as:

$$W_{Fuzzy} = [w_{1_F} \quad \cdots \quad w_{m_F}]^T \qquad (9.40)$$

with

$$w_{i_F} = f_i(x) - \hat{f}_i(x \,|\, \theta_{f_i}^*) + \sum_{j=1}^{p}(g_{ij}(x) - \hat{g}_{ij}(x \,|\, \theta_{g_ij}^*))u_j \quad i = 1, \cdots, m \qquad (9.41)$$

The approximation error w_{i_F} belongs to $L_2[0, t]$, $\forall t \in [0, \infty)$.

First, state vector x must belong to a compact set Ω_x. By applying the universel approximation theorem, we know that:

$$\exists \varepsilon_{1_{ij}} \ such \ that \ \sup_{x \in \Omega_x} | g_{ij}(x) - \hat{g}_{ij}(x \,|\, \theta_{g_ij}^*) | < \varepsilon_{1_{ij}} \qquad (9.42)$$

Thus

$$\exists \varepsilon_1 \in \mathfrak{R}^+ \ such \ that \ \sup_{x \in \Omega_x} | g_{ij}(x) - \hat{g}_{ij}(x \,|\, \theta_{g_ij}^*) | < \max_{ij}(\varepsilon_{1_{ij}}) \forall i, j < \varepsilon_1 \qquad (9.43)$$

Since $x \in \Omega_x$ and U depends on x as well as the desired reference trajectory y_{d_i} and its derivatives, which are supposed bounded, it is obvious to say that $U \in \Omega_u$ (compact set).

Thus:

$$\exists \varepsilon_u \in \mathbb{R}^+ \ such \ that \ \sup_{U \in \Omega_U} | U | < \varepsilon_u \qquad (9.44)$$

It is also proved that

$$\exists \varepsilon_{2_i} \ and \ \varepsilon_2 \in \mathbb{R}^+ \ such \ that \ \sup_{x \in \Omega_x} | f_i(x) - \hat{f}_i(x \,|\, \theta_{f_i}^*) | < \max_i(\varepsilon_{2_i}) \forall i < \varepsilon_2 \qquad (9.45)$$

So

$$\| w_{i_F} \|_2 < \| f_i(x) - \hat{f}_i(x \mid \theta_{f_i}^*) \|_2 + \sum_{j=1}^{p} \| (g_{ij}(x) - \hat{g}_{ij}(x \mid \theta_{g_{ij}}^*) \|_2) \| u_j \|_2 \qquad (9.46)$$

It well know that

$$\| x \|_2 \leq n^{1/2} \| x \|_\infty \ \ if \ \ x \in \mathbb{R}^n \qquad (9.47)$$

$$\| w_{i_F} \|_2^2 < \| f_i(x) - \hat{f}_i(x \mid \theta_{f_i}^*) \|_2^2 + \sum_{j=1}^{p} \| g_{ij}(x) - \hat{g}_{ij}(x \mid \theta_{g_{ij}}^*) \|_2^2 \| u_j \|_2^2$$

$$+ 2 \| f_i(x) - \hat{f}_i(x \mid \theta_{f_i}^*) \|_2 \ (\sum_{j=1}^{p} \| (g_{ij}(x) - \hat{g}_{ij}(x \mid \theta_{g_{ij}}^*) \|_2) \| u_j \|_2)$$

$$\| w_{i_F} \|_2^2 < \| f_i(x) - \hat{f}_i(x \mid \theta_{f_i}^*) \|_\infty + \sum_{j=1}^{p} \| g_{ij}(x) - \hat{g}_{ij}(x \mid \theta_{g_{ij}}^*) \|_\infty \| u_j \|_\infty$$

$$+ 2 \| f_i(x) - \hat{f}_i(x \mid \theta_{f_i}^*) \|_\infty \ (\sum_{j=1}^{p} (\| g_{ij}(x) - \hat{g}_{ij}(x \mid \theta_{g_{ij}}^*) \|_\infty) \| u_j \|_\infty)$$

By using the previous inequalities:

$$\| w_{i_F} \|_2^2 \leq \varepsilon_2 + \varepsilon_1 \varepsilon_u + 2\varepsilon_2 \varepsilon_1 \varepsilon_u \qquad (9.48)$$

w_{i_F} is uniformly bounded. From the definition of $L_2[0 \quad t]$, we get

$$\int_0^T \| w_{i_F} \|_2^2 \ dt \leq \infty \ \ for \ all \ T \in \mathfrak{R}^+ \qquad (9.49)$$

Thus w_{i_F} as well as W_{Fuzzy} belongs to $L_2[0 \quad t] \ \forall t \in [0 \quad \infty]$. By using the same idea, we may prove that W_{PI} belongs to $L_2[0 \quad t] \ \forall t \in \mathbb{R}^+$, as well as d (because d is supposed to be bounded) (see assumption 1).

Finally,

$$W' \ belong \ to \ L_2[0 \quad t] \Leftrightarrow \int_0^T \| W' \|_2^2 \ dt \leq \infty \qquad (9.50)$$

The time derivative of S can be obtained from (9.28) and (9.29):

$$\dot{S} = F(x) + d - Y_d^{(n_r)} + \sum_{i=2}^{n_r} \Lambda_{(i-2)} E^{(i-1)} + (G(x) - \hat{G}(x \mid \theta_2))U + (I - \varepsilon[\varepsilon I$$

$$+ (\hat{G}(x \mid \theta_2)\hat{G}^+(x \mid \theta_2))(\hat{G}(x \mid \theta_2)\hat{G}^+(x \mid \theta_2))^T]^{-1})(-\hat{F}(x \mid \theta_1)$$

$$+ Y_d^{(n_r)} - \sum_{i=2}^{n_r} \Lambda_{(i-2)} E^{(i-1)} + u_r - \hat{\rho}(S \mid \theta_p) - \hat{u}_{rob})$$

$$\dot{S} = F(x) - \hat{F}(x \mid \theta_1) + (G(x) - \hat{G}(x \mid \theta_2))U - \hat{\rho}(S \mid \theta_p) - \hat{u}_{rob} + d + (I - \varepsilon[\varepsilon I$$

$$+ (\hat{G}(x \mid \theta_2)\hat{G}^+(x \mid \theta_2))(\hat{G}(x \mid \theta_2)\hat{G}^+(x \mid \theta_2))^T]^{-1})u_r - I - \varepsilon[\varepsilon I +$$

$$(\hat{G}(x \mid \theta_2)\hat{G}^+(x \mid \theta_2))(\hat{G}(x \mid \theta_2)\hat{G}^+(x \mid \theta_2))^T]^{-1})$$

$$(-\hat{F}(x \mid \theta_1) + Y_d^{(n_r)} - \sum_{i=2}^{n_r} \Lambda_{(i-2)} E^{(i-1)} + u_r - \hat{\rho}(S \mid \theta_p) - \hat{u}_{rob})$$

$$(9.51)$$

In this case, a approached solution of $\dot{S} = 0$ satisfies the followings expression:

$$u_r = [\varepsilon I + (\hat{G}(x \mid \theta_2)\hat{G}^+(x \mid \theta_2))(\hat{G}(x \mid \theta_2)\hat{G}^+(x \mid \theta_2))^T](\hat{G}(x \mid \theta_2)\hat{G}^+(x \mid \theta_2))[\varepsilon I$$

$$+ (\hat{G}(x \mid \theta_2)\hat{G}^+(x \mid \theta_2))^T (\hat{G}(x \mid \theta_2)\hat{G}^+(x \mid \theta_2))^{-1}](\hat{G}(x \mid \theta_2)\hat{G}^+(x \mid \theta_2))^T[\varepsilon I$$

$$+ (\hat{G}(x \mid \theta_2)\hat{G}^+(x \mid \theta_2))(\hat{G}(x \mid \theta_2)\hat{G}^+(x \mid \theta_2))^T]^{-1}\varepsilon[\varepsilon I + (\hat{G}(x \mid \theta_2)\hat{G}^+(x \mid \theta_2))$$

$$(\hat{G}(x \mid \theta_2)\hat{G}^+(x \mid \theta_2))^T]^{-1}(-\hat{F}(x \mid \theta_1) + Y_d^{(n_r)} - \sum_{i=2}^{n_r} \Lambda_{(i-2)} E^{(i-1)} + u_{PI} - \hat{u}_{rob})$$

$$(9.52)$$

(9.51) can be rewritten from (9.32), (9.36), (9.37) and (9.52):

$$\dot{S} = \hat{F}(\hat{x} \mid \theta_1^*) - \hat{F}(x \mid \theta_1) + [\hat{G}(\hat{x} \mid \theta_2^*) - \hat{G}(x \mid \theta_2)]U + \hat{\rho}(S \mid \theta_{\rho_i}^*)$$

$$- \hat{\rho}(S \mid \theta_p) - \hat{\rho}(S \mid \theta_{\rho_i}^*) - \frac{2}{\rho} S + W_{Fuzzy} + d + \Delta_2(x, \varepsilon)$$

$$\dot{S} = \tilde{\theta}_1^T \xi_f(x) + \tilde{\theta}_2^T \Phi(x)U + \Theta(S)\tilde{\theta}_p - \frac{2}{\rho} S - \eta sgn(S)$$

$$+ W_{PI} + W_{Fuzzy} + d + \Delta_2(x, \varepsilon)$$

where $sgn(S) = [sgn(s_1) \quad \cdots \quad sgn(s_m)]^T$, $\tilde{\theta}_1 = \theta_1^* - \theta_1$, $\tilde{\theta}_2 = \theta_2^* - \theta_2$,
$\tilde{\theta}_\rho = \theta_\rho^* - \theta_\rho$ and $W = W_{Fuzzy} + W_{PI} + d + \Delta_2(x, \varepsilon)$.

Remark 2. State vector x must belong to compact set Ω_x. $\Delta_2(x, \varepsilon) < \varepsilon_3$

Then, consider a Lyapunov function candidate as follows:

$$V = \frac{1}{2}S^T S + \frac{1}{2\gamma_f}tr(\tilde{\theta}_1^T \tilde{\theta}_1) + \frac{1}{2\gamma_g}tr(\tilde{\theta}_2^T \tilde{\theta}_2) + \frac{1}{2\gamma_\rho}tr(\tilde{\theta}_\rho^T \tilde{\theta}_\rho), \quad (9.53)$$

where $\gamma_f > 0, \gamma_g > 0$ and $\gamma_\rho > 0$ are the adaptation parameters.

Then, the time derivative of V is given by:

$$\dot{V} = \dot{S}S^T + \frac{1}{\gamma_f}tr(\tilde{\theta}_1^T \dot{\tilde{\theta}}_1) + \frac{1}{\gamma_g}tr(\tilde{\theta}_2^T \dot{\tilde{\theta}}_2) + \frac{1}{\gamma_\rho}tr(\dot{\tilde{\theta}}_\rho^T \tilde{\theta}_\rho)$$

$$= tr[(\tilde{\theta}_1^T \xi_f(x) + \tilde{\theta}_2^T \Phi(x)U + \Theta(S)\tilde{\theta}_\rho)S^T] + tr(WS^T) + \frac{1}{\gamma_f}tr(\tilde{\theta}_1^T \dot{\tilde{\theta}}_1)$$

$$+ \frac{1}{\gamma_g}tr(\tilde{\theta}_2^T \dot{\tilde{\theta}}_2) + \frac{1}{\gamma_f}\dot{\tilde{\theta}}_\rho^T \tilde{\theta}_\rho - \eta tr(sgn(S)S^T) - \frac{2}{\rho}tr(SS^T)$$

$$= \frac{1}{\gamma_f}tr[\tilde{\theta}_1^T(\gamma_f\xi_f(x)S^T + \dot{\tilde{\theta}}_1)] + \frac{1}{\gamma_\rho}tr[(\gamma_\rho S^T\Theta(S) + \dot{\tilde{\theta}}_\rho^T)\tilde{\theta}_\rho]$$

$$+ \frac{1}{\gamma_g}tr[\tilde{\theta}_2^T(\gamma_\rho\Phi(x)US^T + \dot{\tilde{\theta}}_2)] - \eta tr(sgn(S)S^T) - \frac{2}{\rho}tr(SS^T) + tr(WS^T)$$

$$(9.54)$$

Then:

$$\dot{\tilde{\theta}}_1 = -\dot{\theta}_1, \ \dot{\tilde{\theta}}_2 = -\dot{\theta}_2 \ and \ \dot{\tilde{\theta}}_\rho = -\dot{\theta}_\rho \qquad (9.55)$$

Substituting the adaptive laws (9.31) and (9.55) in (9.54), we obtain:

$$\dot{V} \le -\eta|S| - \frac{2}{\rho}S^T S + S^T W$$

$$\dot{V} \le -[\frac{1}{\sqrt{\rho}}S - \frac{\sqrt{\rho}}{2}W]^T[\frac{1}{\sqrt{\rho}}S - \frac{\sqrt{\rho}}{2}W] - \frac{2}{\rho}S^T S + \frac{1}{\rho}S^T S + \frac{1}{4}\rho W^T W$$

$$\text{Hence, } \dot{V} \le -\frac{1}{\rho} S^T S + \frac{1}{4} \rho W^T W \qquad (9.56)$$

Integrating (9.56) for $t=0$ to T:

$$V(T) - V(0) \le -\frac{1}{\rho} \int_0^T S^T S \, dt + \frac{1}{4} \rho \int_0^T W^T W \, dt \qquad (9.57)$$

Then since $V(T) > 0$ and $0 < \rho < 1$, (9.57) can be rewritten as:

$$\int_0^T S^T S \, dt \le 2V(0) + \rho^2 \int_0^T W^T W \, dt < \infty \qquad (9.58)$$

From (9.58), $\int_0^T S^T S \, dt$ has a finished limit when $T \to \infty$. From (9.56), the function $\dot{S}S^T$ is bounded.

Using Barbalat's lemma [1], we can see that the sliding surfaces asymptotically converge to zero in finite time despite the external disturbances. Thus, the tracking errors converge asymptotically to zero in finite time.

Furthermore, it can be proved from (9.58) that the H_∞ tracking performance in (9.20) is achieved, that guarantees the global stability and the robustness of the closed loop system.

The global stability is achieved under the assumption that all the parameter vectors are within the constraint sets. To guarantee that the parameters are bounded, the adaptive laws (9.32) can be modified by using the projection algorithm [30]. The modified adaptive laws are given as follows:

For θ_{f_i}, we use:

$$\dot{\theta}_{f_i} = \gamma_{f_i} s_i \xi(x) \quad \text{if } (\| \theta_{f_i} \| < M_{f_i}) \text{ or } (\| \theta_{f_i} \| = M_{f_i} \text{ and } s_i \theta_{f_i}^T \xi(x) \le 0)$$

$$P_{f_i} [\gamma_{f_i} s_i \xi(x)] \text{ if } (\| \theta_{f_i} \| = M_{f_i} \text{ and } s_i \theta_{f_i}^T \xi(x) > 0) \qquad (9.59)$$

For $\theta_{g_{ij}}$, we use:

$$\dot{\theta}_{g_{ij}} = \gamma_{g_{ij}} s_i \xi(x) u_j \quad if \ (\|\theta_{g_{ij}}\| < M_{g_{ij}}) \ or \ (\|\theta_{g_{ij}}\| = M_{g_{ij}} \ and \ s_i \theta_{g_{ij}}^T \xi(x) u_j \le 0)$$

$$P_{g_{ij}}[\gamma_{g_{ij}} s_i \xi(x) u_j] \ if \ (\|\theta_{g_{ij}}\| = M_{g_{ij}} \ and \ s_i \theta_{g_{ij}}^T \xi(x) u_j > 0)$$

$$(9.60)$$

For θ_{ρ_i}, we use:

$$\dot{\theta}_{\rho_i} = \gamma_{\rho_i} s_i \Theta(s_i) \quad if \ (\|\theta_{\rho_i}\| < M_{\rho_i}) \ or \ (\|\theta_{\rho_i}\| = M_{\rho_i} \ and \ s_i \theta_{\rho_i}^T \Theta(s_i) \le 0)$$

$$P_{\rho_i}[\gamma_{\rho_i} s_i \Theta(s_i)] \ if \ (\|\theta_{\rho_i}\| = M_{\rho_i} \ and \ s_i \theta_{\rho_i}^T \Theta(s_i) > 0),$$

$$(9.61)$$

where $\| . \|$ is the Euclidean norm.

The projection operator, $P_{f_i}[*], P_{g_{ij}}[*]$ and $P_{\rho_i}[*]$ are defined as:

$$P_{f_i}[\gamma_{f_i} s_i \xi(x)] = \gamma_{f_i} s_i \xi(x) - \gamma_{f_i} s_i \frac{\theta_{f_i} \theta_{f_i}^T \xi(x)}{\|\theta_{f_i}\|^2} \qquad (9.62)$$

$$P_{g_{ij}}[\gamma_{g_{ij}} s_i \xi(x) u_j] = \gamma_{g_{ij}} s_i \xi(x) u_j - \gamma_{g_{ij}} s_i \frac{\theta_{g_{ij}} \theta_{g_{ij}}^T \xi(x) u_j}{\|\theta_{g_{ij}}\|^2} \qquad (9.63)$$

$$P_{\rho_i}[\gamma_{\rho_i} s_i \Theta(s_i)] = \gamma_{\rho_i} s_i \Theta(s_i) - \gamma_{\rho_i} s_i \frac{\theta_{\rho_i} \theta_{\rho_i}^T \Theta(s_i)}{\|\theta_{\rho_i}\|^2} \qquad (9.64)$$

From the above discussion, a design procedure for the H_∞ AFSMC (Adaptive Fuzzy Sliding Mode Control) is presented as follows:

Design procedure

Step 1: off line proceeding

– Specify the design parameters M_{f_i}, $M_{g_{ij}}$ and M_{ρ_i} based on practical constraints;

– Select the learning coefficients γ_f, γ_g and γ_ρ;

– Define membership functions $\mu_{F_i}(x)$ for $i = 1, 2 \cdots M$ and construct the fuzzy basis functions as in (9.24);

– Design the fuzzy systems $\hat{F}(x,t)$ and $\hat{G}(x,t)$ in (9.25), (9.26);

– Select the desired attenuation level ρ such that $0 < \rho < 1$;

– Specify the coefficients α_j^i such that all roots of (9.53) are in the left-half plane. See (9.11).

Step 2: on line adaptation

– Apply the control law (9.29) to the plant (9.1) where $\hat{u}_{rob}$ is given by (9.30);

– Use the adaptive law in (9.59) (9.60) and (9.61) to adjust the parameters θ_{f_i}, $\theta_{g_{ij}}$ and θ_{ρ_i}.

Remark 3. In case when the number of inputs is upper than the number of outputs (i.e. $m < p$) the term u_r satisfying (9.52) is equal to zero. Indeed the control laws is given by the following expression:

$$U = (\hat{G}^+(x \mid \theta_2)(-\hat{F}(x \mid \theta_1) + Y_d^{(n_r)} - \sum_{i=2}^{n_r} \Lambda_{(i-2)} E^{(i-1)} + u_{PI} - \frac{2}{\rho} S)$$

$$U = \hat{G}(x \mid \theta_2)(\hat{G}(x \mid \theta_2)\hat{G}(x \mid \theta_2)^T)^{-1}$$
$$(-\hat{F}(x \mid \theta_1) + Y_d^{(n_r)} - \sum_{i=2}^{n_r} \Lambda_{(i-2)} E^{(i-1)} + u_{PI} - \frac{2}{\rho} S) \qquad (9.65)$$

Remark 4. When considered square system (i.e. $m = p$) the control law can be rewritten as:

$$U = (\hat{G}(x \mid \theta_2)^{-1}(-\hat{F}(x \mid \theta_1) + Y_d^{(n_r)} - \sum_{i=2}^{n_r} \Lambda_{(i-2)} E^{(i-1)} + u_{PI} - \frac{2}{\rho} S) \quad (9.66)$$

with $\hat{G}(x \mid \theta_2)$ of full rank for $m < p$ and $m = p$ by correctly choosing Ω_{θ_2}.

9.4. Simulation Results

Case 1: Upper actuated system: In this section, the performance of the proposed approach is evaluated for a numerical example to illustrate the H_∞ tracking performance of the proposed control design algorithm. Furthermore, we also compare our method with other recent methods proposed in the literature.

Consider the nonlinear differential equation given by:

$$
\begin{pmatrix} \dot{x}_1 \\ \dot{x}_2 \\ \dot{x}_3 \end{pmatrix} = \begin{pmatrix} x_2 \\ x_1 + x_2^2 + x_3 \\ x_1 + 2x_2 + 3x_3 x_1 \end{pmatrix} + \begin{pmatrix} 0 \\ 3u_1 + u_2 + u_3 \\ u_1 + 2(2 + 0.5\sin(x_1))u_2 + 2\sin(x_1)u_3 \end{pmatrix}
$$
$$
+ \begin{pmatrix} 0 \\ 0.5\sin(t) \\ 0.5\sin(t) \end{pmatrix}
$$

(9.67)

$$
\begin{pmatrix} y_1 \\ y_2 \end{pmatrix} = \begin{pmatrix} x_1 \\ x_3 \end{pmatrix},
$$

where relative degree $r = [r_1 \quad r_2] = [1 \quad 2]$. Let, $x = [x_1 \quad x_2 \quad x_3]^T$ is the state vector, $U = [u_1 \quad u_2 \quad u_3]^T$ is the input vector, $y = [x_1 \quad x_3]^T$ is the output vector.

The external perturbations are given by: $D = \begin{pmatrix} d_1 \\ d_2 \end{pmatrix} = \begin{pmatrix} 0.5\sin(t) \\ 0.5\sin(t) \end{pmatrix}$ and

$$
F(x) = \begin{pmatrix} x_1 + x_2^2 + x_3 \\ x_1 + 2x_2 + 3x_3 x_1 \end{pmatrix},
$$

$$
G(x) = \begin{pmatrix} g_{11} & g_{12} & g_{13} \\ g_{21} & g_{22} & g_{23} \end{pmatrix} = \begin{pmatrix} 3 & 1 & 1 \\ 1 & 2(2 + 0.5\sin(x_1)) & 2\sin(x_1) \end{pmatrix}.
$$

Then (9.67) can be rewritten as the following state-space form:

$$\dot{x}_1 = x_2$$

$$\dot{x}_2 = f_1(x) + g_{11}(x)u_1 + g_{12}(x)u_2 + g_{13}(x)u_3 + d_1 \qquad (9.68)$$

$$\dot{x}_3 = f_2(x) + g_{21}(x)u_1 + g_{22}(x)u_2 + g_{23}(x)u_3 + d_2$$

$$y_1 = x_1$$

$$y_2 = x_3$$

The control objective is to force the system outputs y_1 and y_2 to track the desired trajectories $y_{d_1}(t) = (\dfrac{\pi}{30})sin(t)$ and $y_{d_2}(t) = (\dfrac{\pi}{30})cos(t)$ respectively in spite of the disturbances. To approximate the unknown nonlinear functions, seven fuzzy membership functions are chosen:

$$\mu_{F_i^1} = \left(\frac{1}{1 + exp(5(x_i + 0.8))} \right), \mu_{F_i^3} = exp(-(x_i + 0.4)^2)$$

$$\mu_{F_i^7} = \left(\frac{1}{1 + exp(-5(x_i - 0.8))} \right), \mu_{F_i^4} = exp(-x_i^2), \mu_{F_i^5} = exp(-(x_i - 0.4)^2)$$

$$\mu_{F_i^6} = exp(-(x_i - 0.6)^2), \mu_{F_i^2} = exp(-(x_i + 0.6)^2) \qquad i = 1, 2, 3 \quad (9.69)$$

The initial conditions are chosen as: $x(0) = (0.1, 0.1, 0)^T$. We choose $M_{f_i} = 0.5, M_{g_{ij}} = 0.1, M_{\rho_i} = 0.01$ and the adaptation adjusting parameters are $\gamma_{f_i} = 0.2, \gamma_{g_{ij}} = 5$ and $\gamma_{\rho_i} = 1$.

Fig. 9.1 – Fig. 9.11 show the simulation results under the proposed adaptive fuzzy sliding mode algorithm. The tracking curves of states, its reference signals and the tracking errors curves are displayed in Fig. 9.1-Fig. 9.3 and Fig. 9.4 - Fig. 9.6 respectively, which indicate the good performance of our controller in particularly the good tracking of the reference signals: $y_{d_1}, \dot{y}_{d_1}, y_{d_2}$ as well as the robustness of the control system with respect to the disturbance. In Fig. 9.7, it is shown that there is no chattering in the control law. In fact, to evaluate the performance of the proposed controller, a comparison the latter with a conventional sliding mode controller is made. Fig. 9.7- Fig. 9.8 show the simulation results for different control schemes. In addition, in order to have a quantitative comparison of tracking errors, we have reported the Integral Absolute Error (IAE) of different SMC algorithms (see Table 9.1).

Table 9.1 shows that our proposed methodology has better performance rather than the classical SMC. The main contribution of our proposed method comparing to other methods is that not only the chattering phenomenon is reduced but also the asymptotical stability of the system is guaranteed as well. The evolution of the surfaces trajectories are shown in Fig. 9.9- Fig. 9.10 and we see the convergence to zero of the system and the attractiveness of the sliding surfaces. Fig. 9.11- Fig. 9.12 illustrate the behavior of adaptation parameters and the convergence of θ_{ρ_1} and θ_{ρ_2} .

Table 9.1. Integral Absolute Error (IAE) values for various methods.

IAE	Method		
	Proposed method	**SMC with saturation function**	**Integral SMC[19]**
Tracking error e_1	0.0.323	0.187	0.081
Tracking error e_2	0.542	0.73	0.65

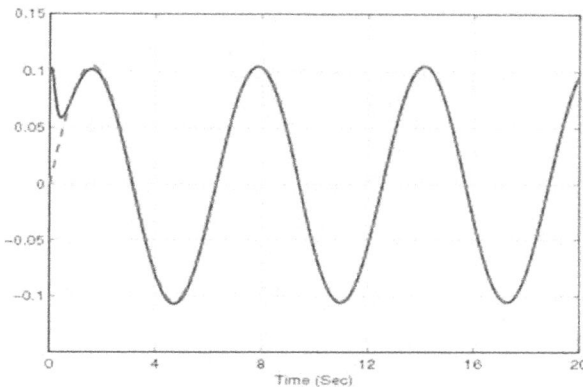

Fig. 9.1. Trajectories of the state variable x_1 (solide line) and the reference state variable y_{d_1} (dashed line).

Case 2: Under actuated system: Inverted pendulum.

In this case we shall demonstrate that the control law defined by (9.29) is applicable to the inverted pendulum system to support the theoretical development. The structure of an inverted pendulum is illustrated in Fig. 9.13, described by the following dynamic equations [11].

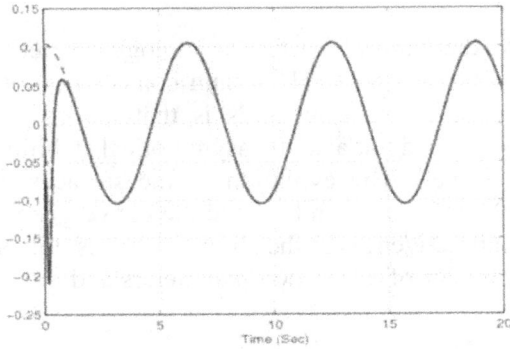

Fig. 9.2. Trajectories of the state variable x_2 (solide line) and the reference state variable $\dot{y}_{d_1}$ (dashed line).

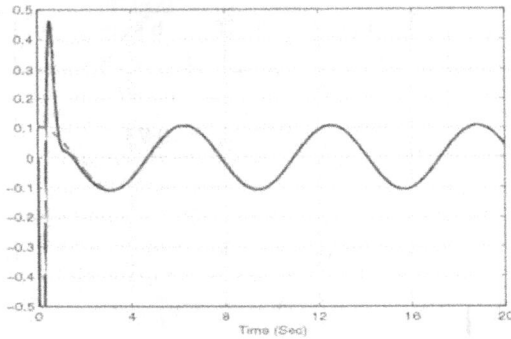

Fig. 9.3. Trajectories of the state variable x_3 (solide line) and the reference state variable y_{d_2} (dashed line).

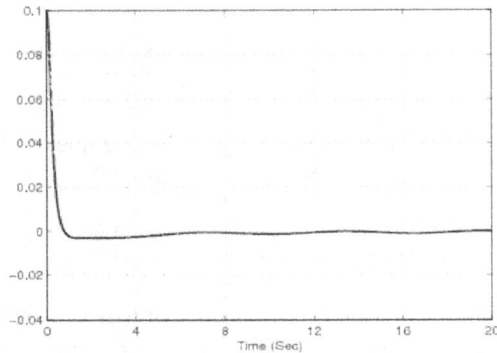

Fig. 9.4. Evolution of the tracking error: $e_{11} = x_1 - y_{d1}$.

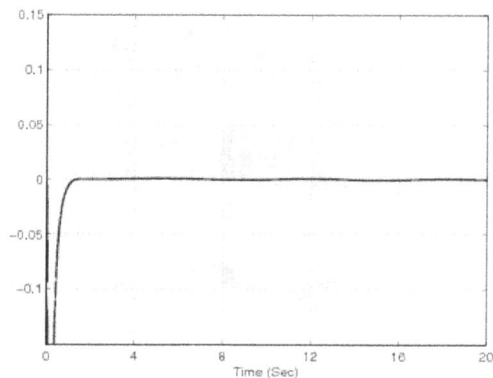

Fig. 9.5. Evolution of the tracking error: $e_{12} = \dot{x}_1 - \dot{y}_{d1}$.

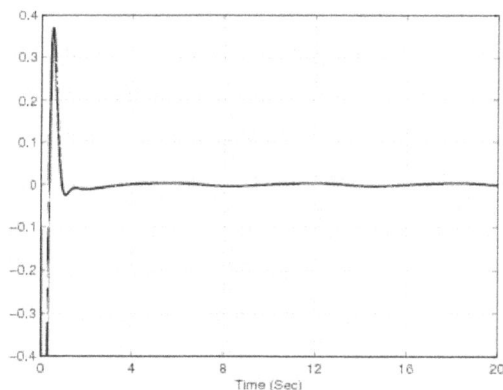

Fig. 9.6. Evolution of the tracking error: $e_{21} = x_3 - y_{d2}$.

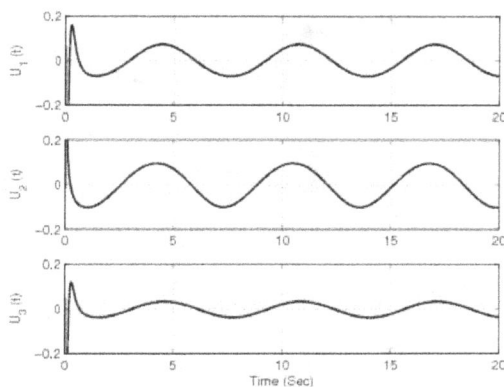

Fig. 9.7. Control input signals with proposed method.

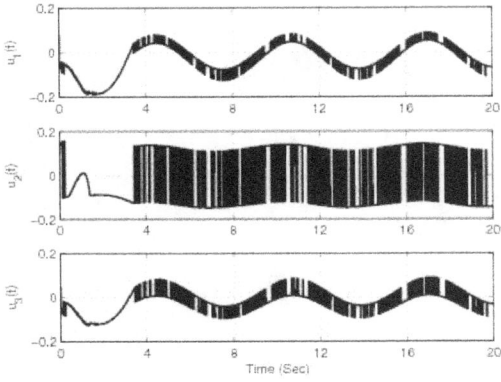

Fig. 9.8. Control input signals with the classical SMC.

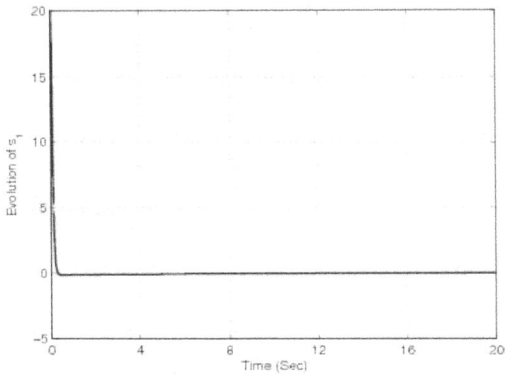

Fig. 9.9. Evolution of surface s_1.

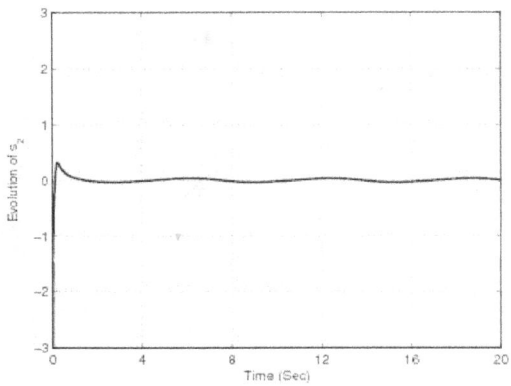

Fig. 9.10. Evolution of surface s_2.

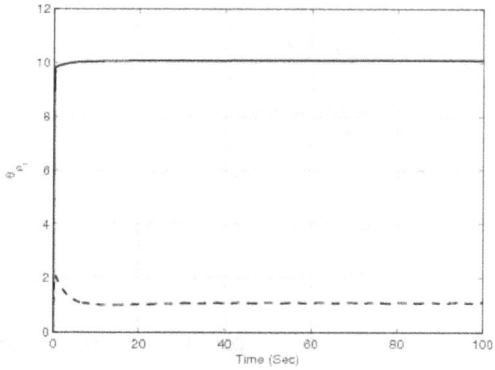

Fig. 9.11. Adaptation parameters θ_{ρ_1} .

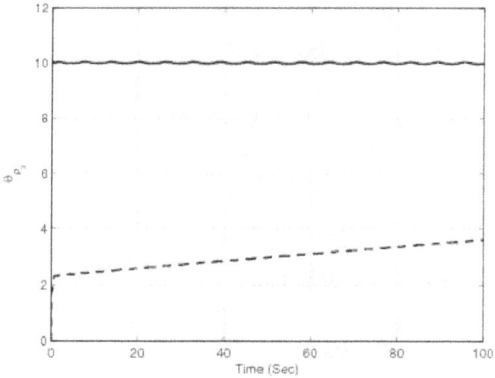

Fig. 9.12. Adaptation parameters θ_{ρ_2} .

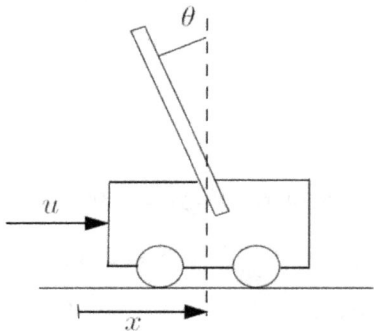

Fig. 9.13. Structure of an inverted pendulum system.

345

$$\begin{cases} \dot{x}_1 = x_2 \\ \dot{x}_2 = f_1 + b_1 u + d_1 \\ \dot{x}_3 = x_4 \\ \dot{x}_4 = f_2 + b_2 u + d_2 \\ y_1 = x_1 \\ y_2 = x_3 \end{cases} \tag{9.70}$$

where:

$x_1 = \theta$ is the angle of the pole with respect to the vertical axis;

$x_2 = \dot{\theta}$ is the angular velocity of the pole with respect to the vertical axis;

$x_3 = x$ is the position of the cart;

$x_4 = \dot{x}$ is the velocity of the cart;

u is the applied force to move the cart

$$f_1 = \frac{m_t g \sin x_1 - m_p L \sin x_1 \cos x_1 x_2^2}{L(\frac{4}{3} m_t - m_p \cos^2 x_1)} \; ; \; b_1 = \frac{\cos x_1}{L(\frac{4}{3} m_t - m_p \cos^2 x_1)}$$

$$f_2 = \frac{-\frac{4}{3} m_p L x_2^2 \sin x_1 + m_p g \sin x_1 \cos x_1}{\frac{4}{3} m_t - m_p \cos^2 x_1} \; ; \; b_2 = \frac{4}{3(\frac{4}{3} m_t - m_p \cos^2 x_1)},$$

where

g : is the acceleration of gravity

L : is the angular velocity of the pole with respect to the vertical axis;

m_c :is the cart mass;

m_p : is the pole mass;

Define $m_t = m_p + m_c$

The control objective is to force system outputs y_1 and y_2 to track desired trajectories $y_{d_1}(t) = (\frac{pi}{30}) sin(t)$ and $y_{d_2}(t) = sin(t)$ respectively in spite of the disturbances. Fig. 9.14 - Fig. 9.20 show the simulation results under the proposed adaptive fuzzy sliding mode algorithm.

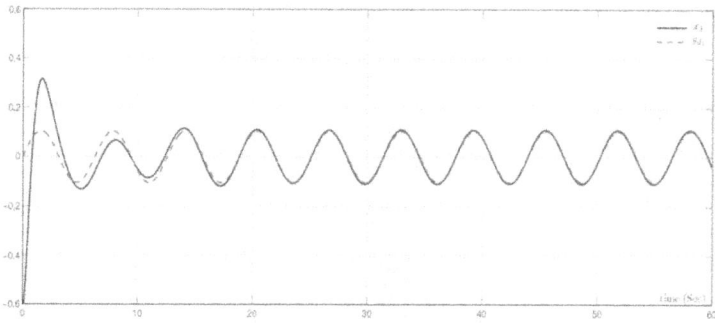

Fig. 9.14. Trajectories of the angle of the pole (solide line) and the reference signal y_{d_1} (dashed line)(rad).

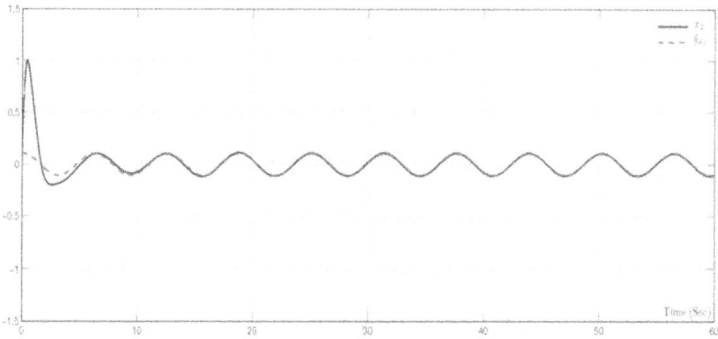

Fig. 9.15. Trajectories of the angular velocity of the pole (solide line) and the reference signal $\dot{y}_{d_1}$ (dashed line)(rad / s).

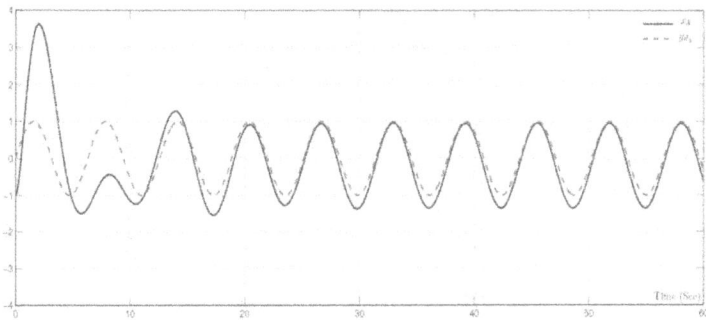

Fig. 9.16. Trajectories of the position of the cart (solide line) and the reference signal y_{d_2} (dashed line)(m).

347

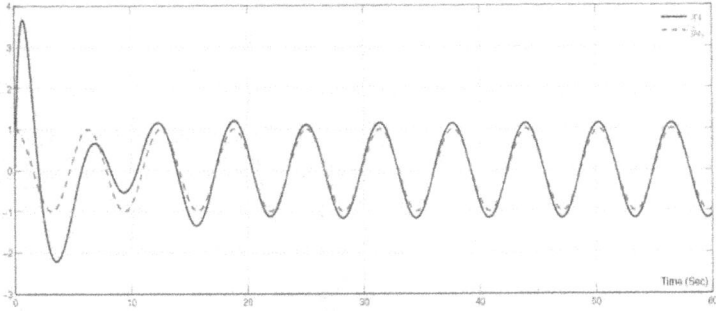

Fig. 9.17. Trajectories of the velocity of the cart (solide line) and their reference signal $\dot{y}_{d_2}$ (dashed line)(m / s).

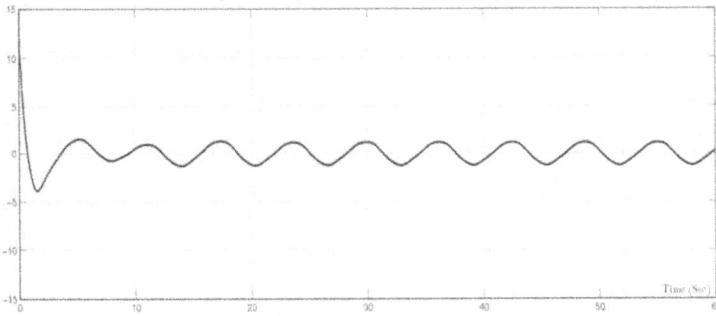

Fig. 9.18. Control signal $N.m$.

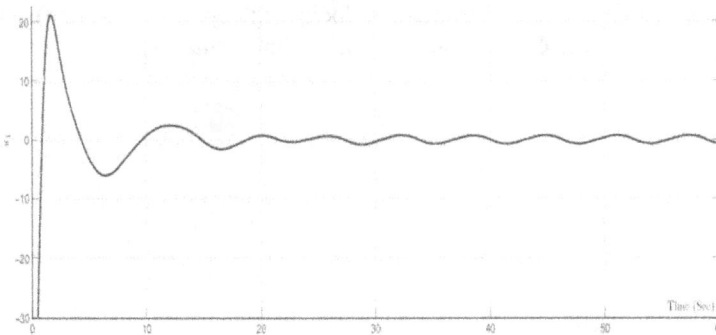

Fig. 9.19. Evolution of surface s_1.

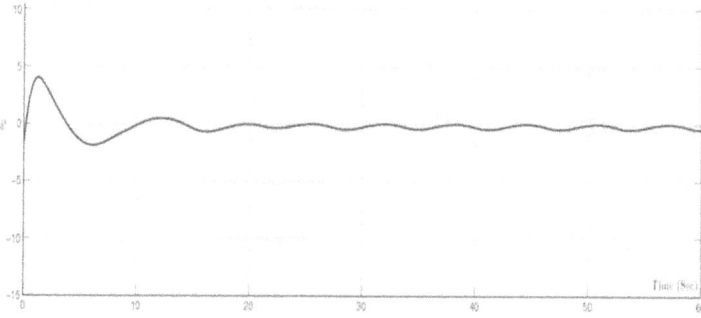

Fig. 9.20. Evolution of surface s_2 .

The tracking curves of states, its reference signals are displayed in Fig. 9.14 - Fig. 9.17 respectively, which indicate the good performance of our controller in particularly the good tracking of the reference signals: $y_{d_1}, \dot{y}_{d_1}, y_{d_2}, \dot{y}_{d_2}$ as well as the robustness of the control system with respect to the disturbance. In Fig. 9.18, it is shown that there is no chattering in the control law. The evolution of the surfaces trajectories are shown in Fig. 9.19 - Fig. 9.20 and we see the convergence to zero of the system and the attractiveness of the sliding surfaces.

Case 3: square system: To illustrate the effectiveness of the proposed adaptive fuzzy controllers, we will study, in simulation, its performances on a robot manipulator described by [15] (Fig. 9.21):

The dynamic equation of the two-link robotic manipulator is given as follows:

$$\ddot{q} = -M^{-1}(q)(C(q,\dot{q}) + G(q)) + M^{-1}(q)(\tau + D) \qquad (9.71)$$

$$C(q,\dot{q}) = \begin{pmatrix} -m_2 l_1 l_2 \sin(q_2)\dot{q}_2 - 2m_2 l_1 l_2 \sin(q_2)\dot{q}_1 \dot{q}_2 \\ m_2 l_1 l_2 \sin(q_2)\dot{q}_1^2 \end{pmatrix}$$

$$G(q) = \begin{pmatrix} m_2 l_2 \cos(q_1 + q_2) + (m_1 + m_2) l_1 g \cos(q_1) \\ m_2 l_2 g \cos(q_1 + q_2) \end{pmatrix}$$

$$M(q) = \begin{pmatrix} M_{11} & M_{12} \\ M_{21} & M_{22} \end{pmatrix},$$

where:

$$M_{12} = m_2 l_2^2 + m_2 l_1 l_2 cos(q_2), \ M_{21} = M_{12},$$
$$M_{11} = (m_1 + m_2)l_1^2 + m_2 l_2^2 + 2m_2 l_1 l_2 \cos(q_2), \ M_{22} = m_2 l_2^2$$

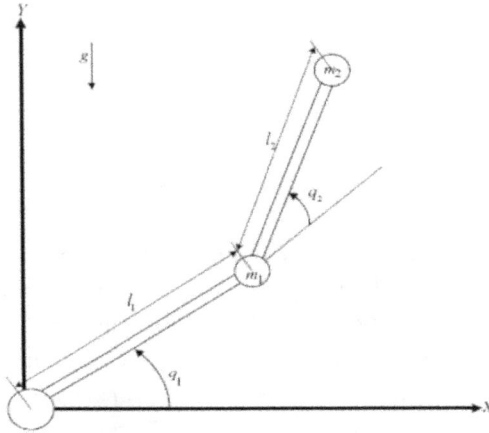

Fig. 9.21. Two degrees of robot manipulator.

The parameters of the robot used for simulation are: $l_1 = 1m$, $l_2 = 0.8m$, $m_1 = m_2 = 1kg$ and $g = 9.8m/s^2$. $q = (q_1 \ q_2)^T$ is the output vector $\tau = [\tau_1 \ \tau_2]^T$ is the input vector. The control objective is to force system outputs $y_1 = q_1$ and $y_2 = q_2$ to track desired trajectories $y_{d_1}(t) = \sin(t)$ and $y_{d_2}(t) = 0.8\cos(t)$ respectively. Moreover the external disturbances $D = (d_1 \ d_2)^T$ with $d_1 = d_2 = 0.5\sin(2t) + 0.1\cos(t)$.

The system (9.71) can be rewritten as followings:

$$\begin{cases} \dot{x}_1 = x_2 \\ \dot{x}_2 = f_1(x) + g_{11}\tau_1 + g_{12}\tau_2 + d_1 \\ \dot{x}_3 = x_4 \\ \dot{x}_4 = f_2(x) + g_{21}\tau_1 + g_{22}\tau_2 + d_2 \\ y_1 = x_1 \\ y_2 = x_3 \end{cases} \qquad (9.72)$$

We define fuzzy memberships functions as follows, to approximate the nonlinear functions:

$$\mu_{Positif} = exp\left(-\frac{1}{2}(\frac{x_i - 1.25}{0.6})^2\right)$$

$$\mu_{Zero} = exp\left(-\frac{1}{2}(\frac{x_i}{0.6})^2\right)$$

$$\mu_{Negatif} = exp\left(-\frac{1}{2}(\frac{x_i + 1.25}{0.6})^2\right) \quad i = 1...4.$$

The robot initial conditions are: $x(0) = (0.5, 0, -0.5, 0)^T$. We choose, $M_{f_i} = 40$, $M_{g_{ij}} = 80$, $M_{\rho_i} = 50$ and adaptation parameters as $\gamma_{f_i} = 50$, $\gamma_{g_{ij}} = 0.5$ and $\gamma_{\rho_i} = 1$. Let $k_{c_1} = 3$, $k_{c_2} = 70$, $k_{c_3} = 8$, $k_{c_4} = 45$, $k_{0_1} = 100$, $k_{0_2} = 65$, $k_{0_3} = 100$, $k_{0_4} = 65$. The sliding surface is selected as $S = \Psi\tilde{e}$, where $\Psi_1 = [10 \quad 1]$ and $\Psi_2 = [10 \quad 1]$.

Fig. 9.22 - Fig. 9.29 show the simulation results under the proposed adaptive fuzzy sliding mode algorithm. The tracking curves of states, its reference signals and the tracking errors curves are displayed in Fig. 9.22 - Fig. 9.23 and Fig. 9.24 - Fig. 9.25 respectively, which indicate the goodperformances of our controller in particularly the good tracking of the reference signals: $y_{d_1}, \dot{y}_{d_1}, y_{d_2}, \dot{y}_{d_2}$ as well as the robustness of the control system with respect to the disturbances. In Fig. 9.26, it is shown that there is no chattering in the control law. In fact, to evaluate the performances of the proposed controller, we propose to compare our approach with a conventional sliding mode controller. Fig. 9.26 - Fig. 9.27 show the simulation results for different control schemes. The main contribution of our proposed method comparing to other method is that not only the chattering phenomenon is reduced but also the asymptotical stability of the system is guaranteed as well. The evolution of the surfaces trajectories are shown in Fig. 9.28 - Fig. 9.29 and we see the convergence to zero of the system and the attractiveness of the sliding surfaces.

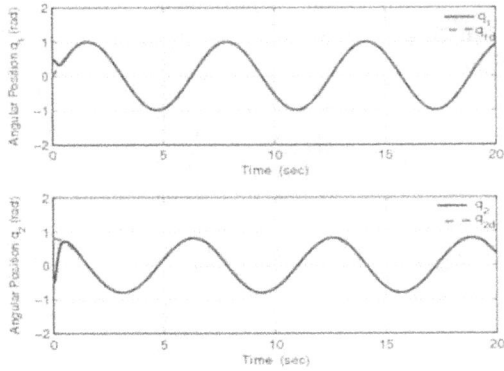

Fig. 9.22. Trajectories of angular position (rad).

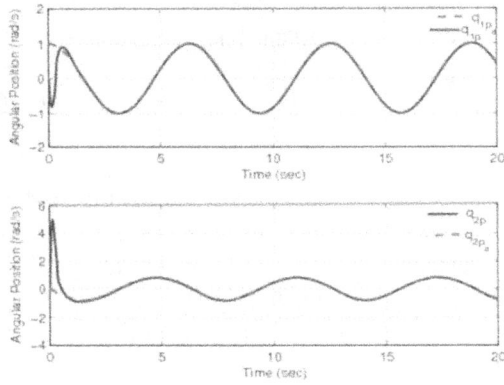

Fig. 9.23. Trajectories of angular velocity $(rad \, / \, sec)$.

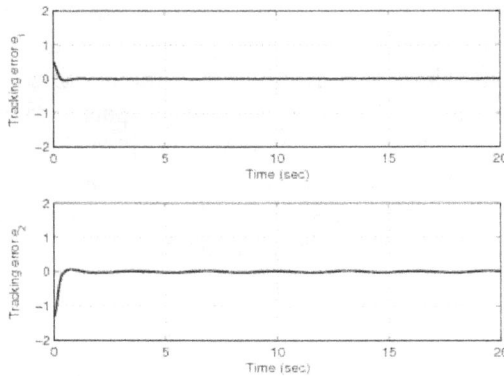

Fig. 9.24. Evolution of tracking errors of position (rad).

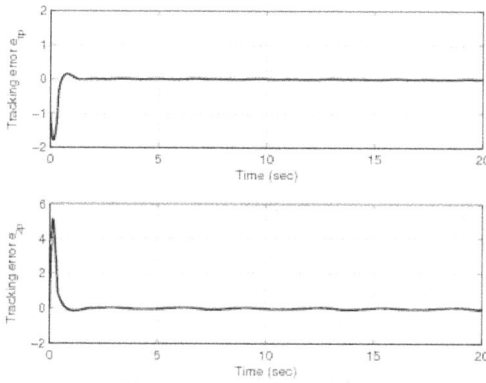

Fig. 9.25. Evolution of tracking errors of velocity (rad / sec).

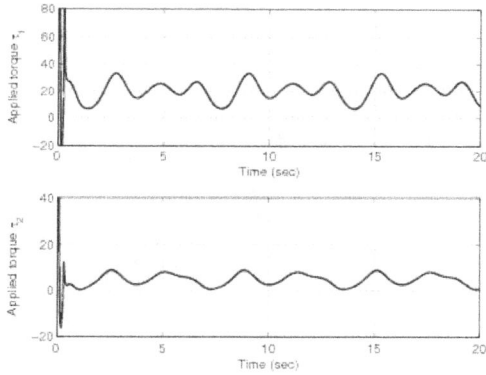

Fig. 9.26. Applied torques $(N.m.)$ with the proposed method.

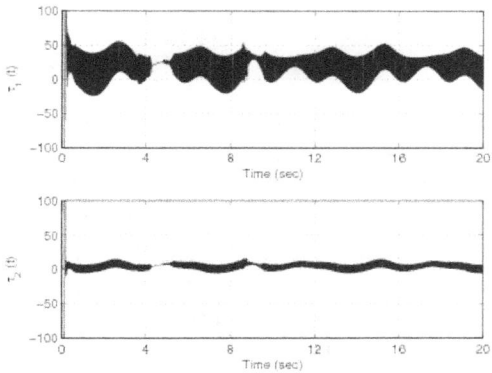

Fig. 9.27. Applied torques $(N.m.)$ with the classical SMC (function Sign).

353

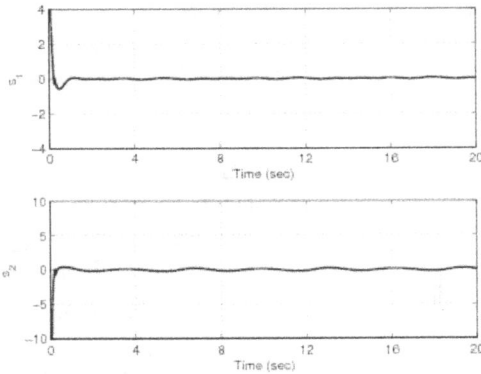

Fig. 9.28. Evolution of two sliding surfaces.

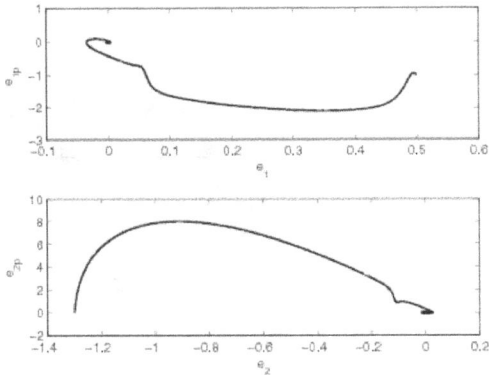

Fig. 9.29. System phase plan: evolution of trajectories.

9.5. Conclusion

In this chapter, an adaptive fuzzy controller using sliding mode approach is developed for a class of MIMO non-square nonlinear disturbed systems. The main contribution of our proposed method in comparison with other methods the system should be controlled in their original non square form. Due to the fact that the pseudo inverse of the gain matrix is used, the introduced error is taken into account in the control law. To obtain the control objectives (stability and robustness), the fuzzy approximator is used to approximate unknown nonlinear functions. In addition, this work has proposed to combine a sliding mode control and an adaptive PI controller into a single fuzzy sliding mode controller in order to overcome the chattering problem and to ensure the tracking of desired trajectories. Finally, based on the Lyapunov stability approach, the proposed adaptive fuzzy sliding mode control scheme can guarantee the global stability and the robustness of the closed loop system with respect to disturbance. Nonlinear system simulation results are presented to verify the effectiveness of the proposed method and present better performances in comparison with other recent methods proposed in literature.

References

[1]. J. J. E. Slotine, W. Li, Applied Nonlinear Control, *Prentice-Hall, Englewoods Cliffs,* New-Jersey, 1991.
[2]. A. Isidori, Nonlinear Control Systems, *Springer-Verlag,* Berlin, 1989.
[3]. V. Utkin, J. Guldner, J. Shi, Sliding Mode Control in Electromechanical Systems, *CRC Press LLC,* 2000.
[4]. G. C. Goodwin, R. S. Long, Generalization of results on multivariable adaptive control, *IEEE Transactions on Automatic Control,* Vol. 25, Issue 6, 1980, pp. 1241-1245.
[5]. G. C. Goodwin, B. C. McInnis, J. C. Wang, Model reference adaptive control for systems having non-square transfer functions, in *Proceedings of the 21st IEEE Conference on Decision and Control,* Orlando, FL, 1982, pp. 744-749.
[6]. S. Aloui, O. Pagès, A. El Hajjaji, A. Chaari, Y. Koubaa, Improved Sliding Mode control for a class of MIMO Nonlinear Uncertain and Perturbed Systems, *International Journal Applied Soft Computer,* Vol. 11, 2011, pp. 820-826.
[7]. S. Aloui, O. Pagès, A. El Hajjaji, A. Chaari, Y. Koubaa, Robust Fuzzy tracking control for a class of perturbed Nonsquare nonlinear systems, in *Proceedings of the American Control Conference,* USA, 2010, pp. 4788-4793.
[8]. S. Aloui, O. Pagès, A. El Hajjaji, A. Chaari, Y. Koubaa, Generalized Fuzzy Sliding Mode Control for MIMO Nonlinear Uncertain and Perturbed Systems, in *Proceedings of the 18th Mediterranean Conference on Control and Automation Marrakech,* 2010, pp. 1164-1169.
[9]. S. Aloui, O. Pagès, A. El Hajjaji, Y. Koubaa, A. Chaari, Robust Adaptive Fuzzy Sliding Mode Control Design for a class of MIMO underactuated system, in *Proceedings of the 18th IFAC World Congress (IFAC'11),* Milano, Italy, 2011, pp. 11127-11132.
[10]. S. Aloui, O. Pagès, A. El Hajjaji, Y. Koubaa, A. Chaari, Improved observer based adaptive fuzzy tracking control for MIMO nonlinear systems, in *Proceedings of the IEEE International Conference on Fuzzy System,* Jeju, Korea, 2009, pp. 2154-2159.

[11]. C.-C. Kung, T.-H. Chen, L.-C. Hung, Adaptive fuzzy sliding mode control for a class of underactuated systems, in *Proceedings of the IEEE International Conference on Fuzzy System*, Jeju, Korea, 2009, pp. 1791-1796.

[12]. W. Wang, J. Yi, D. Zhao, D. Liu, Design of a stable sliding-mode controller for a class of second-order underactuated systems, in *Proceedings of the IEE Control Theory Appl.*, Vol. 151, Issue 6, 2004, pp. 683-690.

[13]. J. Yi, W. Wang, *et al.*, Cascade sliding mode controller for a large-scale underactuated systems, in *Proceedings of the IEEE/RSJ International Conference on Intelligent Robots and Systems*, 2005, pp. 301-306.

[14]. W. Wang, J. Yi, D. Zhao, D. Qian, Robust control using incremental sliding mode for underactuated systems with mismatched uncertainties, in *Proceedings of the American Control Conference*, 2008, pp. 532-537.

[15]. W. Wang, X. -D. Liu, J. -Q. Yi, Structure design of two types of sliding-mode controllers for a class of underactuated mechanical systems, *IET Control Theory Applications*, Vol. 1, Issue 1, 2007, pp. 163–172.

[16]. A. G. Bondarev, S. A. Bondarev, N. -E. Kostyleva, V. I. Utkin, Sliding modes in systems with asymptotic state observers, *Automation and Remote Control*, 1985, pp. 49–64.

[17]. J. Phuah, J. Lu, T. Yahagi, Chattering free sliding mode control in magnetic levitation system, *IEE J. Trans. Electron. Inf. Syst*, Vol. 125, Issue 4, 2005, pp. 600-606.

[18]. J. Wang, A. B. Rad, P. T. Chan, Indirect adaptive fuzzy sliding mode control: Part I: fuzzy switching, *Fuzzy Sets and Systems*, Vol. 122, Issue 1, 2001, pp. 21-30.

[19]. H. F. Ho, Y. K. Wong, A. B. Rad, Robust fuzzy tracking control for robotic manipulators, *Simulation Modelling Practice and Theory*, Vol. 15, 2007, pp. 801-816.

[20]. A. Hamzaoui, N. Manamami, N. Essoumboul, J. Zaytoon, Switching Controller's Synthesis: Combination of Sliding Mode and H_∞ control by a Fuzzy Supervisor, in *Proceedings of the IFAC Conference on Analysis and Design of Hybrid Systems*, 2003, pp. 283-288.

[21]. N. Essounbouli, A. Hamzaoui, K. Benmahammed, Adaptation algorithm for robust fuzzy controller of nonlinear uncertain systems, *International Journal of Computer Applications in Technology*, Vol. 27, Issue 2-3, 2006, pp. 174-182.

[22]. H. F. Ho, Y. K. Wong, A. B. Rad, Adaptive Fuzzy Sliding Mode Control with Chattering Elimination for Nonlinear SISO Systems, *Simulation Modeling Practices and Theory*, Vol. 17, No. 7, 2009, pp. 1199-1210.

[23]. S. Labiod, M. S. Boucherit, T. M. Guerra, Adaptive fuzzy control of a class of MIMO nonlinear systems, *Fuzzy Sets and Systems*, Vol. 151, 2005, pp. 59-77.

[24]. J. Zhang, D. Cheng, Y. Liu, G. Zhu, Adaptive Fuzzy Sliding Mode Control for Missile Electro-hydraulic Servo Mechanism, in *Proceedings*

of the 7th Wold Congress on Intelligent Control and Automation, 2008, pp. 5197-5202.

[25]. X. Yao, Y. Liu, C. Guo, Adaptive fuzzy sliding mode control in variable speed adjustable pitch wind turbine, in *Proceedings of the IEEE International Conference on Automation and Logistics*, Jinan, China, 2007, pp. 313-318.

[26]. T. Tao, C. Y. Huang, Adaptive fuzzy sliding mode control scheme for tracking time-various trajectories, in *Proceedings of the IEEE International Fuzzy Systems Conference, International Conference on Automation and Logistics*, Jinan, China, 2007, pp. 1-5.

[27]. L. M. Capisani, A. Ferrara, L. Magnani, Second Order Sliding Mode Motion Control of Rigid Robot Manipulators, in *Proceedings of the 46th IEEE Conference on Decision and Control*, New Orleans, USA, 2007, pp. 3691-3696.

[28]. S. Purwar, Higher Order Sliding Mode Controller for Robotic Manipulator, in *Proceedings of the 22nd IEEE International Symposium on Intelligent Control (Part of IEEE Multi-Conference on Systems and Control)*, Singapore, 2007, pp. 556-361.

[29]. A. G. Ak, G. Cansever, Fuzzy Sliding Mode Controller with Neural Network for Robot Manipulators, in *Proceedings of the 10th Intl. Conf. on Control, Automation, Robotics and Vision*, Hanoi, Vietnam, 2008, pp. 1556-1561.

[30]. C. -H. Chen, C. -M. Lin, T. -Y. Chen, Intelligent adaptive control for MIMO uncertain nonlinear systems, *Expert Systems with Applications*, Vol. 35, 2008, pp. 865-877.

[31]. M. Roopaei, M. Zalghadri, S. Meshkar, Enchanced adaptive fuzzy sliding mode control for uncertain nonlinear systems, *Commun Nonlinear Sci Numer Simulat*, Vol. 14, 2009, pp. 3670-3681.

[32]. N. Noroozi, M. Roopaei, M. Z. Jahromi, Adaptive fuzzy sliding mode control scheme for uncertain systems, *Communications in Nonlinear Science and Numerical Simulation*, Vol. 14, Issue 11, 2009, pp. 3978-3992.

[33]. Y. C. Chang, Robust tracking control for nonlinear MIMO system via fuzzy approaches, *Automatica*, Vol. 36, 2000, pp. 1535-1545.

[34]. S. Aloui, O. Pagès, A. El Hajjaji, Y. Koubaa, A. Chaari, Robust tracking based adaptive fuzzy sliding mode controller design for robotic manipulators, in *Proceedings of the CDROOM of 9th International Conference on Science and Techniques of Automatic Control Computer Engineering (STA'08)*, Tunisia, 2008, pp. 20-23.

[35]. L. X. Wang, Stable adaptive fuzzy control of nonlinear systems, *IEEE Trans. Fuzzy Syst*, Vol. 1, Issue 2, 1993, pp. 146-155.

[36]. F. Castanos, L. Fridman, Analysis and design of integral sliding manifolds for systems with unmatched perturbations, *IEEE Trans. on Automatic Control*, Vol. 51, Issue 5, 2006, pp. 853-858.

[37]. A. Boulkroune, M. Tadjine, M. M'Saad, M. Farza, Fuzzy adaptive controller for MIMO nonlinear systems with known and unknown control direction, *Fuzzy Sets and Systems*, Vol. 161, Issue 6, 2010, pp. 797-820.

[38]. S. Tong, C. Li, Y. Li, Fuzzy adaptive observer backstepping control for MIMO nonlinear systems, *Fuzzy Sets and Systems*, Vol. 160, 2009, pp. 2755-2775.

[39]. Y. J. Liu, S. C. Tong, W. Wang, Adaptive fuzzy output tracking control for a class of uncertain nonlinear systems, *Fuzzy Sets and Systems*, Vol. 160, Issue 19, 2009, pp. 2727-2754.

[40]. S. Tong, Y. Li, Observer-based fuzzy adaptive control for strict-feedback nonlinear systems, *Fuzzy Sets and Systems*, Vol. 160, Issue 12, 2009, pp. 1749-1764.

Chapter 10

Analytical Solution of Optimized Energy Consumption of Induction Motor Operating in Transient Regime

Riadh Abdelati

10.1. Minimization of a Cost-to-go Function under Constraints

In this section we will introduce the control variables at the level of the functional in order to take into account any possible constraints imposed by the machine's power quantity during the experimental implementation. In this framework, our optimization problem is formulated using the Pontryaguin minimum principle.

10.1.1. Statement of Optimal Control Problem with a Final Free State

Knowledge of the final and initial values of the rotor flux has always been a requirement to solve the Euler-Lagrange equation developed in [27]. The reformulation of the optimization problem that reveals the Euler-Lagrange equation, their limitations and the constraints of the problem can leave free the final value of the state variable. In addition, the resolution uses the Pontryaguin minimum principle which involves intermediate variables called co-state variables that depend on the final state of the system. The optimal control problem envisaged is of type C characterized by a free final state and a fixed final time [3]. The statement of such a problem is as follows:

Find the u continuous optimal controlling the interval $[t_a, t_b]$ such as:

$u: [t_a, t_b] \rightarrow \Gamma \subseteq \mathcal{R}^m$ verifying the constraints (10.1) to (10.3):

Riadh Abdelati
University of Monastir, National School of Engineers of Monastir, Tunisia

$$x(t_a) = x_a \tag{10.1}$$

$$\dot{x}(t) = f(x(t), u(t), t), \text{ for every } t \in [t_a, t_b] \tag{10.2}$$

$$x(t_b) \text{ is free} \tag{10.3}$$

And that the functional $J(u)$, defined by:

$$J(u) = K\big(x(t_b)\big) + \int_{t_a}^{t_b} L(x(t), u(t), t)dt \tag{10.4}$$

should be minimal.

In the literature [3], there are two versions of type C optimal control:

1. The first assumes that t_b time is fixed;

2. The second leaves t_b time free $(t_b > t_a)$.

The t_a time and $x_a \in \mathcal{R}^n$ initial state variable are specified.

For this part, we will reformulate problem into space of the control and we will investigate at the level of type C first version.

10.1.1.1. Induction Motor's Current Magnitudes and Stator Voltages limits

In the case of a real application, it is essential to protect the actuator and the machine by imposing maximum limits on voltages and currents. Therefore, it is recommended an analysis of the performance of the machine should be carried out taking into account the constraints imposed on these two parameters.

As regards the constraint imposed on the stator voltage, it verifies:

$$u_{sd}^2 + u_{sq}^2 < u_{max}^2 \tag{10.5}$$

with $u_{max} = \dfrac{V_{DC}}{\sqrt{3}}$ and V_{DC} as the voltage of the inverter DC bus.

The limitation imposed on the stator current is based on:

$$I_{sd}^2 + I_{sq}^2 < I_{max}^2 \qquad (10.6)$$

with I_{max} as the maximal current delivered by the inverter.

10.1.2. Reformulation of the Optimal Control Problem

We are aiming at an optimal control that allows the minimization of the functional under predefined constraints by the relations (10.5) and (10.6). For this, we can use the Λ: Lagrange multipliers vector in order to rewrite the functional as follows:

$$J(U) = K(X(t_b), t_b) + \int_{t_a}^{t_b} \{L(X(t), U(t), t) +$$
$$\Lambda^{tr}(t)(f(X(t), U(t), t) - \dot{X}(t))\} \, dt \qquad (10.7)$$

with the following constraint:

$$\dot{X}(t) = f(X(t), U(t), t) \qquad (10.8)$$

We can thus describe in a more concise way by defining the Hamiltonian function or Pontryaguin state function by:

$$\boldsymbol{H}(X(t), U(t), \Lambda(t), t) = L(X(t), U(t), t) + \Lambda^{tr}(t)(f(X(t), U(t), t) \qquad (10.9)$$

We integrate by parts the functional given by Equation (10.7) and we obtain:

$$J(U) = K(X(t_b), t_b) - \Lambda^{tr}(t_b)X(t_b) + \Lambda^{tr}(t_a)X(t_a) +$$
$$\int_{t_a}^{t_b} \{\boldsymbol{H}(X(t), U(t), \Lambda(t), t) + \dot{\Lambda}^{tr}(t)X(t)\} \, dt \qquad (10.10)$$

The variation of the functional due to the variation of the δU controlwithin a given time interval $[t_a, t_b]$ is given by:

$$\delta J = \left[\left(\frac{\partial K}{\partial X} - \Lambda^{tr}\right)\delta X\right]_{t=t_b} + [\Lambda^{tr}\delta X]_{t=t_a}$$
$$+ \int_{t_a}^{t_b} \left[\left(\frac{\partial \boldsymbol{H}}{\partial X} + \dot{\Lambda}^{tr}\right)\delta X + \frac{\partial \boldsymbol{H}}{\partial U}\delta U\right] dt \qquad (10.11)$$

It is generally difficult to quantify the variation of the δX state variables from the variation of the δU control variables. That's why we chose $\Lambda(t)$ LaGrange multiplier functions, also referred to as co-state variables, so that the δX coefficients would disappear. Then:

$$\dot{\Lambda}^{tr} = -\frac{\partial H}{\partial X} = -\frac{\partial L}{\partial X} - \Lambda^{tr}\frac{\partial f}{\partial X} \qquad (10.12)$$

with the boundary conditions defined by:

$$\Lambda^{tr}(t_b) = \frac{\partial K}{\partial X(t_b)}(X(t_b), t_b) \qquad (10.13)$$

Combining Equations (10.12) and (10.13), the variation of the functional thus becomes:

$$\delta J = \Lambda^{tr}(t_a)\delta X(t_a) + \int_{t_a}^{t_b}\frac{\partial H}{\partial U}\delta U dt \qquad (10.14)$$

with:

− $\Lambda^{tr}(t_a)$ denotes J gradient according to initial conditions respecting the constraint defined in (10.12) and keeping U(t) constant.

− $\Lambda^{tr}(t)$ denotes the influence function of the X(t) variation on the J functional when the t_a time is arbitrary.

− $\frac{\partial H}{\partial U}$ are called response functions to a pulse when each $\frac{\partial H}{\partial U}$ component represents the J variation due to each unit pulse of the corresponding ∂U componentatt timewhen we keep $X(t_a)$ constant while verifying constraint (10.8).

In the case of an extremum, δJ is forced toward zero for an arbitrary $\partial U(t)$ variation. This is true only when:

$$\frac{\partial H}{\partial U} = 0 \text{ with } t_a \le t \le t_f \qquad (10.15)$$

To conclude, in order to find a U*(t) control vector that minimizes a stationary performance measure of the J functional, the following equations must be solved:

$$\dot{X}^*(t) = f(X^*(t), U^*(t), t) \text{ or}$$

$$\dot{X}^*(t) = \frac{\partial H}{\partial \Lambda}(X^*(t), U^*(t), t) \tag{10.16}$$

By replacing the Hamiltonian expression given by the relation (10.9) in (10.12) we obtain:

$$\dot{\Lambda}^{tr^*} = -\frac{\partial H}{\partial X}(X^*(t), U^*(t), \Lambda^*(t), t) = -\frac{\partial L}{\partial X}(X^*(t), U^*(t), t) -$$
$$\left[\frac{\partial f}{\partial X}(X^*(t), U^*(t), t)\right]^{tr} \Lambda^*(t) \tag{10.17}$$

with $U^*(t)$ determined by equation:

$$\frac{\partial H}{\partial U}(X^*(t), U^*(t), \Lambda^*(t), t) = 0 \text{ or}$$

$$\frac{\partial L}{\partial U}(X^*(t), U^*(t), t) + \left[\frac{\partial f}{\partial X}(X^*(t), U^*(t), t)\right]^{tr} \Lambda^*(t) = 0 \tag{10.18}$$

The initial conditions at both ends are defined by:

-For the vector of state variables:

$X(t_a) = X_a$ et $X(t_b) = X_b$ (if it exists in case of type A problem)

-For the vector of co-state variables or Lagrange multiplier functions:

$$\Lambda^{tr}(t_b) = \frac{\partial K}{\partial X(t_b)}(X(t_b), t_b) \tag{10.19}$$

10.2. Optimal Control via the Hamilton-Jacobi-Bellman Equation

10.2.1. Determination of the HJB Equation

The condition described in Equation (10.15) can be interpreted as a requirement to minimize $H(X(t), U(t), t)$ in relation to U. The partial differentiation operation is possible for t, X et Λ variables provided the U control is continuous and has an unbounded admissible range. However, if we assume that the U admissible control is piecewise continuous and has a boundedrange: $U \in \Omega$. it has been shown [3] that the U^*control can still minimize the Hamiltonian. This is the principle of minimum Pontryaguin.

In this paragraph, we will deal with the problem of type C optimization.

The principle is stated as follows [3]:

If $\boldsymbol{H}$ is the Hamilton function: $\boldsymbol{H} \colon \mathcal{R}^n \times \Omega \times \mathcal{R}^n \times \{0,1\} \times [t_a, t_b] \mathcal{R}$,

$$
\begin{aligned}
\boldsymbol{H}(X(t), U(t), \Lambda(t), t) = {} & \Lambda_0 L(X(t), U(t), t) + \\
& \Lambda^{tr}(t) f(X(t), U(t), t)
\end{aligned}
\tag{10.20}
$$

with $\Lambda(t)$: is the co-state variable or Lagrange multiplier function. $\Lambda(t)$ represents the influence function of the $X(t)$ variation on the $J(u)$ functional when the t_a time is arbitrary. Λ_0 as a coefficient having as optimal value

$$
\Lambda_0^* = \begin{cases} 1 \text{ if the control is non singular} \\ 0 \text{ if the control is singular} \end{cases}
$$

The $U^* \colon [t_a, t_b] \rightarrow \Gamma$ is optimal, there is anon trivial vector

$$
\begin{bmatrix} \Lambda_0^* \\ \Lambda^* \end{bmatrix} \neq 0 \in \mathcal{R}^{n+1}
$$

in such a way that the following conditions are met:

$$
\dot{X}^*(t) = \nabla_\Lambda \boldsymbol{H}^* = f(X^*(t), U^*(t), t);
\tag{10.21}
$$

$$
X^*(t_a) = X_a;
\tag{10.22}
$$

$$
\begin{aligned}
\dot{\Lambda}^*(t) = -\nabla_X \boldsymbol{H}^* = {} & -\Lambda_0^* \, \nabla_X L(X^*(t), U^*(t), t) - \\
& \left[\frac{\partial f}{\partial X}(X^*(t), U^*(t), t) \right]^{tr} \Lambda^*(t) ;
\end{aligned}
\tag{10.23}
$$

$$
\Lambda^*(t_b) = \nabla_X K(X^*(t_b), t_b)
\tag{10.24}
$$

For any $t \in [t_a, t_b]$, the Hamiltonian $\boldsymbol{H}(X^*(t), U, \Lambda^*(t), t)$ has a global minimum with $U = U^*(t)$ with $U \in \Gamma$; implying that there exists a U^*optimal control such as:

$$
\boldsymbol{H}(X^*(t), U^*(t), \Lambda^*(t), t) \leq \boldsymbol{H}(X^*(t), U, \Lambda^*(t), t)
$$

$$
\text{for any } U \in \Gamma \text{ and } t \in [t_a, t_b]
\tag{10.25}
$$

Besides, if the t_b *time is free* (2nd case of type C optimal control), then:

$$H(X^*(t_b), U^*(t_b), \Lambda^*(t_b), t_b) = -\frac{\partial K}{\partial t}(X^*(t_b), t_b) \quad (10.26)$$

The optimal control problem considered here is of type C. We denote by V the cost-to-go function instead of J which characterize the specificity of the HJB equation.

The $f(x(t), u(t), t)$ function can be arbitrary, while both $L(x(t), u(t), t)$ and $K(x(t_b))$ functions must be twice differentiable with respect to their arguments and not negative in order to accomplish the minimization requirements of the cost-to-go function V. Indeed, these two conditions are verified in the present study because the energy balance that is involved in $L(x(t), u(t), t)$ is determined from the dynamic AD model and the weighting factors are calculated in such a way that all the arguments of the functional are non-negative.

As shown in Equation (10.4) inserted in paragraph 10.2.1, $V(x(t), u(.), t)$ depends on initial conditions of the $x(t_a)$ state vector, the t_a initial time and the $u(t)$ control vector for any $t \in [t_a, t_b]$.

In this section, the task to be performed consists in substituting the control that minimizes $V(x(t_a), u(.), t_a)$ denoted $V^*(x(t_a), t_a)$ by $V(x(t), u(.), t)$ $\forall t$ and he $x(t)$ assessment. The resolution of the associated optimal control problem is formulated as follows:

$$V(x(t), u(.), t) = K(x(t_b)) + \int_t^{t_b} L(x(\tau), u(\tau), \tau)d\tau \quad (10.27)$$

Note that by $u[t_a, t_b]$ there is *a restriction on the* $[t_a, t_b]$ interval of the continuous and bounded piecewise u(.) function. In addition, if the system considers -instead of the initial $x(t_a)$ state vector - one corresponding to the x(t) vector at t time, then we get the optimal condition defined by:

$$V^*(x(t), t) = \min_{u[t_a, t_b]} V(x(t), u(.), t) \quad (10.28)$$

The relation (10.28) is independent of $V^*(x(t), t)$. Therefore, we will solve the problem defined in the previous paragraph by replacing t by t_a.

Indeed, for any $t \in [t_a, t_b]$ and $\tau \in [t, t_b]$, we can write:

$$V^*(x(t), t) = \min_{u[t,t_b]} \left[K(x(t_b)) + \int_{t_a}^{t_b} L(x(\tau), u(\tau), \tau) d\tau \right] =$$

$$\min_{u[t,t_a]} \left\{ \min_{u[t_a,t_b]} \left[\int_t^{t_a} L(x(\tau), u(\tau), \tau) d\tau + K(x(t_b)) + \int_{t_a}^{t_b} L(x(\tau), u(\tau), \tau) d\tau \right] \right\} =$$

$$\min_{u[t,t_a]} \left\{ \int_t^{t_a} L(x(\tau), u(\tau), \tau) d\tau + \min_{u[t_a,t_b]} \left[K(x(t_b)) + \int_{t_a}^{t_b} L(x(\tau), u(\tau), \tau) d\tau \right] \right\}$$

yields

$$V^*(x(t), t) = \min_{u[t,t_a]} \left\{ \int_t^{t_a} L(x(\tau), u(\tau), \tau) d\tau + V^*(x(t_a), t_a) \right\} \quad (10.29)$$

At the level of Equation (10.29), we replace t_a time *by* $t + \Delta t$; while assuming that the Δt variation is rather small.

We apply the Taylor theorem and we obtain the following relation:

$$V^*(x(t), t) = \min_{u[t,t+\Delta t]} \left\{ \Delta t L(x(t + \alpha \Delta t), u(t + \alpha \Delta t), t + \alpha \Delta t) + V^* \frac{dx(t)}{dt} \Delta t + \frac{\partial V^*}{\partial t}(x(t), t) \Delta t + 0(\Delta t)^2 \right\} \quad (10.30)$$

with $0 < \alpha < 1$, which leads to:

$$\frac{\partial V^*}{\partial t}(x(t), t) = -\min_{u[t,t+\Delta t]} \left\{ L(x(t + \alpha \Delta t), u(t + \alpha \Delta t), t + \alpha \Delta t) + \left[\frac{\partial V^*}{\partial x}(x(t), t) \right]^{tr} f(x(t), u(t), t) + 0(\Delta t) \right\} \quad (10.31)$$

when $\Delta t \to 0$, we get:

$$\frac{\partial V^*}{\partial t}(x(t), t) = -\min_{u(t)} \left\{ L(x(t), u(t), t) + \left[\frac{\partial V^*}{\partial x}(x(t), t) \right]^{tr} f(x(t), u(t), t) \right\} \quad (10.32)$$

In this equation, L and f functions are known, but the V^* function is not. This is why the preceding relation (10.32) is generally described in the literature by the following expression:

$$\frac{\partial V^*}{\partial t} = -\min_{u(t)} \left\{ L(x(t), u(t), t) + \left[\frac{\partial V^*}{\partial x}\right]^{tr} f(x(t), u(t), t) \right\} \quad (10.33)$$

As $u(t)$ optimal control value is derived from the $\frac{\partial H}{\partial u} = 0$ stationary condition and by analogy to the Hamilton equation, we can generate, in these conditions, a new expression of the Hamiltonian given by:

$$H\left(x(t), u(t), \frac{\partial V}{\partial x}, t\right) = L(x(t), u(t), t) + \left[\frac{\partial V}{\partial x}\right]^{tr} f(x(t), u(t), t) \quad (10.34)$$

The $u(t)$ value that minimizes the straight expression of Equation (10.34) depends on the values taken by $x(t), \frac{\partial V^*}{\partial x}$ and t. In other words, $u(t)$ optimal value is an instantaneous function of these three variables that will be denoted as $u^*(x(t), \frac{\partial V^*}{\partial x}, t)$. Thus the Equation (10.34) is written as follows:

$$\frac{\partial V^*}{\partial t} = -L\left(x(t), u^*\left(x(t), \frac{\partial V^*}{\partial x}, t\right), t\right) - \left[\frac{\partial V^*}{\partial x}\right]^{tr} f\left(x(t), u^*\left(x(t), \frac{\partial V^*}{\partial x}, t\right), t\right) \quad (10.35)$$

The relation (10.35) reflects of the Hamilton-Jacobi-Bellman (HJB) Equation.

From the $V(x(t_a), u(.), t_a)$ expression governed by Equation (10.27), we can deduce that $V(x(t_b), u(.), t_b) = K(x(t_b))$

At t_b time, the cost-to-go function's minimum value is always equal to the penalty function $K(x(t_b))$, or:

$$V^*(x(t_b), t_b) = K(x(t_b)) \quad (10.36)$$

10.3. Determination of the HJB Equation in the Case of IM Minimum Energy Control

In this context, we applied the HJB equation in the case of IM minimal energy control using the lossless balance in iron. The HJB equation is very difficult to solve analytically. This led us to develop hypergeometric solutions generated in particular for cases characterized by mathematical arguments that do not have any physical meaning (complex argument values, or exponential values as a function of time. However, in our application, these arguments must be in accordance with the MA parameters)

In the case of the lossless balance in iron, the functional is presented by the optimal control with closed loop as presented in the previous paragraph and resulting in an HJB equation given by the relation (10.35). This led us to state the problem of minimal energy control in the following way:

Finding the optimal control variable $u_1^*(t, \Phi_r)$ which generates the $\Phi_r^*(t)$ corresponding state variable while taking into account the following condition:

$$V(t, \Phi_r^*) = \min_{u_1^*} \left\{ \frac{\chi_2}{L_R} \left(\Phi_r^2(0) - \Phi_r^2(T) \right) + \int_t^T \left(r_1 (u_1^2 + u_2^2) + q_1 \Phi_r^2 + q_2 \Phi_r u_2 \Omega \right) d\tau \right\}$$

(10.37)

Which transfers $\Phi_r(0) = \Phi_{r0}$ initial state into a final admissible state $\Phi_r(T)$ under the following constraint:

$$\dot{\Phi}_r = -a\Phi_r + bu_1$$

(10.38)

This directly determines the following HJB equation:

$$\frac{\partial V(t, \Phi_r^*)}{\partial t} = -\min_{u_1^*} H(t, \Phi_r^*, u_1^*, \frac{\partial V(t, \Phi_r)}{\partial \Phi_r})$$

(10.39)

Note:

-We omitted the u_2 and state Ω control variables in Equation (10.39) since the latter will be later replaced by an appropriate law of evolution. The expression of rotation speed will be introduced in the IM mechanical equation given by a proposed closed-cycle motor drive process (i.e. This cycle define a specific profile of the motor speed) and the u_2 control variable will be deducted immediately according to Ω and Φ_r.

To define a closed-cycle process, let us introduce the following motor speed profile:

Then the motor speed can be chosen as follows:

$$\Omega = c_0 t + c_1 \tag{10.40}$$

with

$$\begin{cases} c_0 > 0 \text{ in mode (1): The acceleration phase} \\ c_0 < 0 \text{ in Mode (2): The deceleration phase} \end{cases}$$

-In the remainder of this problem of minimal energy control, we will simplify the writing by replacing the Ω and Φ_r state variables by their respective notations: x_2 and x_1.

We will consider the expression of speed according to two modes of transient engine operation:

The Ω expression is imposed by Equation (10.40). The u_2 control command is deduced from the second equation of the system (10.18) taking account the relation (10.40). Then we introduce these two quantities into the Hamiltonian of the problem, which leads, for any $x_1 > 0$, at:

$$H(t, x_1, u_1) = r_1 u_1^2 + r_1 \frac{1}{x_1^2}(At + B)^2 + q_1 x_1^2 + q_2(At + B)(c_0 t + c_1)$$
$$+ \frac{\partial V(t, x_1)}{\partial x_1}(-ax_1 + bu_1) \tag{10.41}$$

with: $A = K_I \dfrac{c_0}{c}$ and $B = \dfrac{J_m}{c}(c_0 + K_I c_1)$

After developing and arranging the relation (10.41), we obtain:

$$H(t,x_1,u_1) = \lambda_1 u_1^2 + (\frac{\lambda_2}{x_1^2} + \lambda_3)t^2 + (\frac{\lambda_4}{x_1^2} + \lambda_5)t + \lambda_6 x_1^2 + \frac{\lambda_7}{x_1^2}$$
$$+ \lambda_8 + \frac{\partial V(t,x_1)}{\partial x_1}(-ax_1 + bu_1) \tag{10.42}$$

with $\lambda_1 = r_1$, $\lambda_2 = r_1 A^2$, $\lambda_3 = q_2 c_0 A$, $\lambda_4 = 2r_1 AB$, $\lambda_5 = q_2(c_0 B + c_1 A)$, $\lambda_6 = q_1$, $\lambda_7 = r_1 B^2$ and $\lambda_8 = q_2 c_1 B$.

The optimal control variable $u_1^*(t,x_1)$ is generated from the equation:
$\dfrac{\partial H(t,x_1,u_1)}{\partial u_1} = 0$; which gives:

$$u_1^*(t,x_1) = -\frac{b}{2\lambda_1}\frac{\partial V(t,x_1)}{\partial x_1} \tag{10.43}$$

Using the HJB Equation (10.42), the relation (10.39) is equivalent to:

$$\frac{\partial V(t,x_1^*)}{\partial t} =$$
$$-\min_{u_1}\left(\lambda_1 u_1^2 + (\frac{\lambda_2}{x_1^2} + \lambda_3)t^2 + (\frac{\lambda_4}{x_1^2} + \lambda_5)t + \lambda_6 x_1^2 + \frac{\lambda_7}{x_1^2} + \lambda_8 + \frac{\partial V(t,x_1)}{\partial x_1}(-ax_1 + bu_1)\right)$$
$$\tag{10.44}$$

Replacing the optimal control $u_1^*(t,x_1^*)$ given by Equation (10.43) in Equation (10.44), we deduce that:

$$\frac{\partial V(t,x_1^*)}{\partial t} = \frac{b^2}{4\lambda_1^2}(\frac{\partial V(t,x_1^*)}{\partial x_1^*})^2 + ax_1^*(\frac{\partial V(t,x_1^*)}{\partial x_1^*}) - (\frac{\lambda_2}{x_1^2} + \lambda_3)t^2$$
$$- (\frac{\lambda_4}{x_1^2} + \lambda_5)t - \lambda_6 x_1^2 - \frac{\lambda_7}{x_1^2} - \lambda_8 \tag{10.44a}$$

10.3.1. Solving the HJB Equation

The relation (10.44a) resulting from the HJB equation constitutes a nonlinear differential equation with the presence of squared terms. Our task is to find the $V^*(t, x_1^*)$ optimal functional, solution to Equation (10.44a):

We therefore propose the following $V^*(t, x_1^*)$ expression:

$$V(t, x_1^*) = \alpha_1(t) + \alpha_2(t) Logx_1^* + \alpha_3(t) x_1^{*2} \tag{10.45}$$

Its partial derivative with respect to time gives:

$$\frac{\partial V(t, x_1^*)}{\partial t} = \overset{\bullet}{\alpha}_1(t) + \overset{\bullet}{\alpha}_2(t) Logx_1^* + \overset{\bullet}{\alpha}_3(t) x_1^{*2} \tag{10.46}$$

and its partial derivative with respect to the rotor flux gives:

$$\frac{\partial V(t, x_1^*)}{\partial x_1^*} = \frac{\alpha_2(t)}{x_1^*} + 2\alpha_3(t) x_1^* \tag{10.47}$$

Using the expressions of partial derivatives of the functional (10.46) and (10.47) in (10.45), we obtain the following equation:

$$\overset{\bullet}{\alpha}_1(t) + \overset{\bullet}{\alpha}_2(t) Logx_1^* + \overset{\bullet}{\alpha}_3(t) x_1^{*2} = \frac{b^2}{4\lambda_1^2}(\frac{\alpha_2(t)}{x_1^*} + 2\alpha_3(t)x_1^*)^2$$

$$+ ax_1^*(\frac{\alpha_2(t)}{x_1^*} + 2\alpha_3(t)x_1^*) - (\frac{\lambda_2}{x_1^2} + \lambda_3)t^2 - (\frac{\lambda_4}{x_1^2} + \lambda_5)t - \lambda_6 x_1^2 - \frac{\lambda_7}{x_1^2} - \lambda_8 \tag{10.48}$$

which –after factorization- leads to:

$$\frac{1}{x_1^{*2}}\left(\frac{b^2}{4\lambda_1}\alpha_2^2(t) - \lambda_2 t^2 - \lambda_4 t - \lambda_7\right) + x_1^{*2}\left(-\overset{\bullet}{\alpha}_3(t) + \frac{b^2}{\lambda_1}\alpha_3^2(t) + 2a\alpha_3(t) - \lambda_6\right)$$

$$+ \left(a\alpha_2(t) - \overset{\bullet}{\alpha}_1(t) - \alpha_2(t)Logx_1^* - \lambda_3 t^2 - \lambda_5 t - \lambda_8 + \frac{\alpha_2\alpha_3}{\lambda_1}b^2\right) = 0 \tag{10.49}$$

To meet relation (10.49) for any $x_1^* > 0$, we have to solve the following sets of equations:

$$\begin{cases} \dfrac{b^2}{4\lambda_1}\alpha_2^2(t) - \lambda_2 t^2 - \lambda_4 t - \lambda_7 = 0 \\[2ex] -\dot{\alpha}_3(t) + \dfrac{b^2}{\lambda_1}\alpha_3^2(t) + 2a\alpha_3(t) - \lambda_6 = 0 \\[2ex] a\alpha_2(t) - \dot{\alpha}_1(t) - \alpha_2(t)Logx_1^* - \lambda_3 t^2 - \lambda_5 t \\[1ex] \quad - \lambda_8 + \dfrac{\alpha_2\alpha_3}{\lambda_1}b^2 = 0 \end{cases} \qquad (10.50)$$

Given the $u_1^*(t, x_1)$ optimal control produced by Equation (10.43) and the expression (10.47) of $\dfrac{\partial V(t, x_1^*)}{\partial x_1^*}$, it would be enough to solve only the first two equations of the system (10.50), i.e., to determine the expression of $\alpha_2(t)$ and $\alpha_3(t)$.

According to the first Equation of (10.50), we have two solutions for $\alpha_2(t)$: a positive one and a negative one. We tried to solve the problem according to both cases, and we finally retained the negative expression of $\alpha_2(t)$ governed by:

$$\alpha_2(t) = -\frac{2}{b}\sqrt{\lambda_1\lambda_2 t^2 + \lambda_1\lambda_4 t + \lambda_1\lambda_7} \qquad (10.51)$$

with $\lambda_1\lambda_2 t^2 + \lambda_1\lambda_4 t + \lambda_1\lambda_7 = Ar_1^2(t + t_1)^2$.

which eventually leads to:

$$\alpha_2(t) = -\frac{2A}{b}r_1(t + t_1) \qquad (10.52)$$

and $t_1 = \dfrac{B}{A} = \dfrac{1}{\dfrac{J_{mt}}{K_l} + \dfrac{c_1}{c_0}}$.

The second relation of the system (10.50) constitutes a first order differential equation with the presence of a squared term in $\alpha_3(t)$.

$$\overset{\bullet}{\alpha}_3(t) - \frac{b^2}{\lambda_1}\alpha_3^2(t) - 2a\alpha_3(t) + \lambda_6 = 0 \qquad (10.53)$$

This type of equation is mentioned in the literature of Riccati [And95].

Replacing $\alpha_3(t)$ in (10.53) by the following expression:

$$\alpha_3(t) = -\frac{\lambda_1}{b^2}\frac{1}{g(t)}\frac{dg(t)}{dt} \qquad (10.54)$$

with $g(t) \subset \Re^*$, $\alpha_3(t)$ partial derivative with respect to time gives:

$$\overset{\bullet}{\alpha}_3(t) = \frac{\lambda_1}{b^2}\frac{1}{g^2(t)}\left(\frac{du(t)}{dt}\right)^2 - \frac{\lambda_1}{b^2}\frac{1}{g(t)}\frac{d^2g(t)}{dt^2} \qquad (10.55)$$

Thus, the relation (10.53) changes into a second order differential equation:

$$\frac{d^2g(t)}{dt^2} - 2a\frac{dg(t)}{dt} - \lambda_6\frac{b^2}{\gamma_1}g(t) = 0 \qquad (10.56)$$

Furthermore, we get

$$a^2 + \frac{b^2}{\lambda_1}\lambda_6 = \left(\frac{R_r}{L_r}\right)^2\left(1 + \frac{q_1}{r_1}M^2\right)$$

$$= \left(\frac{R_r}{L_r}\right)^2\left(1 + \frac{\dfrac{3}{4}\dfrac{\alpha_{11}}{L_r} - \dfrac{3}{2}\dfrac{R_r}{L_r^2}\alpha_{12}}{\dfrac{3}{4}\sigma L_s\alpha_{11} + \dfrac{3}{2}(R_s + R_r\left(\dfrac{M}{L_r}\right)^2)\alpha_{12}}M^2\right) > 0$$

$$(10.57)$$

with α_{11} and α_{12} being the parameters that verify in Equation (26) in [26]

We can therefore deduce the $g(t)$ solution of Equation (10.56), as follows:

$$g(t) = g_1 e^{\left(a+\sqrt{\delta_1}\right)t} - g_2 e^{\left(a-\sqrt{\delta_1}\right)t} \qquad (10.58)$$

The derivative of $g(t)$ function gives:

$$\frac{dg(t)}{dt} = g_1\left(a+\sqrt{\delta_1}\right)e^{\left(a+\sqrt{\delta_1}\right)t} - g_2\left(a-\sqrt{\delta_1}\right)e^{\left(a-\sqrt{\delta_1}\right)t} \qquad (10.59)$$

with $\delta_1 = a^2 + \dfrac{b^2}{\lambda_1}\lambda_6$;

The relation (10.59) introduced in Equation (10.54) allows to deduce the $\alpha_3(t)$ variable:

$$\alpha_3(t) = -\frac{\lambda_1}{b^2}\frac{m_1\left(a+\sqrt{\delta_1}\right)e^{\left(a+\sqrt{\delta_1}\right)t} - m_2\left(a-\sqrt{\delta_1}\right)e^{\left(a-\sqrt{\delta_1}\right)t}}{m_1 e^{\left(a+\sqrt{\delta_1}\right)t} - m_2 e^{\left(a-\sqrt{\delta_1}\right)t}} \qquad (10.60)$$

Which —when posing the $z_1 = \dfrac{m_2}{m_1}$ constant- gives:

$$\alpha_3(t) = -\frac{\lambda_1}{b^2}\frac{\left(a+\sqrt{\delta_1}\right)e^{\left(\sqrt{\delta_1}\right)t} - z_1\left(a-\sqrt{\delta_1}\right)e^{-\left(\sqrt{\delta_1}\right)t}}{e^{\left(\sqrt{\delta_1}\right)t} - z_1 e^{-\left(\sqrt{\delta_1}\right)t}} \qquad (10.61)$$

or:

$$\alpha_3(t) = \frac{\lambda_1}{b^2}\left(-a+\sqrt{\delta_1} - \frac{2\sqrt{\delta_1}e^{\sqrt{\delta_1}\,t}}{e^{\sqrt{\delta_1}\,t} - z_1 e^{-\sqrt{\delta_1}\,t}}\right) \qquad (10.62)$$

The $\alpha_3(t)$ expression is rearranged as follows:

$$\alpha_3(t) = \frac{\lambda_1}{b^2}\left(-a+\sqrt{\delta_1} + \frac{2\sqrt{\delta_1}}{z_1 e^{-2\left(\sqrt{\delta_1}\right)t} - 1}\right) \qquad (10.63)$$

Finally, $\alpha_3(t)$ is given by:

$$\alpha_3(t) = \frac{\lambda_1}{b^2}\left(-a + \sqrt{a^2 + \frac{b^2}{\lambda_1}\lambda_6} + \frac{2\sqrt{a^2 + \frac{b^2}{\lambda_1}\lambda_6}}{-1 + z_1 e^{-2\left(\sqrt{a^2 + \frac{b^2}{\lambda_1}\lambda_6}\right)t}}\right) \qquad (10.64)$$

With z_1 : a constant that depends on the initial conditions.

Using the expression introduced in the relation (10.47) and the relations obtained in (10.51) and (10.64), the term $\dfrac{\partial V(t, x_1^*)}{\partial x_1^*}$ is expressed by:

$$\frac{\partial V(t, x_1^*)}{\partial x_1^*} = \frac{2\lambda_1}{b^2}\left(-a + \sqrt{\delta_1} + \frac{2\sqrt{\delta_1}}{-1 + z_1 e^{-2(\sqrt{\delta_1})t}}\right)x_1^* - \frac{\left(\frac{2}{b}Ar_1(t+t_1)\right)}{x_1^*} \qquad (10.65)$$

with $\delta_1 = a^2 + \dfrac{b^2}{\lambda_1}\lambda_6$.

Finally, from the relation (10.43), the u_1^* optimal control is expressed by:

$$u_1^*(t, x_1^*) = \frac{1}{b}\left(a - \sqrt{\delta_1} - \frac{2\sqrt{\delta_1}}{-1 + z_1 e^{-2(\sqrt{\delta_1})t}}\right)x_1^*(t) + \frac{A}{x_1^*(t)}(t + t_1) \qquad (10.66)$$

Using the flux dynamic Equation and replacing the u_1^* control in (10.66), we obtain the rotor flux differential equation:

$$\dot{x}_1^*(t) = -\left(\sqrt{\delta_1} + \frac{2\sqrt{\delta_1}}{-1 + z_1 e^{-2(\sqrt{\delta_1})t}}\right)x_1^*(t) + \frac{bA}{x_1^*(t)}(t + t_1) \qquad (10.67)$$

Replacing $x_1^*(t)$ by $s(t)$ such as: $s(t) = \left(x_1^*(t)\right)^2$, Equation (10.67) is written:

$$\dot{s}(t) = -2\left(\sqrt{\delta_1} + \frac{2\sqrt{\delta_1}}{-1 + z_1 e^{-2\left(\sqrt{\delta_1}\right)t}}\right)s(t) + 2bA(t + t_1) \qquad (10.68)$$

The conditions at both ends verify $s(t_0) = \left(x_1^*(t_0)\right)^2$ and $s(T) = \left(x_1^*(T)\right)^2$

The relation (10.68) represents a first order differential equation.

We devote by: $\eta(t) = 2\left(\sqrt{\delta_1} + \dfrac{2\sqrt{\delta_1}}{-1 + z_1 e^{-2\left(\sqrt{\delta_1}\right)t}}\right)$ and $\phi(t) = +2bA(t + t_1)$,

which leads to writing:

$$\dot{s}(t) + \eta(t)s(t) = \phi(t) \qquad (10.69)$$

The term to the left of (10.69) represents the first order linear differential equation [Ack03]. We can then deduce the final solution of the differential Equation (10.69) using the following expansion:

$$\frac{d}{dt}\left(s(t)e^{\int_{t0}^{t}\eta(t)dt}\right) = \frac{ds}{st}e^{\int_{t0}^{t}\eta(t)dt} + s(t)\eta(t)e^{\int_{t0}^{t}\eta(t)dt}$$

$$= \left[\frac{ds}{st} + s(t)\eta(t)\right]e^{\int_{t0}^{t}\eta(t)dt} = \phi(t)e^{\int_{t0}^{t}\eta(t)dt}$$

which consequently gives:

$$s(t)e^{\int_{t0}^{t}\eta(t)dt} - s(t_0)\left(e^{\int_{t0}^{t}\eta(t)dt}\right)_{t=t0} = \int_{t_0}^{t}\phi(t)e^{\int_{t0}^{t}\eta(t)dt}dt \qquad (10.70)$$

Finally the solution to Equation (10.69) is summarized by:

$$s(t) = e^{-\int_{t0}^{t}\eta(t)dt}\left[s(t_0)\left(e^{\int_{t0}^{t}\eta(t)dt}\right)_{t=t0} + \int_{t_0}^{t}\phi(t)\, e^{\int_{t0}^{t}\eta(t)dt}\, dt\right] \quad (10.70a)$$

We first start by calculating $\eta(t)$ integral:

$$\int_{t_0}^{t}\eta(t)dt = 2\sqrt{\delta_1}\int_{t_0}^{t}\left(1 + \frac{2}{-1 + z_1 e^{-2\left(\sqrt{\delta_1}\right)t}}\right)dt$$

$$= 2\sqrt{\delta_1}\,(t\text{-}t_0) + 2\mathrm{Log}\left(\frac{z_1 - e^{2\left(\sqrt{\delta_1}\right)t_0}}{z_1 - e^{2\left(\sqrt{\delta_1}\right)t}}\right) \quad (10.71)$$

Which –in exponential terms- leads to:

$$e^{\int_{t0}^{t}\eta(t)dt} = \left(\frac{z_1 - e^{2\left(\sqrt{\delta_1}\right)t_0}}{z_1 - e^{2\left(\sqrt{\delta_1}\right)t}}\right)^2 e^{2\sqrt{\delta_1}\,(t\text{-}t_0)} = \frac{e^{4\left(\sqrt{\delta_1}\right)t_0}}{e^{4\left(\sqrt{\delta_1}\right)t}}\left(\frac{z_1 e^{-2\left(\sqrt{\delta_1}\right)t_0} - 1}{z_1 e^{-2\left(\sqrt{\delta_1}\right)t} - 1}\right)^2 e^{2\sqrt{\delta_1}\,(t\text{-}t_0)}$$

$$= \left(\frac{z_1 e^{-2\left(\sqrt{\delta_1}\right)t_0} - 1}{z_1 e^{-2\left(\sqrt{\delta_1}\right)t} - 1}\right)^2 e^{-2\sqrt{\delta_1}\,(t\text{-}t_0)}$$

$$(10.72)$$

The second integral of Equation (10.70) gives the following expression:

$$\int_{t_0}^{t}\phi(t)e^{\int_{t0}^{t}\eta(t)dt}\, dt =$$

$$2bA\left(z_1 e^{-2\left(\sqrt{\delta_1}\right)t_0} - 1\right)^2 \int_{t_0}^{t}(t + t_1)\frac{1}{\left(z_1 e^{-2\left(\sqrt{\delta_1}\right)t} - 1\right)^2}e^{-2\sqrt{\delta_1}\,(t\text{-}t_0)}dt$$

$$(10.73)$$

or:

$$\int_{t_0}^{t} \phi(t) e^{\int_{t_0}^{t} \eta(t)dt} \, dt = 2bA(z_1 e^{-2\left(\sqrt{\delta_1}\right)t_0} - 1)^2 e^{2\left(\sqrt{\delta_1}\right)t_0} \int_{t_0}^{t} (t+t_1) \frac{e^{-2\left(\sqrt{\delta_1}\right)t}}{(z_1 e^{-2\left(\sqrt{\delta_1}\right)t} - 1)^2} \, dt$$

$$(10.74)$$

By integrating the last term of Equation (10.74), we obtain:

$$\int_{t_0}^{t} (t+t_1) \frac{e^{-2\left(\sqrt{\delta_1}\right)t}}{\left(z_1 e^{-2\left(\sqrt{\delta_1}\right)t} - 1\right)^2} \, dt = \frac{1}{2z_1\left(\sqrt{\delta_1}\right)} \int_{t_0}^{t} (t+t_1) \left[\frac{-(-2z_1\left(\sqrt{\delta_1}\right))e^{-2\left(\sqrt{\delta_1}\right)t}}{\left(z_1 e^{-2\left(\sqrt{\delta_1}\right)t} - 1\right)^2} \right] dt$$

$$(10.75)$$

or:

$$\int_{t_0}^{t} (t+t_1) \frac{e^{-2\left(\sqrt{\delta_1}\right)t}}{\left(z_1 e^{-2\left(\sqrt{\delta_1}\right)t} - 1\right)^2} \, dt = \frac{1}{2z_1\left(\sqrt{\delta_1}\right)} \left[(t+t_1) \frac{1}{z_1 e^{-2\left(\sqrt{\delta_1}\right)t} - 1} \right]_{t0}^{t}$$

$$- \frac{1}{2z_1\left(\sqrt{\delta_1}\right)} \int_{t_0}^{t} \frac{1}{z_1 e^{-2\left(\sqrt{\delta_1}\right)t} - 1} \, dt$$

$$(10.76)$$

The last term of Equation (10.76) leads to:

$$\int_{t_0}^{t} \frac{1}{z_1 e^{-2\left(\sqrt{\delta_1}\right)t} - 1} \, dt =$$

$$= -\frac{1}{2\left(\sqrt{\delta_1}\right)} \int_{t_0}^{t} \frac{-2\left(\sqrt{\delta_1}\right)e^{2\left(\sqrt{\delta_1}\right)t}}{z_1 - e^{2\left(\sqrt{\delta_1}\right)t}} \, dt = -\frac{1}{2\left(\sqrt{\delta_1}\right)} Log\left(\frac{z_1 - e^{2\left(\sqrt{\delta_1}\right)t}}{z_1 - e^{2\left(\sqrt{\delta_1}\right)t_0}} \right)$$

$$(10.77)$$

Replacing (10.77) in (10.76), we deduce:

$$\int_{t_0}^{t} (t+t_1) \frac{e^{-2(\sqrt{\delta_1})t}}{\left(z_1 e^{-2(\sqrt{\delta_1})t} - 1 \right)^2} dt =$$

$$= \frac{1}{2z_1(\sqrt{\delta_1})} \left[(t+t_1) \frac{1}{z_1 e^{-2(\sqrt{\delta_1})t} - 1} \right]_{t_0}^{t} + \frac{1}{2z_1(\sqrt{\delta_1})} \frac{1}{2(\sqrt{\delta_1})} Log \left(\frac{z_1 - e^{2(\sqrt{\delta_1})t}}{z_1 - e^{2(\sqrt{\delta_1})t_0}} \right)$$

(10.78)

Replacing(10.78) in (10.74), we obtain:

$$\int_{t_0}^{t} \phi(t) e^{\int_{t_0}^{t} \eta(t)dt} dt = 2bA \left(z_1 e^{-2(\sqrt{\delta_1})t_0} - 1 \right)^2 e^{2(\sqrt{\delta_1})t_0} \times$$

$$\times \left[\frac{1}{2z_1(\sqrt{\delta_1})} \left[(t+t_1) \frac{1}{z_1 e^{-2(\sqrt{\delta_1})t} - 1} \right]_{t_0}^{t} + \frac{1}{2z_1(\sqrt{\delta_1})} \frac{1}{2(\sqrt{\delta_1})} Log \left(\frac{z_1 - e^{2(\sqrt{\delta_1})t}}{z_1 - e^{2(\sqrt{\delta_1})t_0}} \right) \right]$$

(10.79)

By integrating the integral-factor represented by the first term of Equation (10.70), we get:

$$e^{-\int_{t_0}^{t} \eta(t)dt} = \left(\frac{z_1 - e^{2(\sqrt{\delta_1})t_0}}{z_1 - e^{2(\sqrt{\delta_1})t}} \right)^{-2} e^{-2\sqrt{\delta_1}(t-t_0)} = \frac{e^{4(\sqrt{\delta_1})t}}{e^{4(\sqrt{\delta_1})t_0}} \left(\frac{z_1 e^{-2(\sqrt{\delta_1})t} - 1}{z_1 e^{-2(\sqrt{\delta_1})t_0} - 1} \right)^2 e^{-2\sqrt{\delta_1}(t-t_0)}$$

$$= \left(\frac{z_1 e^{-2(\sqrt{\delta_1})t} - 1}{z_1 e^{-2(\sqrt{\delta_1})t_0} - 1} \right)^2 e^{2\sqrt{\delta_1}(t-t_0)}$$

(10.80)

with the following condition:

$$s(t_0)(e^{\int_{t_0}^{t} \eta(t)dt})_{t=t_0} = x_1^{*2}(t_0)$$

(10.81)

Taking into account the results illustrated in Equations (10.79) to (10.81) in the relation (10.70), we obtain:

$$s(t) = e^{-\int_{t0}^{t} \eta(t)dt} \left[s(t_0) \left(e^{\int_{t0}^{t} \eta(t)dt} \right)_{t=t0} + \int_{t_0}^{t} \phi(t) e^{\int_{t0}^{t} \eta(t)dt} dt \right]$$

$$= \left(\frac{z_1 e^{-2\left(\sqrt{\delta_1}\right)t} - 1}{z_1 e^{-2\left(\sqrt{\delta_1}\right)t_0} - 1} \right)^2 e^{2\sqrt{\delta_1}(t-t_0)} \left[x_1^{*2}(t_0) + \frac{2bA\left(z_1 e^{-2\left(\sqrt{\delta_1}\right)t_0} - 1 \right)^2 e^{2\left(\sqrt{\delta_1}\right)t_0}}{2z_1(\sqrt{\delta_1})} \times \right.$$

$$\left. \times \left(\frac{t+t_1}{z_1 e^{-2\left(\sqrt{\delta_1}\right)t} - 1} - \frac{t_0+t_1}{z_1 e^{-2\left(\sqrt{\delta_1}\right)t_0} - 1} + \frac{1}{2\left(\sqrt{\delta_1}\right)} Log \left(\frac{z_1 - e^{2\left(\sqrt{\delta_1}\right)t}}{z_1 - e^{2\left(\sqrt{\delta_1}\right)t_0}} \right) \right) \right]$$

$$\tag{10.82}$$

The solution to the differential equation given by Equation (10.69) is as follows:

$$s(t) = \frac{(z_1 e^{-2\left(\sqrt{\delta_1}\right)t} - 1)^2 e^{2\left(\sqrt{\delta_1}\right)t}}{(z_1 e^{-2\left(\sqrt{\delta_1}\right)t_0} - 1)^2 e^{2\left(\sqrt{\delta_1}\right)t_0}} \times \left(\left(x_1^{*}(t_0) \right)^2 + \frac{bA}{z_1 \sqrt{\delta_1}} (z_1 e^{-2\left(\sqrt{\delta_1}\right)t_0} - 1)^2 e^{2\left(\sqrt{\delta_1}\right)t_0} \times \right.$$

$$\left. \times \left(\frac{t+t_1}{z_1 e^{-2\left(\sqrt{\delta_1}\right)t} - 1} - \frac{t_0+t_1}{z_1 e^{-2\left(\sqrt{\delta_1}\right)t_0} - 1} + \frac{1}{2\sqrt{\delta_1}} Log \left(\frac{z_1 - e^{2\left(\sqrt{\delta_1}\right)t}}{z_1 - e^{2\left(\sqrt{\delta_1}\right)t_0}} \right) \right) \right) \tag{10.83}$$

Finally, we reach an optimal analytical solution of the $x_1^{*}(t) = \Phi_r^{*}(t)$ rotor flux that minimizes the energy consumption of the Vector controlled induction motor drives. This solution is established by developing the HJB equation in the case of a minimum-energy problem. The optimal rotor flux analytical expression is summarized by:

$$\Phi_r^{*}(t) = (z_1 e^{-2\left(\sqrt{\delta_1}\right)t} - 1)e^{\left(\sqrt{\delta_1}\right)t} \sqrt{\beta_{z_1}(t_0) + \nu \gamma_{z_1}(t)} \tag{10.84}$$

with:

$$\beta_{z_1}(t_0) = \frac{\Phi_r^{*2}(t_0) e^{-2(\sqrt{\delta_1}) t_0}}{(z_1 e^{-2\left(\sqrt{\delta_1}\right)t_0} - 1)^2}$$

$$\gamma_{z_1}(t) = \frac{1}{z_1}\left(\frac{t+t_1}{z_1 e^{-2\left(\sqrt{\delta_1}\right)t} - 1} - \frac{t_0+t_1}{z_1 e^{-2\left(\sqrt{\delta_1}\right)t_0} - 1} + \frac{1}{2\sqrt{\delta_1}} Log\left(\frac{z_1 - e^{2\left(\sqrt{\delta_1}\right)t}}{z_1 - e^{2\left(\sqrt{\delta_1}\right)t_0}} \right) \right)$$

and $v = \dfrac{bA}{\sqrt{\delta_1}}$

As shown in relation (10.84), the optimal flux expression is a time function dependent on the speed reference. z_1 has to be determined to ensure a unique solution to our problem. This constant depends on the initial and final conditions of the rotor flux. In order to find z_1, the expression of the rotor flux for its nominal value at $t=T$ gives the following equation:

$$(\Phi_r^*(T))^2 - (z_1 e^{-2\left(\sqrt{\delta_1}\right)T} - 1)^2\, e^{2\left(\sqrt{\delta_1}\right)T}\left(\beta_{z_1}(t_0) + v\,\gamma_{z_1}(T) \right) = 0 , \quad (10.85)$$

where T is the time required for the speed to go from its initial state to the final targeted one. It is the duration of Mode (1) or Mode (2), according to the c_0 sign.

$$\beta_{z_1}(t_0) = \frac{\Phi_r^{*2}(t_0)\, e^{-2\left(\sqrt{\delta_1}\right)t_0}}{(z_1 e^{-2\left(\sqrt{\delta_1}\right)t_0} - 1)^2} \qquad (10.86)$$

and

$$\gamma_{z_1}(T) = \frac{1}{z_1}\left(\frac{T+t_1}{z_1 e^{-2\left(\sqrt{\delta_1}\right)T} - 1} - \frac{t_0+t_1}{z_1 e^{-2\left(\sqrt{\delta_1}\right)t_0} - 1} + \frac{1}{2\sqrt{\delta_1}} Log\left(\frac{z_1 - e^{2\left(\sqrt{\delta_1}\right)T}}{z_1 - e^{2\left(\sqrt{\delta_1}\right)t_0}} \right) \right) \quad (10.87)$$

Using Taylor's expansion, the term in $\gamma_{z_1}(T)$ Logarithm [17], page 354, can be approximated as follows:

$$Log\left(\frac{z_1 - e^{2\left(\sqrt{\delta_1}\right)T}}{z_1 - e^{2\left(\sqrt{\delta_1}\right)t_0}} \right) = \qquad (10.88)$$

$$= Log\left(1 + \frac{e^{2\left(\sqrt{\delta_1}\right)t_0} - e^{2\left(\sqrt{\delta_1}\right)T}}{z_1 - e^{2\left(\sqrt{\delta_1}\right)t_0}} \right) \cong \sum_{K=0}^{+\infty}(-1)^K \frac{\left(e^{2\left(\sqrt{\delta_1}\right)t_0} - e^{2\left(\sqrt{\delta_1}\right)T} \right)^{K+1}}{(K+1)\left(z_1 - e^{2\left(\sqrt{\delta_1}\right)t_0} \right)^{K+1}}$$

Substituting the relation (10.86) into (10.85), the z_1 variable can be expressed according to K order as a function of the limit values of $\Phi_r^*(T)$ and $\Phi_r^*(t_0)$ variables:

$$z_1 = \Gamma_{K,c_0,c_1}(\Phi_r^*(T),\Phi_r^*(t_0)) \tag{10.89}$$

with $\Gamma_{K,c_0,c_1}(\Phi_r^*(T),\Phi_r^*(t_0))$ as the solution of Equation (10.85) resulting from Taylor's expansion of the $\gamma_{z_1}(T)$ function. This solution is determined numerically. The calculation of the variable z_1 is followed by relative errors of unavoidable truncations. These uncertainties depend on the choice of the K order of the Taylor expansion of $\gamma_{z_1}(T)$ function.

Note: The v term given by Equations (10.84) and (10.85) $v = bA / \sqrt{\delta_1} = c_0 bK_l / c\sqrt{\delta_1}$ depends on c_0 and δ_1. This last term depends on λ_1 and c_1. c_0 and c_1 are the coefficients of the expression proposed in (10.40) for the reference rotation speed. This confirms that z_1 depends on the load level.

Finally, we can deduce the optimal control law to be implemented in a magnetizing stator current control application. This law deduced from Equations (10.66) and (10.84) is as follows:

$$
\begin{aligned}
u_1^*(t,x_1^*) &= \\
&= \frac{1}{b}\left(a - \sqrt{\delta_1} - \frac{2\sqrt{\delta_1}}{-1+z_1 e^{-2(\sqrt{\delta_1})t}}\right)(z_1 e^{-2(\sqrt{\delta_1})t} - 1)e^{(\sqrt{\delta_1})\,t}\sqrt{\beta_{z_1}(t_0)+v\gamma_{z_1}(t)} \\
&\quad + \frac{A}{(z_1 e^{-2(\sqrt{\delta_1})t} - 1)e^{(\sqrt{\delta_1})\,t}\sqrt{\beta_{z_1}(t_0)+v\gamma_{z_1}(t)}}(t+t_1)
\end{aligned}
\tag{10.90}
$$

10.4. Implementation of the Optimal Solution Obtained by HJB Equation

10.4.1. Determination of the Rotor Flux Optimum Trajectory

Most industrial applications with variable speeds and designed around electrical machines consider velocity profiles that are typically trapezoidal [10], [20] and [23], and triangular [6], [21] and [25].

For instance, this study focuses on the optimization of energy consumption in particular during the IM transitional mode. As indicated in the paragraph (10.4.2), we considered two operation phases: engine in acceleration: Mode (1) and engine in deceleration: Mode (2).

The task consists in determining the rotor flux optimal trajectory from the solution of HJB Equation (10.35). However, for a speed reference illustrated in Fig.10.1 and assuming that the speed controller delivers the reference torque current, it is possible to ensure a response in terms of the rotation speed confused with that of its reference.

Fig. 10.1 illustrates the profile of the selected speed reference with the following coefficient values:

1) Mode (1), for Ω (rev /min): $c_0 = 1433$ and $c_1 = 0$;

2) Mode (2), for Ω (rev/min): $c_0 = -1144$ and $c_1 = 1717$.

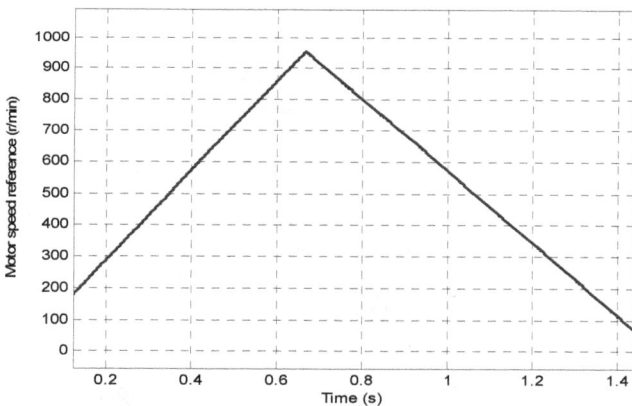

Fig. 10.1. Mechanical speed reference (rev/mn).

The asynchronous motor is accelerated from 181 to 954 rpm for the first phase and decelerates from 954 to 70 rpm for the second phase. From the speed reference trajectory, we can determine for each mode all the parameters that appear in the optimal analytical expression of the rotor flux governed by Equation (10.84). Taking as initial and final conditions as well as the values of the weighting factors those described in Table 10.1, we can obtain the optimum trajectory of the optimal rotor flux for both operation modes. This is illustrated in Fig. 10.2. In addition, we can deduce the trajectory of the optimal control variable given by Equation (10.89), as shown in Fig. 10.3.

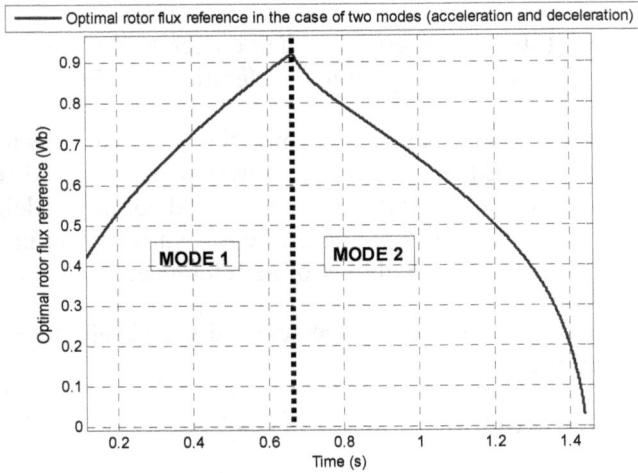

Fig. 10.2. Optimal rotor flux trajectory via the HJB equation.

The initial and final conditions as well as all the parameters that describe the optimal analytical flux equation for the two modes are described in Table 10.1.

Table 10.1. Initial and final conditions as well as all the parameters that describe the flux optimal analytical equation for both modes.

	Mode (1)	Mode (2)
$\Phi_r^*(T)$ **(Wb)**	$\Phi_r^*(T) = 0.924\,(\text{Wb})$	$\Phi_r^*(T) = 0.924\,(\text{Wb})$
$\Phi_r^*(t_0)$ **(Wb)**	Depending on the motor flux level	Depending on the motor flux level
t_0 **and** T **(s)**	$t_0 = 0.12$ (s) and $T = 0.667$ (s)	$t_0 = 0.667$ (s) and $T = 1.44$ (s)
α_{11}, α_{12} **et** α_{13}	$\alpha_{11} = 5,\ \alpha_{12} = 1$ and $\alpha_{13} = 0.1$	$\alpha_{11} = 5,\ \alpha_{12} = 1$ and $\alpha_{13} = 0.1$
$t_1 = (B/A)$	$t_1 = 0.0731$	$t_1 = -1.4269$
K **(Taylor expansion order)**	$K = 6$	$K = 6$
v	$v = 1.2201$	$v = -0.9424$
c_0 **and** c_1	$c_0 = 150$ and $c_1 = 0$	$c_0 = -120$ and $c_1 = 180$
z_1	$z_1 = 34624306450.391$	$z_1 = 1059529196371687.679$

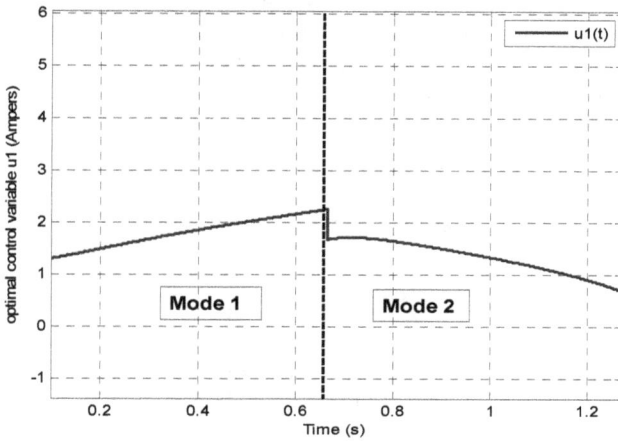

Fig. 10.3. Evolution of the optimal control variable reference.

10.4.2. Implementation of the Flux Optimal Trajectory in the VECTOR CONTROL Structure

The main structural elements of optimal vector-controlled induction motor drives are illustrated in Fig. 10.4. There is in particular the minimization block of the IM energy consumption via the HJB equation. Using the speed reference given in Fig. 10.4 at the speed controller input, the 'A' block constructed around Equation (10.84) delivers the optimum rotor flux trajectory used by the flux controller at finite response time. These two controllers respectively deliver the optimum torque stator currents and reference rotor flux in order to reconstitute the remainder of the vector control structure.

10.4.2.1. Simulation Results of the IM Minimal-energy Control via the HJB Equation

Once the optimal trajectories of the vector control are synthesized, it would be possible to conduct a comparative study between the simulation results of the latters and those obtained for the conventional vector control. In the two control laws, we considered the IM magnetic saturation model in function of the rotor flux. The simulation results of the two control laws are governed by the same speed reference illustrating two operation phases in transient mode: Mode (1) and Mode

(2) described in paragraphs 4.4.2 and 4.4.4.1. Fig. 10.5 illustrates the trajectory variation of the real mechanical rotation speed with that of the reference. The IM conventional VECTOR CONTROL operates at constant nominal rotor flux.

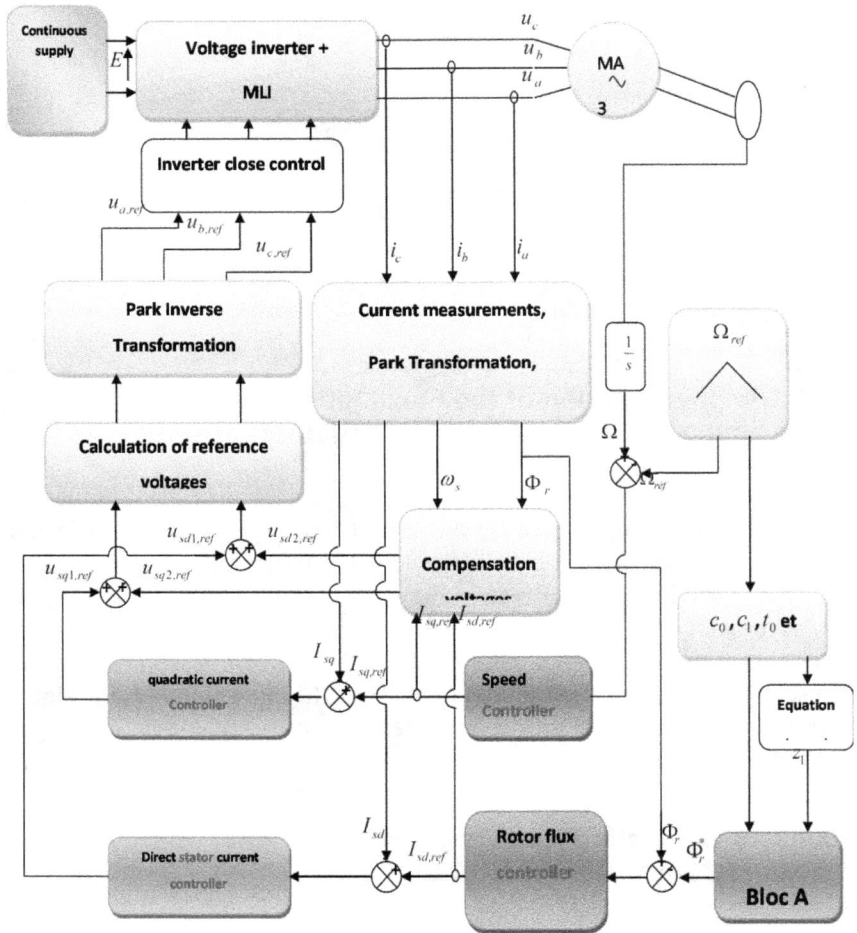

Fig. 10.4. The implementation block diagram of the optimization strategy by the HJB equation.

The VECTOR CONTROL combined with the energy optimization algorithm via the HJB equation uses the optimal flux trajectory of

Fig. 10.2. Fig. 10.6 illustrates the trajectories of the optimal rotor flux. We note that the finite response time of the flux controller ensures a good tracking towards its reference.

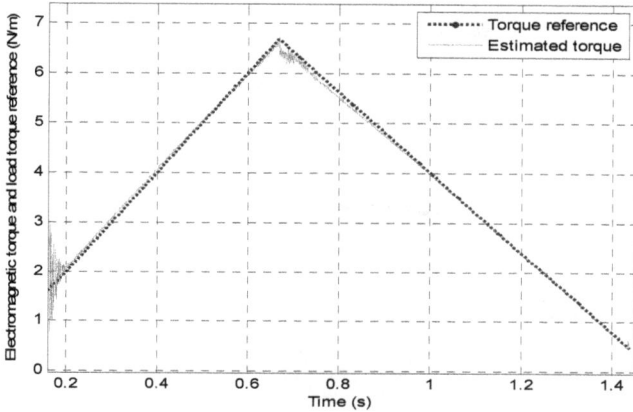

Fig. 10.5. Trajectory of the real mechanical rotation speed with that of the reference.

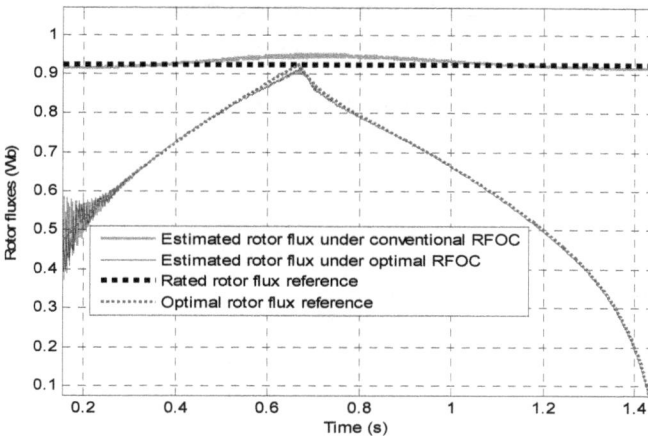

Fig. 10.6. Trajectory of the rated flux used by the conventional VECTOR CONTROL and that of the optimal flux.

Fig. 10.8 gives the direct stator current delivered by the flux controller. This current follows the same law of variation as that given by the optimal control synthesized through Equation (10.89) and illustrated in Fig. 10.3. The level of the quadrature axis current is represented by Fig. 10.7.

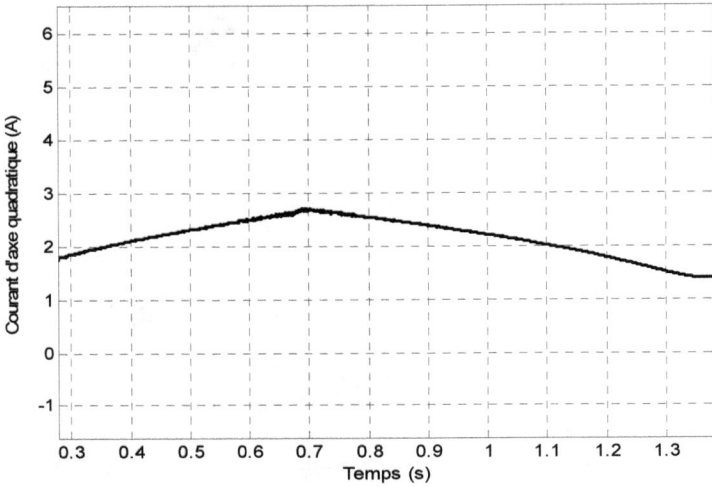

Fig. 10.7. Reference of the quadrature axis current and the estimated electromagnetic torque current.

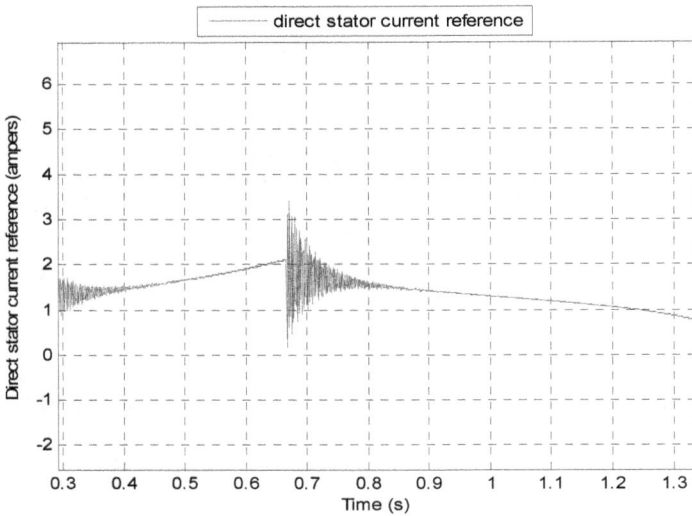

Fig. 10.8. Reference direct stator current.

The simulation results shown in Fig. 10.9 and Fig. 10.10 respectively show the evolution of the stator components of the flux and torque currents for the conventional VECTOR CONTROL and that using the optimization strategy by the HJB equation. Through these two figures, we observe a marked decrease in the current responsible for the flux in the case of the optimal VECTOR CONTROL compared to the conventional one. Therefore, a slight increase in the stator current of the optimal torque is noted compared to that delivered by the conventional vector control. This first observation allows us to say that our approach minimizes energy.

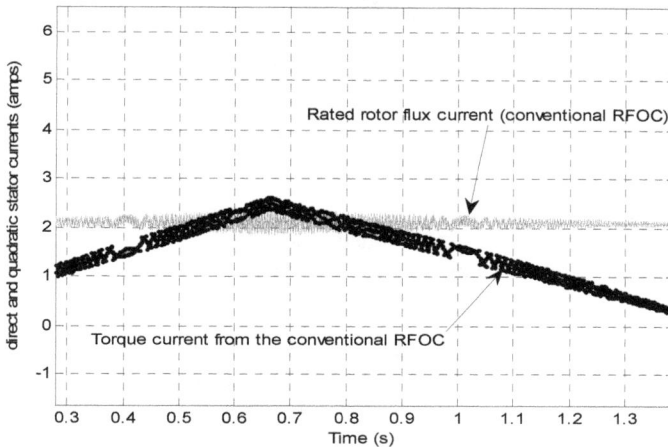

Fig. 10.9. Flux current and torque current in the conventional RFOC.

The finite-time-response rotor flux controller allows a good performance of the reference. Its start up at the beginning of each mode gives rise to short-term deficiencies which are manifested by some oscillations at the level of the reference direct stator current.

Fig. 10.11 shows the temporal evolution of the energy consumed by the IM under the conventional VECTOR CONTROL and the VECTOR CONTROL using the optimization approach via the HJB equation. It can be noted that the optimal solution of the flux allows to reduce the energy consumption throughout the IM acceleration and deceleration phase compared with the same mechanical cycle operating with nominal rotor flux. To further demonstrate the potential of this optimal control via the

HJB equation, the IM performance under optimum VECTOR CONTROL is also found to be better throughout the whole cycle, as shown in Fig. 10.12.

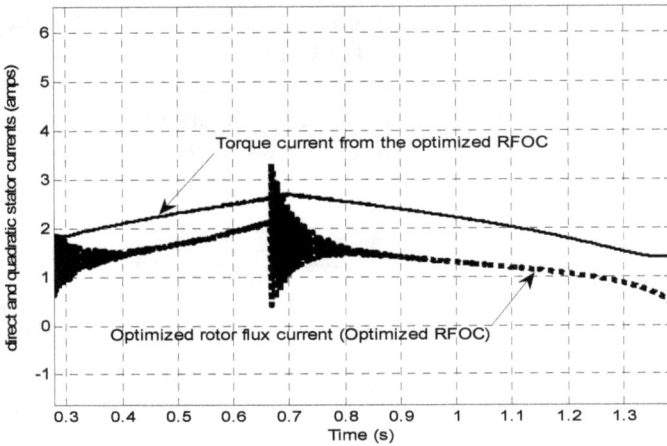

Fig. 10.10. Stator currents responsible for the rotor flux and the torque generated by optimal VECTOR CONTROL.

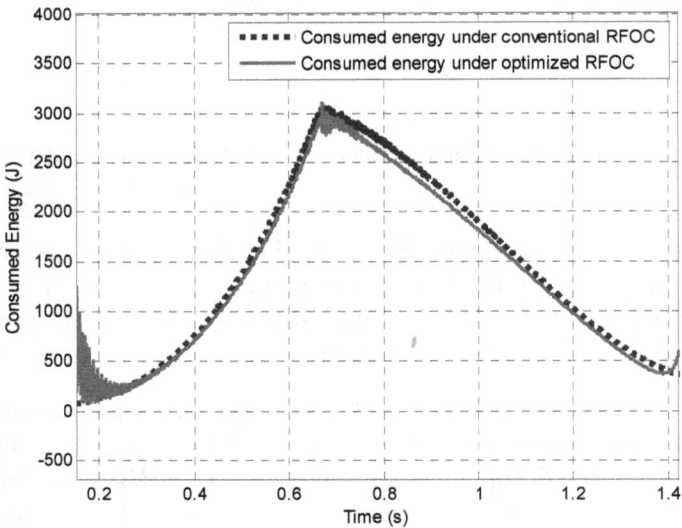

Fig. 10.11 (a). Consumed energy under conventional and optimized RFOC

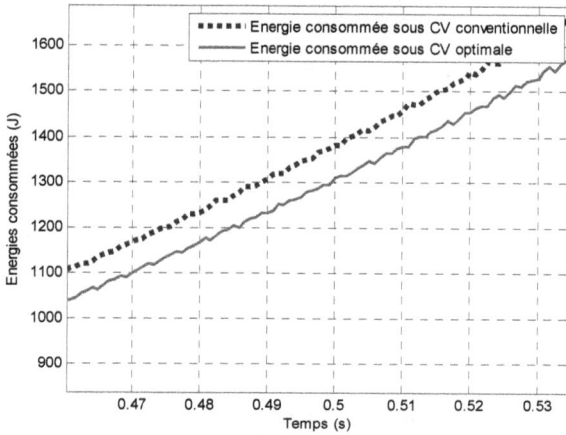

(b) Temporal evolution of the same energy at acceleration mode.

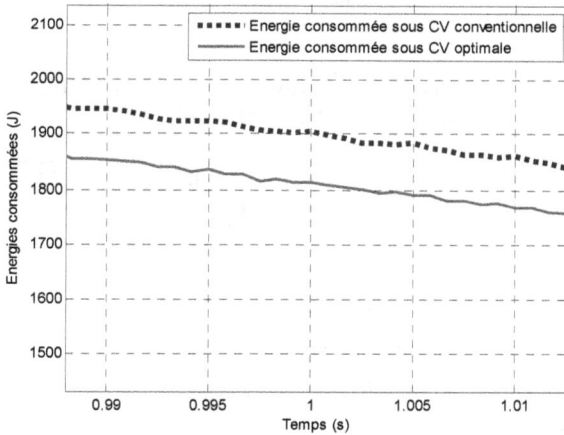

(c) Temporal evolution of the same energy at deceleration mode.

Fig. 10.11 (b-c). Temporal evolution of energy consumed by the two types of vector control.

We reported some results obtained by the two vector controls in Table 10.2. Given the power of our laboratory IM (average energy consumption around 2000J) we can see that the reduction in energy consumption achieved via the optimum VECTOR CONTROL is promising.

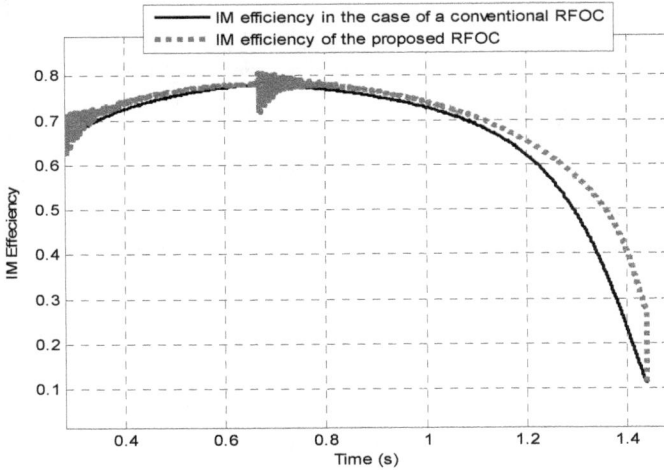

Fig. 10.12. The IM performance with optimal VECTOR CONTROL and that obtained with the conventional.

Table 10.2. Some results obtained with the classical VECTOR CONTROL and the optimal VECTOR CONTROL.

Energy and Performance	Consumed Energy (J)	IM performance
Mode 1 (time 0.5 s and load torque = 5.02 Nm)		
VECTOR CONTROLoperating with optimal solution flux through HJB	1310	0766
VECTOR CONTROL operating with constant nominal flux	1380	0755
Energy reduction (J)	**-70**	
Performancegain		**+1.1%**
Mode 2(time 1.2 s and load torque = 2.41 Nm)		
VECTOR CONTROL operating with optimal solution flux through HJB	974	0652
VECTOR CONTROL operating with nominal flux	1026	0621
Energy reduction (J)	**-52**	
Performancegain		**+3.1%**

10.5. Conclusion

In this chapter, two formulations of the optimal control problem associated with the optimization of the energy consumed by the IM under VECTOR CONTROL were presented. The emphasis was placed on the advantage of limiting the control quantities during a real application in order to protect the actuators and the machine. These limitations essentially relate to the maximum values which can be taken by the current and stator voltages. Through a formulation in a control space which is different from that of the variational calculation, we showed the control variables in the functional as well as the constraints of the problem.

The minimization of the functional or the integral of the energy weighted sum and the losses of the AM under the dynamic stress of the flux and of the rotation speed is realized in a first part by applying the principle of Minimum of Pontryaguin. This principle uses a formulation, other than that of the Euler equation, which hinges on the Hamilton-Jacobi equation and which sets the necessary and sufficient optimality conditions around the control variables, state and co-state. The application of this principle in the case of the IM minimal-energy problem led us to derive differential equations relating to these above-mentioned variables. The expansion of these equations allowed us to find the same result determined in the Section 10.3. The reformulation of the optimization problem via the Hamilton-Jacobi equation is essential to give off the control variables. The crucial problem related to the application of the Pontryaguin minimum principle such as the one encountered when solving the Euler equation, lies at the level of the initial and final values of the flux and its primary and secondary derivatives which must be stopped right from the beginning.

In the second part of this chapter, we overcame this problem of conditions at both ends by considering optimal control in closed loop. Thus, the optimization technique using the Hamilton-Jacobi-Bellman equation allowed an optimal solution of the rotor flux that is independent of these conditions or at the limit converges to the optimal solution for any initial flux value. This obtained solution is the subject of one of the contributions of these theses. It is presented in an analytical form and is dedicated to optimizing a transient regime characterized by a speed ramp. The VECTOR CONTROL using this analytical solution imposed itself by a remarkable reduction in IM energy consumption.

References

[1]. Marlin O. Thurston, Energy-Efficient Electric Motors, *Electrical and Computer Engineering*, Marcel Dekker, New York, 2005.

[2]. John Chiasson, Modelling and High-Performance Control of Electric Machines, *IEEE Press Series on Power Engineering, John Willey &Sons, Inc.*, Hobokon, New Jersey, 2005.

[3]. Hans P. Geering, Optimal Control with engineering application, *Springer-Verlog Berlin Heidelberg*, 2007, pp. 115-117.

[4]. Arturo Locatelli, Optimal control: an introduction, *Birkhauser Verlog*, 2001.

[5]. Robert D. Lorenz, Sheng Ming Yang, Efficiency-Optimized flux trajectories for closed cycle operation of field oriented induction machine drives, in *Proceedings of the IEEE Conference Record of the Industry Application Society Annual Meeting*, 2-7 Oct., Vol. 1, 1988, pp. 457-762.

[6]. Robert D. Lorenz, Sheng Ming Yang, AC Induction Servo Sizing for Motion Control Applications via Loss Minimizing Real-Time Flux Control, in *Proceedings of the IEEE Transaction on Industry Applications*, Vol. 28, No. 3, May-June 1992.

[7]. C. Canudas de Wit, J. Ramirez, Optimal Torque Control for Current-feeded Induction Motors, in *Proceedings of the IEEE Transactions on Automatic Control*, Vol. 44. No. 5, May 1999.

[8]. Didier Georges, Carlos Canudas de Wit, Jose Ramirez, Nonlinear H_2 and H_∞ Optimal controllers for Current-Fed Induction Motors, in *Proceedings of the IEEE Transaction on Automatic Control*, Vol. 44, No. 7, July 1999, pp. 1430-1435.

[9]. Peda V. Medagam, Farzad Pourboghrat, Online H_∞ Speed Control of Sensorless Induction Motors with Rotor Resistance Estimation, in *Proceedings of the IEEE 7th International Conference on Power Electronics and Drive Systems (PEDS'07)*, 27-30 Nov. 2007, pp. 1307-1312.

[10]. I. Ya. Braslavsky, A. V. Kostylev, D. P. Stepanyuk, Optimization of Starting Process of the Frequency Controlled Induction Motor, in *Proceedings of the 13th IEEE International Power Electronics and Motion Control Conference(EPE-PEMC'08)*, 1-3 Sept. 2008, pp. 1050-1053.

[11]. Tr. Munteanu, E. Rosu, M. Gaiceanu, R. Paduraru, T. Dumitriu, M. Culea, C. Dache, The optimal Control for Position Drive System with Induction Machine, in *Proceedings of the IEEE 13th European Conference on Power Electronics and Applications (EPE'09)*, 8-10 Sept. 2009, pp. 1-8.

[12]. Aiyuan Wang, Zhihao Ling, Improved Efficiency Optimization for Vector Controlled Induction Motor, in *Proceedings of the IEEE Asia Pacific Power and Energy Engineering Conference (APPEEC'09)*, 27-31 March 2009, pp. 1-4.

[13]. Hou-Tsan Lee, Li-Chen Fu, Nonlinear Control of Induction Motor with Unknown Rotor Resistance and Load Adaptation, in *Proceedings of the*

American Control Conference, Arlington, VA, Vol. 1, 25-27 June 2001, pp. 155-159.

[14]. H. K. Khalil, Nonlinear systems, *Prentice Hall,* New Jersy, 1996.

[15]. A. C. King, J. Billingham, S. R. Otto, Differential equations linear, non linear, ordinary, partial, *Cambridge University Press*, 2003.

[16]. Andrei D. Polyamin, Valentin F. Zeitsev, Exact solutions For ordinary differential equations, *CRC Press Inc.,* Boca Raton, FL, 1995.

[17]. Andrei D. Polyanin, Alexander V. Manzhirov, Handbook of Mathematics for Engineers and Scientist, *Chapman & Hall/CRC*, 2007.

[18]. M. F. Mimouni, R. Dhifaoui, Modelling and Simulation of Double-Star Induction Machine Vector Control using Copper Losses Minimization and Parameters Estimation, *International Journal of Adaptive Control and Signal Processing*, 2002, Vol. 16, pp. 1-24.

[19]. I. Braslavsky, Z. Ishhmatov, Y. Plotnikov, I. Averbakh, Energy Consumption and Losses Calculation Approach for Different Classes of Induction Motor Drives, in *Proceedings of the International Symposium on Power Electronics, Electrical Drives, Automation and Motion (SPEEDAM'06)*, pp. S15 60 - S15 65.

[20]. K. L. Shi, T. F. Y. K. Wong, S. L. Ho, A Rule-Based Acceleration Control Scheme for an Induction Motor, *IEEE Transactions on Energy Conversion*, Vol. 17, No. 2, June 2002, pp. 254-259.

[21]. Sung-Don Wee, Myoung-Ho Shin, Dong-Seok Hyun, Stator-Flux-Oriented Control of Induction Motor Considering Iron Loss, *IEEE Transactions on Industrial Electronics*, Vol. 48, No. 3, June 2001, pp. 602-608.

[22]. Ian T. Wallace, *et al.*, Verification of Enhanced Dynamic Torque per Ampere Capability in Saturated Induction Machines, *IEEE Transactions on Industry Applications*, Vol. 30, No. 5, September / October 1994, pp. 1193-1201.

[23]. Vladimir V. Pankratov, *et al.*, New Approach to Energy Efficient Control of Induction Motor Drives, in *Proceedings of the 30th Annual Conference of IEEE Industrial Electronics Society*, Eusan, Korea, Vol. 2, 2-6 November 2004, pp. 1400-1404.

[24]. Kouki Matsuse, Shotaro Taniguchi, Tatsuya Yoshizumi, Kazushige Namiki, A Speed-Sensorless Vector Control of Induction Motor Operating at High Efficiency Taking Core Loss into Account, *IEEE Transactions on Industry Applications*, Vol. 37, No. 2, March/April 2001, pp. 548-558.

[25]. Ahmed Rubaai, Oscar Uribina, M. D Kankam, Design of an Adaptive Nonlinear Controller-Based Optimal Control Theory for a Voltage Source Induction Motor Drive System, in *Proceedings of the IEEE Industry Applications Conference Thirty-Sixth IAS Annual Meeting Conference Record*, Vol. 2, 2001, 30 Sept. -4 Oct. 2001, pp. 1279-1284.

[26]. R. Abdelati, M. F. Mimouni, Analytical Solution of Optimized Energy Consumption of Induction Motor Operating in Transient Regime *European Journal of Control*, Vol. 4, 2011, pp. 397-411.

Index

S

T

www.ingramcontent.com/pod-product-compliance
Lightning Source LLC
Chambersburg PA
CBHW050453190326
41458CB00005B/1266